SUSTAINABLE DEGRADATION OF LIGNOCELLULOSIC BIOMASS - TECHNIQUES, APPLICATIONS AND COMMERCIALIZATION

Edited by **Anuj K. Chandel**
and **Silvio Silvério da Silva**

Sustainable Degradation of Lignocellulosic Biomass - Techniques, Applications and Commercialization
http://dx.doi.org/10.5772/1490
Edited by Anuj K. Chandel and Silvio Silvério da Silva

Contributors

Elba Bon, Ayla Sant'Ana da Silva, Viridiana Santana Ferreira-Leitão, Rodrigo da Rocha Olivieri de Barros, Maria Antonieta Ferrara, Ricardo Sposina Sobral Teixeira, Rondinelle de Oliveira Moutta, Masahide Yasuda, Rosa Estela Quiroz- Castañeda, Jorge Luis Folch-Mallol, Vittaya Punsuvon, Tim Dumonceaux, Thomas Canam, Jennifer Town, Lope G. Tabil, Kingsley Iroba, Wagner Rodrigo De Souza, Fernanda Motta, Maria Helena Santana, Cristiane Andrade, Zhijia Liu, Benhua Fei, Larissa Canilha, Rita Rodrigues, Felipe F.A. Antunes, Anuj Chandel, Silvio Silva, Danielle Julie Carrier

Published by InTech

Janeza Trdine 9, 51000 Rijeka, Croatia

Publishing Process Manager Ana Pantar
Technical Editor InTech DTP team
Cover InTech Design team

Additional hard copies can be obtained from orders@intechopen.com

Sustainable Degradation of Lignocellulosic Biomass - Techniques, Applications and Commercialization,
Edited by Anuj K. Chandel and Silvio Silvério da Silva
p. cm.
ISBN 978-953-51-1119-1

For EU product safety concerns:
IN TECH d.o.o., Prolaz Marije Krucifikse Kozulić 3, 51000 Rijeka, Croatia,
info@intechopen.com or visit our website at intechopen.com.

Meet the editors

Dr. Anuj K. Chandel completed doctorate in 2009 from Jawaharlal Nehru Technological University, Hyderabad, India. After his masters from Indian Institute of Technology Roorkee, he joined Dalas Biotech Ltd. Bhiwadi, for the large scale production of penicillin acylase. Subsequently, he worked at University of Delhi in a research project funded by Department of Biotechnology (DBT), Govt. of India. Later he joined Celestial Labs Ltd., Hyderabad as a research associate. After this, he did post doctoral studies at University of Stellenbosch, South Africa. Currently he is working at Engineering School of Lorena, University of São Paulo in a thematic project funded by Bioen-FAPESP. He is the author of 1 book on xylitol, 45 articles in peer-reviewed journals and 12 book chapters.

Professor Silvio Silvério da Silva was born on December 31st, 1959. He is a professor at Department of Biotechnology, Engineering School of Lorena, University of São Paulo (USP), Brazil. He completed his doctorate in Biochemical and Pharmaceutical Technology from USP and GBF German Research Centre for Biotechnology, Germany in 1994. He has published more than 130 papers in peer reviewed international journals and 18 book chapters. He has recorded two patents on technological processes for xylitol production. Prof. Silva has guided 10 doctoral students, 25 masters' dissertations and 55 scientific initiation students. He has successfully completed 19 research projects funded by Brazilian Government. Prof. Silvio has been actively working on various aspects of lignocellulose biotechnology for 25 years.

Contents

Preface

This book is primarily focused on the key aspects of lignocellulosic biomass (LB) degradation and eventually the production of value-added products. LB is the most abundant renewable organic resources (~200 billion tons annually) on earth that are readily available for conversion to biofuels and other value-added products (industrial enzymes, organic acids, pharmaceuticals, commodity chemicals, and food/feed). However, LB has not yet been tapped for the commercial production of these products.

The book content has been divided into four sections. Section 1 deals with the various kinds of potential biomass sources. Section 1 contains 3 chapters. Chapter 1 details about the bamboo as biomass source, the characterization of cell wall components and chemical pretreatment. Chapter 2 comprehensively reviews the characterization of sugarcane residues (sugarcane bagasse and straw) which is important biomass in countries like Brazil, India and China. This chapter is particularly dedicated to hemicellulose fraction of cell wall of these residues which can be converted into several value-added products. Methods of hemicellulose de-polymerization, detoxification of xylose-rich solution, microbial conversion of pentose sugars into various products and the economic aspects of by-products derived during hemicellulose destruction. Chapter 3 exclusively discusses various pretreatment technologies for sugarcane residues and woody biomass. Emphasis has been placed on mechanistic changes in cell wall after pretreatment of biomass sources.

Section 2 covers the important aspects of biomass processing involved in biomass pretreatment, saccharification and ethanol production. There are 4 chapters in this section. Chapter 4 specifically shows the effect of lignin removal on ethanol production via simultaneous saccharification and co-fermentation from napier grass, rice straw, silver grass and bamboo. Minimization of water usage is an important feature for any bioprocess industry. The effect of washing the cellulignin from poplar for ethanol production has been elaborated in chapter 5. Chapter 6 shows the effect of cellulase loadings and other inevitable conditions on biomass saccharification for the clean sugars production. Chapter 7 demonstrates the statistical optimization of delignification and hydrolysis of steam exploded oil palm trunk for ethanol production.

Section 3 provides a thorough analysis on recent progress on microbial pretreatment of biomass in 2 chapters. Microbial pretreatment of biomass has shown promising results for the delignification while retaining maximum carbohydrates in the plant cell wall for the subsequent conversion into fermentable sugars. Chapter 8 provides the technical overview on microbial delignification of biomass in addition to limitations and various process configurations. Microorganism mediated biodelignification, the mechanism involved in biodelignification, enzyme production profile of various microorganisms, and the possible approaches for developing the robust microbial pretreatment strategy have been elaborated in Chapter 9.

Section 4 finally devoted to industrial enzymes production from lignocellulosic biomass employing various process configurations. Chapter 10 comprehensively reviews the xylanase production, microorganisms, fermentation methodologies and xylanolytic machineries for the complete depolymerisation of carbohydrate fraction from cell wall.

We sincerely believe that this book will be a useful source of information for graduate and post graduate students, researchers in the field of life sciences, biochemical engineering, and professionals from both academia and industries. Anuj would like to express sincere thanks to his wife, Meenakshi, and son, Abhay, for their immense support during editing of this book. Silvio Silvério is also thankful to his wife, Deyse, and daughter, Isabela, for their cooperation while editing this book.

We welcome the reader's suggestions to improve future editions. Readers' benefits will be the best reward for the editors and authors.

Dr. Anuj K. Chandel,
University of São Paulo, Brazil

Professor Silvio Silvério da Silva,
University of São Paulo, Brazil

"Learning gives creativity
Creativity leads to thinking
Thinking provides knowledge
Knowledge makes you great."
(Dr. A. P. J. Abdul Kalam)

Section 1

Potential Biomass Sources

Chapter 1

Characteristics of Moso Bamboo with Chemical Pretreatment

Zhijia Liu and Benhua Fei

Additional information is available at the end of the chapter

http://dx.doi.org/10.5772/55379

1. Introduction

The world's present economy is highly dependent on various fossil energy sources such as oil, coal, natural gas, etc [1]. There are several alternative energies which can replace fossil fuels in the future, such as hydro, solar, wind, biomass and ocean thermal energy. Among these energy sources, biomass is the only carbon-based sustainable energy and is utilized by most people around the world [2]. Furthermore, it is also confirmed that the petroleum-based fuels can be replaced by biomass fuels such as bioethanol, bio-diesel, bio-hydrogen, which derives from agricultural residues, forestry residues, municipal solid waste, manufacturing waste, vegetable oils, dedicated energy crops, etc [3]. Recently, it is a growing interest in manufacturing bioethanol using biomass materials. It is well known that production of ethanol from biomass includes three major processes, such as pretreatment, hydrolysis, and fermentation. One of the most important processes is biomass pretreatment in the production of biofuel. Biomass pretreatment can remove lignin and hemicelluloses, which significantly enhance the hydrolysis of cellulose. It is required to alter the biomass macroscopic and microscopic size and structure as well as its submicroscopic structural and chemical composition to facilitate rapid and efficient hydrolysis of carbohydrates to fermentable sugars [4]. Figure 1 shows simplified impact of pretreatment on biomass [5].

Pretreatment technology contains physical pretreatment (mechanical size reduction, pyrolysis, microwave oven and electron beam irradiation pretreatment), physicochemical pretreatment (steam explosion or autohydrolysis, liquid hot water method, ammonia fiber explosion, CO_2 explosion), chemical pretreatment (acid pretreatment, alkaline pretreatment, wet oxidation, organosolv pretreatment), and biological pretreatment [1]. Recently, the information of various pretreatment methods is available. Lin et al. found that the yields of glucose and xylose were improved by adding any of the following dilute chemical reagents, such as H_2SO_4, HCl, HNO_3, CH_3COOH, HCOOH, H_3PO_4, and NaOH, KOH, $CaOH_2$, NH_3 H_2O in the ball milling pretreatment of corn stover [6]. Bjerre et al. studied the wet oxidation process of wheat straw as a pretreatment method. By using a specially constructed autoclave system, the wet oxidation process was optimized with respect to both reaction time and temperature (20 g/L straw, 170℃, 5 to 10 min) and gave about 85% w/w yield of converting cellulose to glucose [7]. Zhu and Pan evaluated the performances of three of the most promising pretreatment technologies, including steam explosion, organosolv, and sulfite pretreatment to overcome lignocelluloses recalcitrance (SPORL) for softwood pretreatment. SPORL was the most efficient process and produced highest sugar yield [8]. Keshwani et al. examined the feasibility of microwave pretreatment to enhance enzymatic hydrolysis of switchgrass. It was found that the application of microwave radiation for 10 minutes at 250 watts to switchgrass immersed in 3% sodium hydroxide solution (w/v) produced the highest yields of reducing sugar [9]. Yu et al. studied a two-step liquid hot water pretreatment (TSLHW) mehod. The first step of pretreatment was temperature range from 180 to 200℃, and the highest yield of total xylose achieved was 86.4% after 20 min at 180℃. The second-step of pretreatment was temperature range from 180 to 240℃ for 0-60 min. The optimum reaction conditions of pretreatment with minimal degradation of sugars were 200℃ for 20 min [10]. Sulfuric acid is widely used for acid pretreatment among various types of acid such as hydrochloric acid, nitric acid and phosphoric acid [11]. Maarten et al. compared the efficiencies of fumaric, maleic, and sulfuric acid in wheat straw pretreatment. At 150℃ and 20-30% (w/w) dry wheat straw, the pretreatment with dilute fumaric or maleic acid could be a serious alternative to dilute sulfuric acid pretreatment [12]. Sun et al. studied the effectiveness of different alkaline solutions by analyzing the delignification and dissolution of hemicelluloses in wheat straw. The optimal process condition was 1.5% NaOH for 144h at 20℃, releasing 60% and 80% lignin and hemicelluloses, respectively [13].

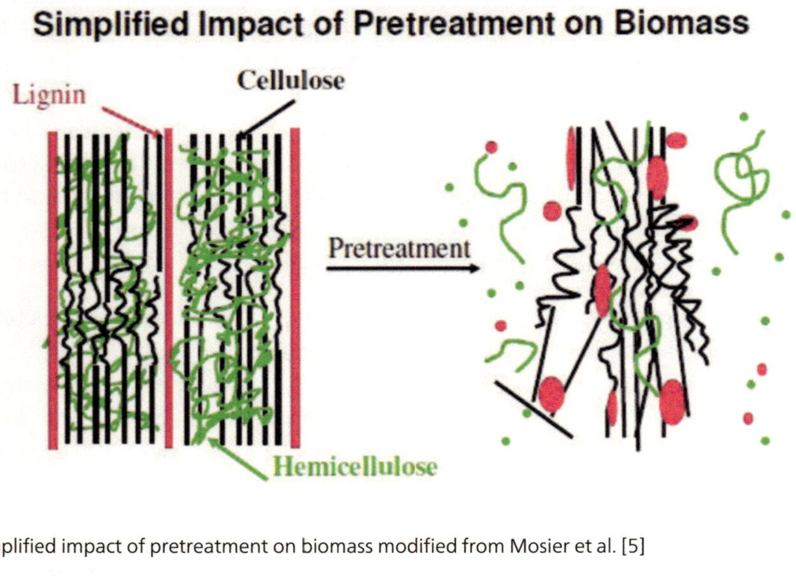

Figure 1. Simplified impact of pretreatment on biomass modified from Mosier et al. [5]

Agricultural and forest residues represent a major fuel source for potential bio-energy projects in many developing countries [14]. Bamboo, like wood and agricultural residue, is mainly composed of hemicelluloses, cellulose and lignin, even though the contents of these compositions are different. The cellulose, hemicelluloses and lignin content in some agriculture, wood and bamboo wastes is showed in Table 1. Bamboo has been widely cultivated in the west and south of China. Currently, bamboo resources are very abundant. The total area of bamboo is about five million hectares and that of moso bamboo is about 3 million hectares in China [15]. Annual yield of moso bamboo is about eighteen million tons, and it is widely used to produce furniture, flooring and interior decoration materials. It has great potential as a bio-energy resource of the future in China. Despite these previous researches are very helpful in understanding the pretreatment of biomass materials, bamboo is a different type of material. To date, no research about bamboo biofuel is available. In this research, moso bamboo was therefore pretreated by 2% of sulfuric acid (w/w bamboo) and 10% sodium hydrate (w/w bamboo). Characteristics of pretreated bamboo were determined by Fourier transform infrared spectroscopy (FTIR) and X-ray diffraction (XRD), respectively. Pretreatment method was evaluated through comparing with characteristics of untreated and pretreated bamboo. The objective of this research is to investigate characteristics of bamboo pretreated by chemical methods and select a pretreatment method for exploring the biofuel using bamboo.

Biomass type	Species	Cellulose (%)	Hemicelluloses (%)	Lignin (%)	Ash (%)
Wood	Pine (hardwood)	40-45	25-30	26-34	----
	Maple (softwood)	45-50	21-36	22-30	----
Bamboo	Moso bamboo	42-50	24-28	24-26	1.3-2.0
Agriculture residues	Rice straw	41-57	33	8-19	8-38
	Rice husk	35-45	19-25	20	14-17
	Baggasse	40-46	25-29	12.5-20	1.5-2.4
	Cotton stalk	43-44	27	27	1.3

Table 1. Chemical composition of bamboo, wood and agriculture residue

2. Materials and methods

2.1. Material

Moso bamboo aging with 4 years was used in this study. They were taken from a bamboo plantation located in Anhui province, China. The initial moisture content of samples was about 6.13%, and the density was about 0.65g/cm3. Bamboo materials were cut off to sample size 40mm (longitudinal) by 3-8mm (radial) by 20-30mm (tangential). Then, they were broken down to particles with a Wiley mill and the size of bamboo particles used in the test was about

250-425 um. Finally, the particles were dried at 105℃ until mass variability of samples was less than 0.2%.

2.2. Pretreatment

Bamboo particles were pretreated using a microwave accelerated reaction system. 2% of sulfuric acid (w/w of H_2SO_4/bamboo) and 10% sodium hydrate (w/w of NaOH/bamboo) were used in the pretreated process. About 3g bamboo particles were mixed with the solvent in a 100 mL vessel. The mass ratio of liquor-to-bamboo was about 8:1. The vessel with samples was positioned at the centre of a rotating circular ceramic plate in the microwave oven for pre-treatment at the power level of 400 W. The temperature was raised to 180 ℃ (target temperature) in about 10 min and maintained for an additional 30min. After the pretreatment, a few minutes were allowed for the temperature to drop down below 50 ℃, and then pretreated substrate and spent liquor were then separated by filtration. The substrate was washed using distilled water until pH value of washed liquor was about 7. Each experiment was carried out in three times and average results were reported.

2.3. Property test

1. FTIR test

The functional group difference of untreated and pretreated samples was analyzed by means of FTIR-spectrometer (Bruker, Bremen, Germany). The concentration of the sample in the tablets was constant of 1 mg/400 mg KBr. Scans were run at a resolution of $4cm^{-1}$ and each sample consisted of 64 scans recorded in absorbance units from 3800 to $750cm^{-1}$. The spectra were ATR and baseline corrected and the spectra analyzed for carbonyl bands using relative indices. A minimum of two samples were tested, and data from the first run was used when it was shown to be in accordance with the second run. The experimental data could be directly obtained though FTIR-spectrometer, but they were analyzed by using Origin 8.0 software.

2. XRD test

XRD test of untreated and pretreated bamboo samples was carried out using an X-ray diffractometer (Diffraktometer D5000, Siemens, Germany) with an X-ray generator and a Co target (λ=0.1729 nm) at a scanning speed of 3º/min, and the data were recorded every 0.02º (2θ) for the angle range of 2θ=5-45º. The cellulose crystallinity (CrI) was calculated based on formula (1):

$$CrI = \left(I_{002} - I_{am}\right)/I002 \quad (1)$$

Where, I_{002} was the overall intensity of the peak at 2θ about 22º and I_{am} was the intensity of the baseline at 2θ about 18º.

http://dx.doi.org/10.5772/55379

3. Results and discussion

3.1. FTIR analysis

The functional groups of untreated and pretreated bamboo were shown in the FTIR spectra presented in Figure 2. For untreated bamboo, there was a strong broad O-H stretching absorbance at 3350cm^{-1}. The absorbance at 2910 cm^{-1} was a prominent C-H stretching. In the fingerprint region (from 1800 to 800 cm^{-1}), some important information on various functional groups presented in untreated bamboo. The absorbance at 1740 cm^{-1} was attributed to C=O stretching vibration in hemicelluloses. The bands from 1600 to 1450 cm^{-1} was due to aromatic skeletal vibration in lignin. The absorbance at 1370cm^{-1} was C-H bending of cellulose or hemicelluloses, and that of 1230cm^{-1} was C-O stretching of phenolic hydroxyl group in the lignin. The absorbance at 1160cm^{-1} and 1030 cm^{-1} was respectively attributed to C-O-C stretching of cellulose or hemicelluloses and C-O stretching of cellulose, hemicelluloses or lignin [16].

The chemical group difference of bamboo-H_2SO_4 and bamboo was showed in Figure 3. It was very obvious that the number of most chemical groups on the bamboo-H_2SO_4 sample surface was more than that of untreated bamboo surface, except for absorption of C=O stretching vibration. The information is very important, which confirms the removal of hemicelluloses of bamboo. The main feature of difference between hemicelluloses and cellulose is that hemicelluloses has branches with short lateral chains consisting of different sugars. These monosaccharides include pentoses (xylose, rhamnose, and arabinose), hexoses (glucose, mannose, and galactose), and uronic acids (e.g., 4-omethylglucuronic, D-glucuronic, and D-galactouronic acids) [29]. Hemicelluloses is known to coat the cellulose microfibrils in the plant cell wall, forming a physical barrier to access by hydrolytic enzymes. Removal of hemicelluloses from the microfibrils is believed to expose the cellulose surface and to increase the enzymatic hydrolysis of cellulose [17, 18]. Dilute-acid pretreatment is a main method for the selective fractionation of hemicelluloses from biomass. Both cellulose and hemicelluloses components can also be hydrolysed using dilute-acid catalysed processes but in this case a two step-hydrolysis is required. The difference between two steps is mainly the operational temperature, which is high in the second step (generally around 230-240 ℃) [19, 20].

The chemical group difference of bamboo-NaOH and bamboo was showed in Figure 4. It was found that the number of all chemical groups on the bamboo-NaOH sample surface was less than that of bamboo surface. This indicated the removal of hemicelluloses and lignin of bamboo. Alkaline pre-treatments are very effective for lignin solubilisation exhibiting only minor cellulose and slightly higher hemicelluloses solubilisation [21]. Generally, alkaline pre-treatments include wet oxidation and the ammonia. The wet oxidation was used in this research. Wet oxidation is defined as pretreatment process including oxygen and water at elevated temperatures and pressure, promoting the oxidation of lignin and decomposing it to CO_2, H_2O and carboxylic acids [22, 23]. The hemicellulosic sugars remain mainly in the oligomeric form, and although there is a low formation of furan-aldehydes, a significant formation of carboxylic acids still exists [24]. It is well known that lignin confers integrity and structural rigidity on the plant cell wall. There is several information that cellulolytic enzymes are adsorbed non-specifically on the lignin fraction of lignocelluloses even in the absence of a carbohydrate-binding module [25, 26, 27]. Therefore, the increase in the initial hydrolysis rate of cellulose and hemicelluloses should be due in part to the decreasing number of non-specific binding sites on lignin, making more enzyme available for hydrolysis [28].

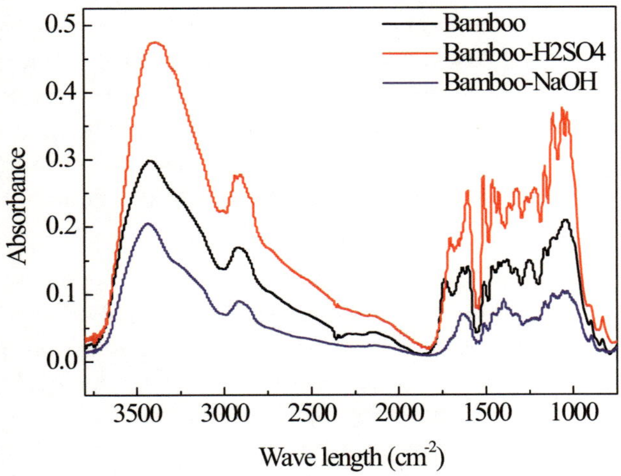

Figure 2. The FTIR spectra of untreated and pretreated bamboo

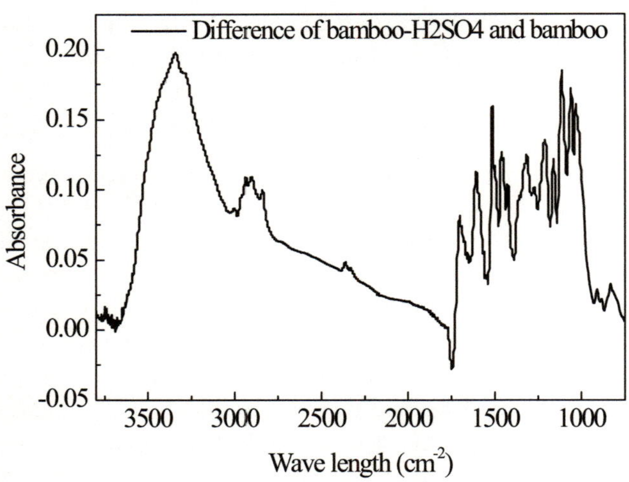

Figure 3. The FTIR spectra difference of untreated and sulfuric acid pretreated bamboo

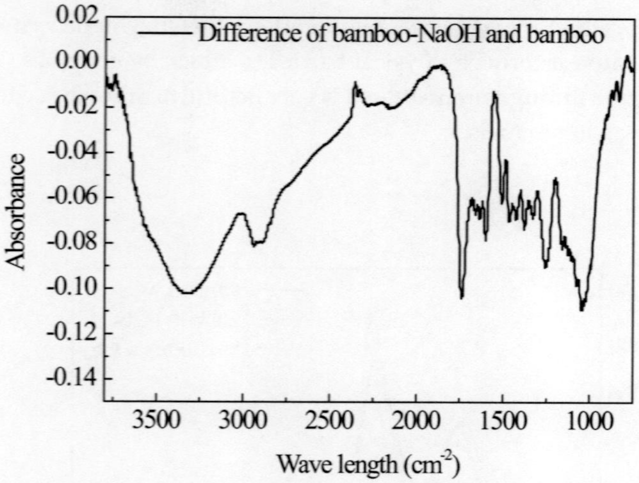

Figure 4. The FTIR spectra difference of untreated and sodium hydrate pretreated bamboo

3.2. XRD analysis

The untreated and pretreated bamboo was determined using an X-ray generator in this study. The XRD results were shown in the Figure 5. The cellulose crystallinity was calculated according to formulas (1). The cellulose crystallinity of bamboo, bamboo-H_2SO_4 and bamboo-NaOH samples were shown in Figure 6. It was obvious that the cellulose crystallinity of pretreated bamboo was greater than that of bamboo. Biomass materials exhibit two types of cellulose crystallinity: absolute and relative crystallinity. The relative crystallinity was used in this research. The relative crystallinity of bamboo, bamboo-H_2SO_4 and bamboo-NaOH samples was 44.4%, 49.8% and 55.2%, respectively. This phenomenon further confirmed that some compositions of bamboo were removed in the pretreatment process. It also expressed that the NaOH pretreatment was better than H_2SO_4 pretreatment. This can be explained based on substrate content of different pretreatment. It was found from Table 2 that the substrate content of NaOH pretreatment (64.1%) was lower than that of H_2SO_4 pretreatment (70.3%).

Bamboo is composed of cellulose, hemicelluloses, lignin and a remaining smaller part comprising extractives and minerals. The cellulose and hemicelluloses typically comprise up to 70% of the bamboo and are the substrates for second generation ethanol production. The microbial conversion of the hemicelluloses fraction, either in the monomeric form or in the oligomeric form, is essential for increasing fuel ethanol yields from bamboo. Unlike cellulose, hemicelluloses are not chemically homogeneous and different hydrolytic technologies and various biological and non-biological pre-treatment options are available both for fractionation or solubilisation of hemicelluloses from lignocellulosic materials [21]. Depending on the process and conditions used during pre-treatment, hemicelluloses sugars may be degraded to weak acids and furan derivatives which potentially act as microbial inhibitors during the fermentation step to ethanol. For fuel ethanol production, hemicelluloses are commonly removed during the initial stage of biomass processing aiming to reduce structural constraints for further enzymatic cellulose hydrolysis. Lignin affects the enzymatic hydrolysis of lignocellulosic biomass because it forms a physical barrier to attack by enzymes. The removal of hemicelluloses and lignin through pretreatment is very helpful to make the cellulose accessible to hydrolysis for conversion to fuels.

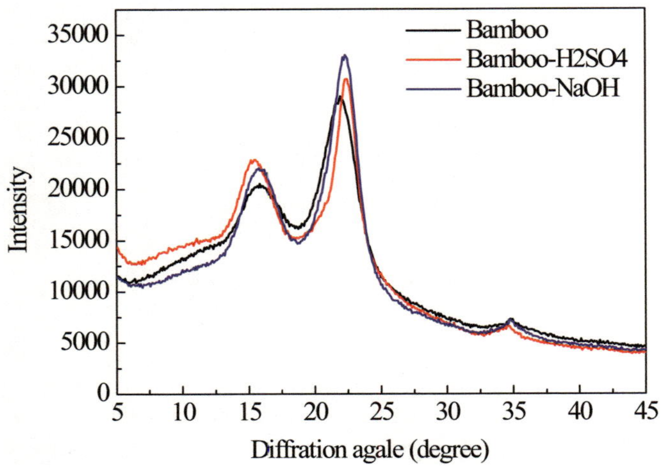

Figure 5. XRD curve of untreated and pretreated bamboo

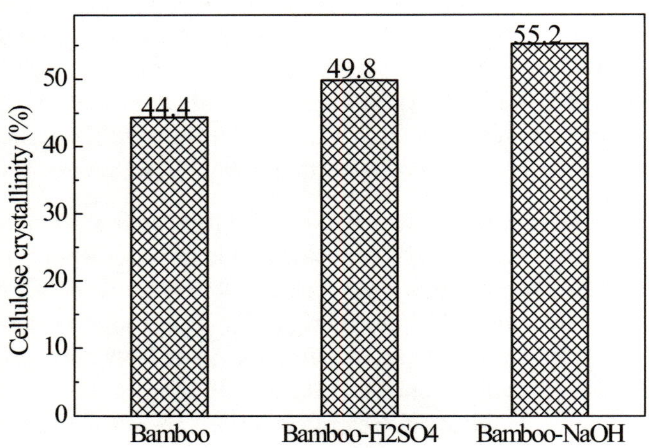

Figure 6. The cellulose crystallinity of untreated and pretreated bamboo

http://dx.doi.org/10.5772/55379

Pretreated method	Bamboo mass (g)		Substrate content (%)
	Untreated	Pretreated	
H_2SO_4	3.015	2.119	70.3
NaOH	3.017	1.934	64.1

Table 2. The substrate content of pretreated bamboo

4. Conclusions

It can be concluded from this research that hemicelluloses and lignin of bamboo are removed by H_2SO_4 and NaOH pretreatment. The number of most chemical groups on the bamboo-H_2SO_4 sample surface is more than that of bamboo surface, except for absorption of C=O stretching vibration, which indicates hemicelluloses removal of bamboo through H_2SO_4 pretreatment. The number of all chemical groups on the bamboo-NaOH sample surface is less than that of bamboo surface, which expresses hemicelluloses and lignin removal of bamboo through NaOH pretreatment. The relative crystallinity of bamboo-H_2SO_4 and bamboo-NaOH samples is 49.8% and 55.2%, respectively. The substrate content of NaOH pretreatment (64.1%) is lower than that of H_2SO_4 pretreatment (70.3%).This phenomenon further confirms that NaOH pretreatment is better than H_2SO_4 pretreatment. This research represents an initial stage in the study of bamboo bioethanol and may provide guidelines for further research, such as substrate hydrolysis, biofuels synthesis, etc.

Acknowledgements

This research was financially supported by '12th Five Years Plan-Study on manufacturing technology of functional bamboo (rattan)-based materials' (Grant No. 2012BAD54G01) and 'Basic Scientific Research Funds of International Centre for Bamboo and Rattan' (Grant No. 1632012002).

Author details

Zhijia Liu and Benhua Fei*

*Address all correspondence to: Feibenhua@icbr.ac.cn

International Centre for Bamboo and Rattan, Beijing, China

References

[1] Sarkar N, Ghosh SK, Bannerjee S, Aikat K. Bioethanol production from agricultural wastes: An overview. Renewable Energy, 2012, 37: 19-27.

[2] Faizal HM, Latiff ZA, Wahid MA, Darus AN. Physical and combustion characteristics of biomass residues from palm oil mills. Heat Transfer Environ., 2010: 34-39.

[3] Filho AP, Badr O. Biomass Resources for Energy in North-Eastern Brazil. Applied Energy, 2004, 77: 51-57.

[4] Zheng Y, Pan ZL, Zhang RH. Overview of biomass pretreatment for cellulosic ethanol production. Int. J. Agric. & Biol. Eng., 2009, 2: 51-68.

[5] Mosier N, Wyman C, Dale B, Elander R, Lee YY, Holtzapple M, Ladisch M R. Features of Promising Technologies for Pretreatment of Lignocellulosic Biomass. Bioresourc. Technol., 2005, 96: 673-686.

[6] Lin Z, Huang H, Zhang H, Zhang L, Yan L, Chen J. Ball milling pretreatment of corn stover for enhancing the efficiency of enzymatic hydrolysis. Appl Biochem Biotechnol. 2010, 162: 1872-1880.

[7] Bjerre AB, Olesen AB, Tomas F, Annette P, Anette SS. Pretreatment of wheat straw using combined wet oxidation and alkaline hydrolysis resulting in convertible cellulose and hemicellulose. Biotechnol. Bioeng. 1996, 49: 568-577.

[8] Zhu JY, Pan XJ. Woody biomass pretreatment for cellulosic ethanol production: Technology and energy consumption evaluation. Bioresourc. Technol., 2010, 101: 4992-5002.

[9] Keshwani DR, Cheng JJ, Burnsz JC, Li LG, Chiangy V. Microwave Pretreatment of Switchgrass to Enhance Enzymatic Hydrolysis. An ASABE Meeting Presentation. Paper Number: 077127.

[10] Yu Q, Zhuang X, Yuan Z, Wang Q, Qi W, Wang W, Zhang Y, Xu J, Xu H. Two-step liquid hot water pretreatment of Eucalyptus grandis to enhance sugar recovery and enzymatic digestibility of cellulose. Bioresour. Technol., 2010, 101:4895-4899.

[11] Cardona CA, Quintero JA, Paz IC. Production of bioethanol from sugarcane bagasse: status and perspectives. Bioresourc. Technol., 2009, 101:4754-4766.

[12] Maarten A, Kootstra J, Beeftink HH, Scott EL, Sanders JPM. Comparison of dilute mineral and organic acid pretreatment for enzymatic hydrolysis of wheat straw. Biochem. Eng. J., 2009, 46: 126-131.

[13] Sun RC, Lawther JM, Banks WB. Influence of alkaline pretreatments on the cell-wall components of wheat-straw. Industrial Crops and Products, 1995, 4:127-145.

[14] Jiang ZH. World bamboo and rattan. Liaoning Science & Technology Press 2002; China. 43 pp.

[15] Mcmullen J, Fasina O, Wood W, Feng YC, Mills G. Physical characteristics of pellets from poultry litter. ASABE Paper No. 046005. St. Joseph, MI: ASABE 2004.

[16] Li J. Spectroscopy of wood. Chinese Science & Technology Press 2003; China. 110 pp.

[17] Yang B, Wyman CE. Effect of xylan and lignin removal by batch and flow through pretreatment on the enzymatic digestibility of corn stover cellulose. Biotechnol. Bioeng., 2004, 86: 88-95.

[18] Ohgren K, Bura R, Saddler J, Zacchi G. Effect of hemicellulose and lignin removal on enzymatic hydrolysis of steam pretreated corn stover. Bioresour. Technol., 2007, 98: 2503-2510.

[19] Lee, YY, Iyer P, Torget RW, In: Tsao, G.T. (Ed.), Dilute-acid Hydrolysis of Lignocellulosic Biomass, vol. 65. Springer-Verlag, Berlin, 1999a, pp. 93-115.

[20] Wyman, CE. Potential synergies and challenges in refining cellulosic biomass to fuels, chemicals, and power. Biotechnol. Prog. 2003, 19: 254-262.

[21] Gírio FM, Fonseca C, Carvalheiro F, Duarte LC, Marques S, Bogel-Łukasik R. Hemicelluloses for fuel ethanol: A review. Bioresourc. Technol., 2010, 101: 4775-4800.

[22] Bjerre, AB, Olesen AB, Fernqvist T, Ploger A, Schmidt AS. Pretreatment of wheat straw using combined wet oxidation and alkaline hydrolysis resulting in convertible cellulose and hemicellulose. Biotechnol. Bioeng., 1996, 49: 568-577.

[23] Klinke HB, Ahring BK, Schmidt AS, Thomsen AB. Characterization of degradation products from alkaline wet oxidation of wheat straw. Bioresour. Technol., 2002, 82: 15-26.

[24] Martin C, Marcet M, Thomsen AB. Comparison between wet oxidation and steam explosion as pretreatment methods for enzymatic hydrolysis of sugarcane bagasse. Bioresources, 2008, 3: 670-683.

[25] Sutcliffe R, Saddler JN. The role of lignin in the adsorption of celluloses during enzymatic treatment of lignocellulose material. Biotechnol. Bioeng. Symp., 1986, 17: 749-762.

[26] Chernoglazov VM, Ermolova OV, Klyosov AA. Adsorption of high-purity endo-1,4-β-glucanases from Trichoderma reesei on components of lignocellulosic materials: cellulose, lignin, and xylan. Enzyme Microb. Technol., 1988, 10: 503-507.

[27] Converse AO, Ooshima H, Burns DS. Kinetics of enzymatic hydrolysis of lignocellulosic materials based on surface area of cellulose accessible to enzyme and enzyme adsorption on lignin and cellulose. Appl. Biochem. Biotechnol., 1990, 24-25: 67-73.

[28] Yoshida M, Liu Y, Uchida S, Kawarada K, Ukagami Y, Ichinose H, Kaneko S, Fukuda K. Effects of Cellulose Crystallinity, Hemicelluloses, and Lignin on the Enzymatic

Hydrolysis of Miscanthus sinensis to Monosaccharides Biosci. Biotechnol. Biochem., 2008, 72: 805-810.

[29] Kumar P, Barrett DM, Delwiche MJ, Stroeve P. Methods for Pretreatment of Lignocellulosic Biomass for Efficient Hydrolysis and Biofuel Production. Ind. Eng. Chem. Res., Article ASAP.DOI: 10.1021/ie801542g.

Chapter 2

Bioconversion of Hemicellulose from Sugarcane Biomass Into Sustainable Products

Larissa Canilha,
Rita de Cássia Lacerda Brambilla Rodrigues,
Felipe Antônio Fernandes Antunes,
Anuj Kumar Chandel,
Thais Suzane dos Santos Milessi,
Maria das Graças Almeida Felipe and
Silvio Silvério da Silva

Additional information is available at the end of the chapter

http://dx.doi.org/10.5772/53832

1. Introduction

Sugarcane is main crop cultivated in countries like Brazil, India, China, etc. It plays a vital role in the economy of these countries in addition to providing employment opportunities [1]. Only in the 2012/13 Brazil harvest, for example, it was estimated that more than 602 million tons of sugarcane will be processed by the sugar-alcohol mills [2].

During the processing of sugarcane, the sugarcane straw (SS) is remained on field and do not presents suitable use. After the juice extraction from sugarcane stem, the fraction that is left over is called sugarcane bagasse (SB) [3]. Both residues (SB and SS) represent a sizeable fraction of agro-residues collected annually. The annual world production of sugarcane is ~1.6 billion tons, which yields approximately 279 million metric tons (MMT) of SB and SS [1, 4].

SB is used as a source of heat and electricity in sugar producing mills while SL is openly burnt on the fields causing environmental pollution. The harnessing of both residues via biotechnological routes into value-added products (xylitol, organic acid, industrial enzymes, ethanol, etc) is much more likely to be complimentary than competitive in the near term without jeopardizing the food requirements [5, 6, 7]. Both residues (SB and SS) are principally constituted of cellulose, hemicellulose and lignin. Among these constituents, hemicellulose is of particular interest because of its unique properties and composition. In the last two decades of research has been witnessed the technological development for the hemicellulose depolymerization into its monomeric constituents, mainly xylose, and their subsequent conversion into value-added products via microbial fermentation [8, 9, 10]. Dilute acid hydrolysis is a well established process for hemicellulose depolymerization, however, inhibitory compounds of microbial metabolism are also formed and should be reduced/eliminated prior to using the liquid in the fermentation process [8, 9]. On the other hand, enzymatic conversion of hemicellulose, that requires cocktail of enzymes for its breakdown, is slow, costly and requires combinatorial mixture of specialized enzymes [9]. The recovered sugar solution after hemicellulose hydrolysis contains primarily pentose sugars and the fermentation of these pentosans is problematic. Only limited numbers of microorganisms that use pentose are known and the fermentation of pentose sugars at industrial scale is not established yet [10, 11]. Generally, pentose utilizing microorganisms have slow growth rate, low osmotolerance and have poor resistance against inhibitors. The microorganisms that use pentose more extensively explored in laboratories are *Candida shehatae, Pichia stipitis, Pachysolen tannophilus* (for ethanol production), *C. utilis, C. intermedia, C. guilliermondii* (for xylitol production) and *Klebsiella oxytoca* ATCC 8724, *Bacillus subtilis, Aeromonas hydrophilia* (for 2, 3-butanediol production) [8, 9, 12].

Rather than summarizing all the literature on hemicellulose bioconversion from sugarcane agro-residues, we aim to highlight in this chapter technological developments focusing hemicellulose hydrolysis, detoxification of hydrolysates and microbial fermentation of sugars into sustainable products.

2. Sugarcane, bagasse and straw

2.1. Structure of sugarcane

The sugarcane basically consists of stem and straw. Stem is the part normally associated with sugarcane (cleaned cane). It is the piece of cane plant between plantation level and end node from last stem. The sugarcane stem are crushed to obtain cane juice, which is subsequently used for sugar (sucrose) or alcohol (ethanol) production. Sugarcane bagasse (SB) is the left over residue from stems after extraction of juice. It is normally burned to supply all the energy required in the process [13]. Sugarcane straw (or trash) (SS) is composed by fresh leaves, dry leaves and tops available before harvesting. Fresh (green and yellow) leaves and tops are the part of cane plant between the top end and the last stalk node. Dry leaves are normally in brownish color [14]. The SS is also normally burnt in the field after the harvest of the crop [15]. Potential applications of the leaves include: 1) as a fuel for direct combustion; 2) as a raw material for conversion by pyrolysis to char, oil and/or gas; and 3) as a raw material for conversion by gasification and synthesis to methanol. Potential applications of the tops include: 1) as a ruminant feed, either fresh or dried; 2) as a substrate for anaerobic fermentation to methane pro-

http://dx.doi.org/10.5772/53832

duction; and 3) after reduction in water content, for the three energy uses listed for cane trash. Figure 1 presents the scanning electronic microscopy (SEM) of SS and SB before pretreatment. In the Figures 1A and 1B the SS was amplified 500 and 10.000x which reveal the presence of some vacuoles in the structure, which is not common in SB (Figure 1C).

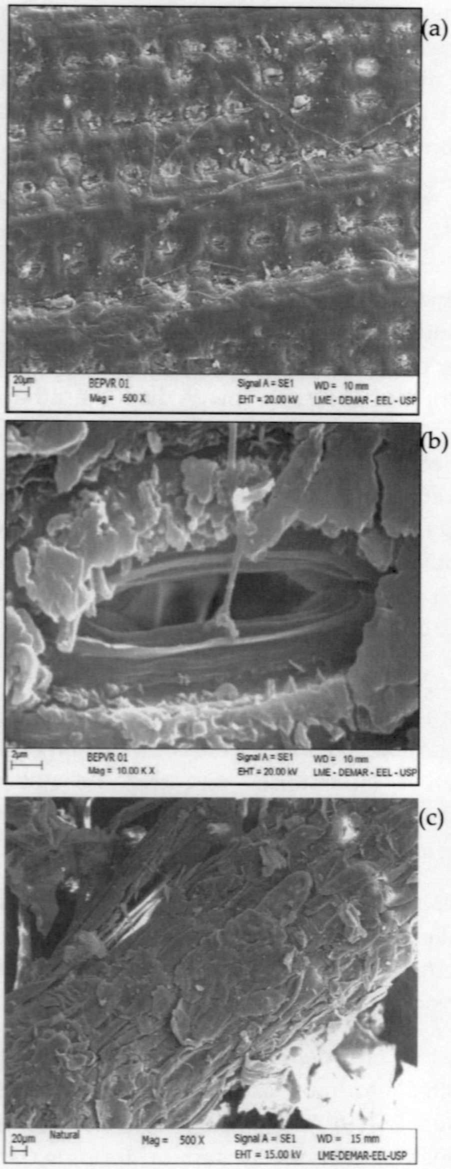

Figure 1. SEM of sugarcane straw (A) 500x and (B) 1000x [16] and sugarcane bagasse (C) 500x (Chandel et al., unpublished work).

2.2. Physical and chemical compositions of sugarcane

Physically, sugarcane is constituted by four fractions, whose relative magnitude depends on the agro industrial process: fiber, non-soluble solids, soluble solids and water. The fiber is composed of the whole organic solid fraction, non-soluble in water, originally found in the cane's stalk, and characterized by its marked heterogeneity from the morphological point of view. The non-soluble solids, or the fraction that cannot be dissolved in water, are constituted mainly by inorganic substances (rocks, soil and extraneous materials) and it is greatly influenced by the conditions of the agricultural cane processing and harvesting type. Soluble solids, fraction that can be dissolved in water, are composed basically of sucrose as well as other small chemical components such as waxes [17].

Bagasse and straw (trash), which are the focus of second generation ethanol production, are lignocellulosic materials chemically composed by cellulose, hemicelluloses and lignin. According to some works in the literature, sugarcane bagasse of the Brazilian territory is quantitatively composed by 38.8-45.5% cellulose, 22.7-27.0% hemicellulose and 19.1-32.4% lignin (Table 1). Non-structural components of biomass namely ashes (1.0-2.8%) and extractives (4.6-9.1%) are the other substances that are part of the chemical compositional of bagasse. The ash content of bagasse is lower than the others crop residues, like rice straw and wheat straw (with approximately 17.5 and 11.0% of this compound, respectively) and the bagasse is considered a rich solar energy reservoir due to its high yields and annual regeneration capacity (about 80 t/ha) in comparison with others agricultural residues, like wheat, grasses and tree (1, 2 and 20 t/ha, respectively) [3]. The bagasse also can be used as a raw material for cultivation of microorganisms for the production of value-added products such as xylitol and ethanol. Due to these and others advantages the bagasse is not only a sub-product of sugar industry, but it is a co-product with high added-value [3].

As can be seen in the Table 1, the chemical compositions of sugarcane bagasse samples varied widely. In fact, it is impossible to compare the composition of samples from different origins, performed by different laboratories and that do not use the same methods. Furthermore, factors like plant genetics, growth environment and processing conditions also influence the compositional analysis [18].

The large variation in the values of chemical components also is observed for the sugarcane straw, that it is composed approximately by 33.3-36.1% cellulose, 18.4-28.9% hemicellulose, 26.1-40.7% lignin (Table 2). Ashes (2.1-11.7%) and extractives (5.3-11.5%) are also present on the sugarcane straw composition.

When mechanically harvested, and depending on the harvesting technology applied, the range of straw that is collected and transported to the mill together with the stalks is 24% to 95% of the total trash available [19]. The amount of trash from sugarcane harvesting depends on several factors such as: harvesting system, topping, height, cane variety, age of crop (stage of cut), climate, soil and others. The average stalks yield per hectare was estimated to 83.23 tons/ha over an average of 5 seasons (cuts), resulting average availability of trash of 11.98 tons/ha (dry basis) [20].

Component (%)	Reference						
	[22]1* Brazil	[23]² Brazil	[24]3* Brazil	[25]4* Brazil	[26]5* India	[27]6 Cube	[28]7* USA
Cellulose	41.1	38.8	45.0	45.5	43.0	43.1	39.6
Hemicellulose	22.7	26.0	25.8	27.0	24.0	31.1	29.7
Lignin	31.4	32.4#	19.1	21.1	20.0	11.4	24.7
Ash	2.4	2.8	1.0	2.2	-	5.5	4.1
Extractives	6.8	-	9.1	4.6	-	-	14.3
Others	-	-	-	-	-	8.5	-

*Extractives-free basis; #Lignin and others

Extracting solvents: [1]dichloromethane, ethanol: toluene (1:2), ethanol, hot water; [2]none; [3]water and ethanol; [4]ethanol; [5]none; [6]none; [7]not described

Table 1. Chemical composition (% w/w, dry basis) of Brazilian and worldwide sugarcane bagasse samples reported in the literature

Component (%)	Reference					
	[16]1* Brazil	[22]2* Brazil	[29]3* Brazil	[23]4 Brazil	[30]5 Brazil	[15]6 India
Cellulose	36.1	34.4	36.1	33.6	33.3	45.0
Hemicellulose	28.3	18.4	26.9	28.9	27.4	25.0
Lignin	26.2	40.7	26.2	31.8#	26.1	18.0
Ash	2.1	11.7	2.1	5.7	2.6	-
Extractives	5.3	11.5	5.3	-	-	-
Others	-	-	-	-	10.6	-

*Extractives-free basis; #Lignin and others

Extracting solvents: [1]ethanol; [2]dichloromethane, ethanol: toluene (1:2), ethanol, hot water; [3]water; [4]none; [5]none; [6]none

Table 2. Chemical composition (% w/w, dry basis) of Brazilian and Indian sugarcane straw samples reported in the literature

From the technological viewpoint, sugars that are present in the cellulosic (glucose) and hemicellulosic (xylose, arabinose, glucose, mannose and galactose) fractions representing the substrates that can be used in fermentative process for production of some sustainable products such as xylitol, butanediol, single cell protein, ethanol and xylitol. However, the close association between the three major fractions (cellulose, hemicellulose and lignin) of the lignocelulosic materials, like bagasse and straw, causes difficulties for the recovery of these substrates in the form of monomers with high purity. Therefore, to use these three constituents it is required a selective separation of each fraction by pretreatment techniques, delignification and hydrolysis, involving the breakdown of hemicellulose-lignin-cellulose complex [21].

3. Methods of separation of hemicellulose from cellulignin complex

The lignocellulosic materials are renewable resources which can be used to obtain sustainable products as well as value-added biomolecules [31]. However, cellulose, hemicellulose and lignin are arranged to form a highly recalcitrant structure [32], hindering the availability of carbohydrates for fermentation processes, representing a high barrier for the bioconversion of lignocellulosic materials [33]. Through a pretreatment process, the biomass components can be separated, releasing fermentable sugars such as xylose, arabinose and glucose and making the cellulose more accessible to the action of cellulolytic enzymes [34, 35]. This step is one of the most expensive step of biomass processing, thus, studies to lower the cost are extremely important [35].

According to Brodeur et al. [36], the typical characteristics that must be attained in a pretreatment process are: production of highly digestible solids that enhances sugar yields during enzyme hydrolysis; avoid the degradation of sugars; minimize the formation of inhibitors; recover the lignin for conversion into valuable co-products. Pretreatment process should be cost effective and environment friendly. All these features are considered in order that pretreatment results balance against their impact cost on downstream processing steps and the trade-off with operational cost, capital cost and biomass cost [37]. The pretreatments methods can be divided into physical, chemical, physic-chemical and biological [38]. Some methods of pretreatments as well as their advantages and disadvantages are shown in the Table 3.

Different types of biomass (woody plants, grasses, agricultural crops, etc) has different contents and proportions of cellulose, hemicellulose and lignin which determine the digestibility of the biomass [37]. There is not a universal pretreatment process for all biomass. Depending on the process and conditions used, hemicellulose sugars may be degraded to weak acids, furan derivates and phenolics that inhibit the fermentation process, leading to lower yields and productivities of the desired product [8]. Thus, the method of pretreatment used will depend on the type of raw material used, the objective of the process (the constituent to be degraded) and the product to be obtained, which will directly affect the cost benefit.

http://dx.doi.org/10.5772/53832

	Process	Advantages	Disadvantages	Solubilized Fraction			References
				Hemicel.	Cel.	Lignin	
			Physical Pretreatments				
Milling	ball milling provides the reduction of particles size and breaks down the structure of lignocellulosic materials	- environment friendly - chemical addition is not required - inhibitors are not produced	- high power - high energy costs	Alteration in structure			[23] [39]
Pyrolysis	treatment with temperatures higher than 300ºC	- fast degradation of cellulose into H_2, CO and residual char	- high temperature - ash production		X		[40]
			Physic-Chemical Pretreatments				
Steam Explosion or hydrothermal	structure compounds breakdown by heat addition in form of steam and forces by the moisture expansion	- needs few or no chemical addition - environment friendly	- inhibitors production	X			[41] [42]
Ammonia Fiber Explosion (AFEX)	expose the lignocellulosic material with ammonia to high temperature and pressure followed by a fast pressure release	- inhibitors are not produced - simple process -short time process	- cost of ammonia - ammonia recovery - depending of lignin content			X	[38] [40]
CO_2 Explosion	formation of carbonic acid and increase the hydrolysis rate of substrates	- more cost effective - inhibitors are not produced	- hard operation method		X	X	[40] [43]
			Chemical Pretreatments				
Acid Pretreatment	dilute-acid hydrolysis of the lignocellulosic material	- low and medium temperatures	- equipment corrosion - inhibitors production	X	X		[10] [24] [35]
Alkaline Pretreatment	delignification process employing bases such as sodium hydroxide, calcium hydroxide (lime), etc	- low temperature - low pressure	- long processing time - environment pollution			X	[40] [44]

	Process	Advantages	Disadvantages	Solubilized Fraction			References
				Hemicel.	Cel.	Lignin	
Ozonolysis	the ozone incorporates conjugated double bonds and functional groups with high electron densities	- removal of lignin - inhibitors are not produced - ambient temperature and pressure	- large amount of ozone is required - expensive process	X		X	[45] [46]
Organosolv	simultaneous process of hydrolyses and delignification catalyzed by solvents and diluted acid solution	- facilitates the enzyme access - needs few chemical addition - low waste generation	- expensive process			X	[47] [48]
Wet Oxidation	occurs in the presence of oxygen or catalyzed air, sodium carbonate is the preferred catalyst	- released sugars without generation of inhibitors	- expensive process - high pressure	X		X	[9]
Biological Pretreatments							
Microorganism	modification of the chemical composition and/or structure of lignocellulosic materials employing microorganisms	- environment friendly	- low efficiency - considerable loss of carbohydrates - long processing time			X	[15] [38] [44]

Hemicel.: hemicellulosic fraction; Cel.: cellulosic fraction; X: solubilized fraction by the pretreatment

Table 3. Advantages and disadvantages of different methods of pretreatment

4. Hemicellulosic fraction

4.1. Structure of hemicellulose

Hemicellulose differs substantially from cellulose to be amorphous, which makes it more easily hydrolyzed than cellulose [49]. The hemicellulosic fraction reaches 40% of lignocellulosic material and acts as substance of reserve and support. This fraction presents branched structure composed by pentoses (D-xylose and L-arabinose), hexoses (D-galactose, D-mannose and D-glucose) and small amounts of acetic and uronic (D-glucuronic, D-4-O-methyl-glucuronic and D-galacturonic acids) acids [8, 21]. Other sugars such as L-rhamnose and L-fucose may also be present in small amounts. Xylose is the main carbohydrate present in the hemicellulosic fraction, representing about 80% of total sugars [35, 50].

The heterogeneous structure of hemicellulose with a low polymerization degree makes it interesting fraction for fermentation process. The open three-dimensional conformation of hemicellulose favors the diffusion of the catalyst in the molecule, providing a better yield of hydrolysis in mild conditions [51, 52].

The hemicellulosic fraction can be removed of lignocellulosic materials by some type of pretreatments, summarized in the Table 3, liberating sugars, mainly xylose, that subsequently can be fermented to sustainable products such as xylitol, butanediol, single cell protein and ethanol [24, 27].

4.2. Methods of detoxification of hemicellulosic hydrolysates

When the lignocellulosic matrix is breakdown by different types of pretreatments, particularly by dilute acid process, undesired compounds that are toxic for microbial metabolism are liberated and/or formed in addition of sugars. These products can be divided into three groups according to their origin: derived from sugars (furfural and 5-hydroxymethylfurfural), lignin derivatives (phenolics i.e. vanillin, *p*-hydroxybenzaldehyde, lignans, etc.) and weak acids (acetic, formic and levulinic) [53]. Several studies have shown that these byproducts generated during the hydrolysis of the hemicellulose fraction from different materials affect negatively the microbial metabolism, hindering the conversion of sugars in some products of interest [53, 54, 55].

Several chemical, physical and biological methods have been used for removing these byproducts present in the hemicellulosic hydrolysates. Some detoxification methods as well as their advantages and disadvantages are summarized in the Table 4.

4.3. Products from hemicellulose

Hemicelluloses have a wide variety of applications. They can be hydrolyzed into hexoses (glucose, galactose, and mannose) and pentoses (xylose and arabinose), can be transformed into fuel ethanol and other value-added products such as 5-hydroxymethylfurfural (HMF), xylitol, ethanol, butanediol, butanol, etc. In addition, hemicelluloses also can be converted into various biopolymers, like polyhydroxyalkanoates (PHA) and polylactates (PLA).

In industrial applications, hemicelluloses are used to control water and the rheology of aqueous phases. Thus, they may be used as food additives, thickeners, emulsifiers, gelling agents, adhesives and adsorbents [71]. According to Peng et al. [72], hemicelluloses have also been investigated for their possible medical uses such as ulcer protective [73], antitussive [74], immunostimulatory [75] and antitumor properties [76]. For example, xylooligosaccharides have been shown to have economic utilization in the pharmaceutical industry for applications such as treating viral and cancer processes in the human body [77, 78].

	Process	Advantages	Disadvantages	References
		Physical Methods		
Evaporation/ Concentration	removes toxic compounds by evaporation in a vacuum concentrator based on the volatility	- reduces volatile compounds as acetic acid, furfural, and vanillin	- increasing the nonvolatile toxic compounds as extractives	[56] [57]
Membrane	membranes have surface functional groups attached to their internal pores, which may eliminate metabolic inhibitors	- avoids the need to disperse one phase and minimize the entrainment of small amounts of organic phase	- high cost - selective removal of inhibitors	[58] [59]
		Physic-Chemical Methods		
Ion Exchange Resin	resins change undesirable ions of the liquid phase to be purified by saturating of functional groups of resins	- can be regenerated and reused - remove lignin-derived inhibitors, acetic acid and furfural - does not cause high sugars loss	- high pressure - long processing time - possible degradation of fragile biological product molecules - difficult to scale-up	[59] [60] [61] [62]
Overlimming	increase of the pH followed by reduction	- precipitate toxic compounds	- high sugars loss - filtration complexity	[63]
Activated Charcoal	adsorption of toxic compounds by charcoal which is activated to increase the contact surface	- low cost - remove phenolics and furans - minimizes loss of sugars	- filtration complexity	[60] [64] [65]
Extraction with Organic Solvents	mix of liquid phase to be purified and a organic solvent. The liquid phase is recovered by separation of two phases (organic and aqueous)	- recycling of solvents for consequent cycles - remove acetic acid, furfural, vanillin, 4-hydroxybenzoic acid and low molecular weight phenolics	- high cost - long processing time	[66] [67]
Vegetable Polymer	biopolymers are composed by tannins with astringent properties that flocculate inhibitors compounds	- low cost - biodegradable - minimizes loss of sugars - reducing toxic compounds	- cell death when the tannin content is high - significant volume loss	[68]
		Biological Methods		
Microorganism	specific enzymes or microorganisms that act on the inhibitors compounds present in hydrolysates and change their composition	- low waste generation - environmental friendly - less energy requirements	- long processing time	[56] [59] [69] [70]

Table 4. Advantages and disadvantages of different detoxification methods of hemicellulosic hydrolysate

Arabinoxylans are used as emulsifiers, thickeners, or stabilizers in the food, cosmetic, or pharmaceutical industries. Glucomannans are used in the food industry (as caviar substituent), whereas arabinogalactans have applications in the mining (for processing of iron and copper ores) or pharmaceutical industry (as a tablet binder or emulsifier). 4-O-methylglucuronoxylan is a water absorption agent and also presents antitumor activity [71].

4.3.1. Xylose and glucose

The D-xylose ($C_5H_{10}O_5$) is the main carbohydrate found in the hemicellulose fraction of sugarcane bagasse and straw. It is used as a sweetener for diabetics [79], as non-cariogenic sweetener [80], to enhance the flavor of food made from beef and poultry [81], to prepare marinades and baked [81] and as substrate in fermentation processes to produce different products, such as penicillin, biodegradable polymers and xylitol [50, 82]. Monomeric xylose from hemicellulose has a selling price of ~$1.2/kg [83]. It is known that in industrial scale, xylose is obtained from lignocellulosic materials rich in xylan. These materials are hydrolyzed in the presence of dilute acids. Then, the hemicellulose hydrolysates are purified, in order to remove the byproducts generated during the hydrolysis of hemicellulose. After the purification steps, xylose is recovered of purified media by crystallization [84].

D-glucose is also found in the hemicellulose fraction of sugarcane bagasse and straw and can be obtained by hydrolysis of cellulosic materials. Some compounds that are obtained from glucose fermentation are alcohols (ethanol, isopropanol, butanol, 2,3-butanediol, glycerol), carboxylic acids (acetic acid, propanoic acid, lactic acid, gluconic acid, malic acid, citric acid) and other products such as acetone, amino acids, antibiotics, enzymes and hormones [85].

4.3.2. 5-Hydroxymethylfurfural and levulinic acid

5-Hydroxymethylfurfural (HMF) ($C_5H_4O_2$), which is derived from the hexoses (6-carbon sugars) present in the hemicellulose, is produced by steam treatment followed by dehydration [85, 86, 87]. HMF is an intermediate in the production of levulinic acid from 6-carbon sugars in the biofinery process. HMF is very useful not only as intermediate for the production of the biofuel, dimethylfuran (DMF) and other molecules, but also for important molecules such as levulinic acid, 2,5-furandicarboxylic acid (FDA), 2,5-diformylfuran (DFF), dihydroxymethylfuran and 5-hydroxy-4-keto-2-pentenoic acid [88]. Glucose is still utilized in industry for the preparation of HMF because of its price lower than fructose [89].

Levulinic acid (4-oxopentanoic acid) ($C_5H_8O_3$) is a valuable platform chemical due to its specific properties. It has two highly reactive functional groups that allow a great number of synthetic transformations. Levulinic acid can react as both a carboxylic acid and a ketone. The carbon atom of the carbonyl group is usually more susceptible to nucleophilic attack than that of the carboxyl group. Due to the spatial relationship of the carboxylic and ketone groups, many of the reactions proceed, with cyclisation, to form heterocyclic type molecules (for example methyltetrahydrofuran). Levulinic acid is readily soluble in water, alcohols, esters, ketones and ethers. The worldwide market has estimated the price of $ 5/kg for pure levulinic acid [86].

4.3.3. *Furfural and formic acid*

Furfural (2-furaldehyde) and its derivatives, furfuryl alcohol, furan resins, and tetrahydrofuran, are produced in many countries from corn cobs, wheat and oat hulls, and many other biomass materials [90]. Furfural, which is derived from the pentoses (five-carbon sugars) present in hemicellulose, is produced by steam treatment followed by dehydration with hydrochloric or sulfuric acid [87, 90]. The market price of furfural was approximately $1/kg compared with prices in 1990 of $1.74/kg for furfural and $1.76/kg for furfuryl alcohol [83, 86]. The most important furfuryl alcohol is used to produce furan resins for foundry sand binders. Tetrahydrofuran is made by the decarbonylation of furfural with zinc-chromium-molybdenum catalyst followed by hydrogenation. It is also made by the dehydration of 1,4-butanediol [87]. Other uses for furfural, such as production of adiponitrile, might be found if furfural prices were reduced by expanded production [90].

Formic acid (methanoic acid) is an important organic chemical which is widely used in industries. Recently, it received renewed attraction to be used as environmentally benign storage and transportation medium for hydrogen, the clean energy in future. Extensive studies have shown that hydrogen and CO_2 could be quickly and efficiently generated by the decomposition of formic acid by hydrothermal reaction or catalyst reaction. Also, some researchers have demonstrated that formic acid has the potential to direct power fuel cells for electricity generation and automobiles [91]. It is used extensively as a decalcifier, as an acidulating agent in textile dyeing and finishing, and in leather tanning. It is also used in the preparation of organic esters and in the manufacture of drugs, dyes, insecticides, and refrigerants. Formic acid can also be converted to calcium magnesium formate which can be used as a road salt. The current market price of formic acid is $0.16/liter [86].

4.3.4. *Xylitol*

Xylitol is a polyol of five carbons ($C_5H_{12}O_5$) easily found in nature in many fruits and plants. Among them, the yellow plum is the vegetable that contains highest level of xylitol [92]. This polyol is an intermediate metabolite in the carbohydrate metabolism in mammals, with an endogenous production followed by assimilation of 5-15g per day in a normal adult. Xylitol is widely used as a sweetener by diabetics due to its slow adsorption and entrance in pathways which is independent of the insulin and does not contribute in rapid change of blood glucose levels [93].

This polyol, with anti-cariogenic properties, is employed in foods, dental applications, medicines and surfactants [94, 95]. In the dental applications, the use of xylitol reduces the salivary flow, reduce gingivitis, stomatitis and lesions to poorly fitted dentures. If used in toothpaste, its action enhances the action of sodium fluoride and chlorhexidine, increasing the concentrations of xylitol 5-P [96]. For human diet, the Food and Drug Administration classifies this product as "GRAS" - "Generally Recognized as Safe" [97].

At large scale, xylitol is produced by chemical reduction of xylose derived mainly of wood hydrolysates. This process consists in steps of acid hydrolysis of the vegetal material, hydrolysate purifications and crystallization of xylitol [92]. However, there are disadvantages in the chemical production such as the use of high temperature and pressure in the process and the purification steps with low efficiency and productivity [52]. In this context, the biotechnological production of xylitol from hemicellulosic hydrolysates is a promising process with great economic interest. This process can add value to the lignocellulosic residues, like sugarcane bagasse and straw, promoting a complete utilization of these materials, using the cellulosic and hemicellulosic fractions to obtain xylitol and others value-added bioproducts [98]. Among the microorganisms that produce xylitol, yeast, particularly the genus *Candida, Pichia, Debaryomyces* are the most employed due to their ability to convert xylose to xylitol, with significant yields [99]. Xylitol can be produced through microbial transformation reactions by yeast from D-xylose, or by both yeast and bacteria from D-glucose [100]. D-xylose can also be directly converted into xylitol by NADPH-dependent xylose reductase [101].

Considering xylose fermentation by yeasts, the main factors that should be controlled are: substrate concentration, cellular concentration, the presence of inhibitors, aeration flow, adaptation of the microorganism to the hydrolysate, temperature and pH [102, 103]. It can be found many works in the literature where these factors were studied extensively using the hemicellulosic hydrolysates obtained from different lignocellulosic materials. From sugarcane bagasse hemicellulosic hydrolysate, for example, Carvalho et al. [61] reported the production of 19.2 g/L of xylitol by *Candida guilliermondii*; Santos et al. [104] achieved 18 g/L of xylitol with a bioconversion yield of 0.44 g/g employing a fluidized bed reactor operated in semi-continuous mode with the same yeast; and recently, Prakash et al. [105] produced xylitol, with a yield and volumetric productivity of 0.69 g/g and 0.28 g/L.h, respectively, using *Debaryomyces hansenii*.

The world xylitol production exceeds 10,000 tons per year and is directed mainly to the food, pharmaceutical and cosmetics [106]. The American xylitol market is estimated at $159 million for 2012 while it expected $400 million to $500 million for global market [107]. From the Figure 2, it can be seen the average annual prices of xylitol from 1995 to 2007. Xylitol price has decreased over last decades until 2007 (Figure 2), however since 2009, the price of xylitol has increased to $4-5/kg [108].

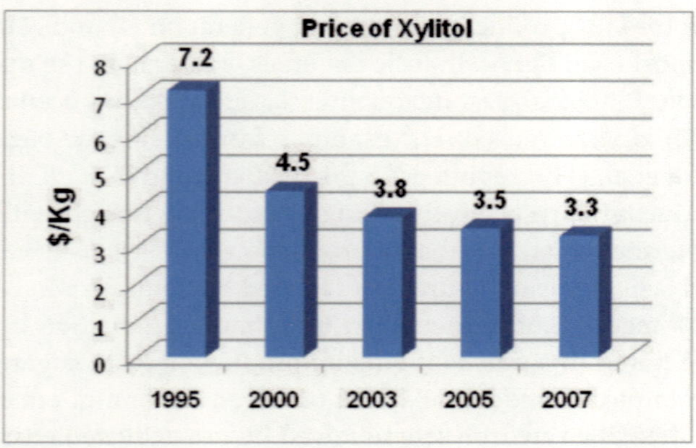

Figure 2. Xylitol price profile from 1995 to 2007 (Source: adapted from reference [109]).

4.3.5. Ethanol

Currently, the world has the prospect of a significant increase in demand for ethanol. The use of this fuel is concentrated on a global scale in power generation, in its mixture with gasoline or simply dehydrated, being a considerable product in the global energy matrix [110].

The first generation ethanol production consists in conversion of hexose sugars to ethanol, and it is relatively simple and usually performed in three steps: acquisition of fermentable sugars, fermentation of sugars by microorganisms and separation and purification of ethanol, usually carried out by distillation, rectification and dehydration [111]. Microorganisms such as *Saccharomyces cerevisiae* consumes directly the sucrose present in sugarcane juice producing ethanol. However, in the long scenario, the use of juice or molasses to produce ethanol will not be able to supply the increasing demand.

Biofuels from renewable sources, such as second generation ethanol production from lignocellulosic materials (bagasse and straw), may represent a sustainable alternative to environmental and social problems caused due to the extensive use of fossil fuels [112]. The process for second generation ethanol requires three steps: pretreatment of lignocellulosic materials, to make the hemicellulose sugars and cellulose more accessible, fermentation of sugars and separation and purification of ethanol [111]. Although it is an eminent perspective, the development of this technology requires some additional challenges. The production of ethanol from lignocellulosic biomass can increase the productivity of ethanol per hectare of sugar cane planted [113], without increasing the cultivated area in the same proportions, not competing with food production for land use [112].

S. cerevisiae is the most common microorganism used for ethanol production from hexose sugars, but it is unable to produce ethanol from pentoses such as xylose. Among the microorganism that can assimilate pentose sugars such as xylose, yeasts have shown more ethanol yield and productivity than bacteria and fungi [114]. There are some naturally yeast which ferments xylose to ethanol, among them, *Pichia stipitis* [116] and *Candida shehatae* [117] are the most employed in bioprocess.

Considering the process for production of second generation ethanol, sugarcane bagasse is reported as one of most used lignocellulosic materials, and among the microorganisms used for xylose conversion, *P. stipitis* yeast (taxonomic classification has been changed to *Scheffersomyces stipitis* [118]) is widely used. For example, from sugarcane bagasse hemicellulosic hydrolysate, Canilha et al. [119] reported 7.5 g/L, 0.30 g/g and 0.16 g/L.h of ethanol production, yield and productivity, respectively, using hydrolysate treated with ion exchange resins as a medium of fermentation for ethanol production by *P. stipitis* DSM 3651 while Hande et al. [120] obtained 0.45 g/g using hydrolysate treated by neutralization and activated charcoal adsorption as a medium of fermentation for ethanol production by *Pichia* strain BY2. Other yeasts can be found in studies for ethanol production from sugarcane hemicellulosic hydrolysate. For example, Chandel et al. [121] observed maximum ethanol yield (0.48 g/g) from ion exchange detoxified hydrolysate followed by use activated charcoal, by *C. shehatae* NCIM 3501 and Cheng et al. [122] obtained 19 g/L ethanol, yield of 0.34 g/g and productivity of 0.57 g/L.h when used a batch culture with pretreated hydrolysate as substrate for *Pachyso-*

http://dx.doi.org/10.5772/53832

len tannophilus DW06. For sugarcane straw, the ethanol production is only from cellulosic fraction. For example, Krishnan et al. [4] verified an ethanol production about 34–36 g/L using the recombinant *S. cerevisiae* (424A LNH-ST) from the bagasse and straw pretreated by ammonia fiber expansion method (AFEX). Sindhu et al. [123] observed the ethanol production of 11.365 g/L using *S. cerevisae* yeast from leaves pretreated with dilute acid hydrolysis followed by enzymatic saccharification with cellulases.

Regarding the world ethanol scenario, a regular increase in the production has been observed (Figure 3). The Americas are the largest producer continent of ethanol. The United States of America is the largest producer country of ethanol with production levels over 51 billion liters (13.5 U.S. gallons) in 2011 [124].

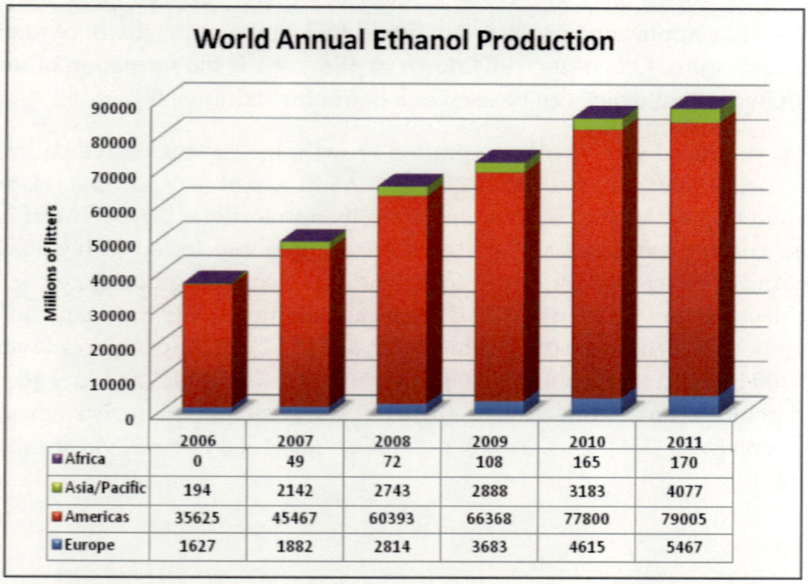

Figure 3. World Annual Ethanol Production since 2006 (Source:[125]).

4.3.6. Butanol

Biobutanol, a four carbon primary alcohol (butyl alcohol-$C_4H_{10}O$), is second generation alcoholic fuel with a higher energy density and lower volatility as compared to ethanol in addition to its existing applications as a solvent [126, 127]. The primary use of butanol is as an industrial solvent in the manufacturing of products such as lacquers and enamels. Butanol can be used directly in any gasoline engine without modification and/or substitution, because it has several similar characteristics to gasoline, besides being compatible with ethanol blending may improve the blending of gasoline with ethanol [128]. It can be produced through processing of domestically grown crops, such as corn and sugar beets, and other biomass residues [128].

Production of butanol by using fermentation to replace the chemical process depends largely on the availability of inexpensive and abundant raw materials and efficient bioconversion of these materials. The producers strains of biobutanol which have been extensively studied are *Clostridium sp.* [126, 127] and genetically engineered *E. coli* [129, 130]. Studies to determined the recovery of biobutanol from fermentation broth (dry corn and wet corn milling) whey permeate and molasses) by distillation showed that it was not economical when compared with butanol derived from the current petrochemical route [131]. The use of lignocellulosic substrates in combination with developed process technologies is expected to make the production of biobutanol economically viable [132].

4.3.7. Butanediol

2,3-Butanediol (2,3-BDL), also known as 2,3-butylene glycol, is a valuable chemical feedstock because of its application as a solvent, liquid fuel, and as a precursor of many synthetic polymers and resins. One of its well known applications is the formation of methyl ethyl ketone, by dehydration, which can be used as a liquid fuel additive [8].

Butanediol is produced during oxygen-limited growth, by a fermentative pathway known as the mixed acid-butanediol pathway [133]. The 2,3-BDL pathway and the relative proportions of acetoin and butanediol serve to maintain the intracellular NAD/NADH balance under changed culture conditions. All of the sugars commonly found in hemicellulose and cellulose hydrolysates can be converted to butanediol, including glucose, xylose, arabinose, mannose, galactose, and cellobiose. The theoretical maximum yield of butanediol from sugar is 0.50 kg per kg. With a heating value of 27,200 J/g, 2,3-BDL compares favorably with ethanol (29,100 J/g) and methanol (22,100 J/g) for use as a liquid fuel and fuel additive [134]. Hexose and pentose can be converted to 2,3-BDL by several microorganisms including *Klebsiella* [135], *Aeromonas* [136], *Bacillus* [137], *Paenibacillus* [138], *Serratia, Aerobacter* [139] and *Enterobacter* [140].

4.3.8. Biopolymers

The use of plastics is consistently increasing in the society due to its advantages such as low cost and durability, and the replacement of conventional materials such as paper and glass [141]. However, these materials have xenobiotic and recalcitrant nature, having an extremely long degradation rate [142, 143]. Besides its slow degradation, the accumulation of plastic is a major risk to marine animals. When is landfilled, is more difficult to occurs the process of decomposition and when it is incinerated, causes the release of several toxic compounds [144]. Due to the increasing demand of plastics and its incorrect disposal, these materials have become a major environmental problem. An alternative to trying to solve this rising problem is the replacement of conventional plastics for biodegradable plastics. Biodegradable plastics are natural biopolymers that are synthesized and catabolized by microorganisms and are made from renewable resources and do not lead to the depletion of finite resources [145, 146]. Among bioplastics, polyhydroxyalkanoates (PHA) and polylactates (PLA) got significant attraction. The PHA's are typically accumulated by bacterial via intra or extracellular while PLA is produced by polymerizing lactic acid via microbial fermentation [147].

For replacement of the conventional plastic, biodegradable plastic is a feasible option. However, it is necessary that the price of biopolymers should be competitive [148]. The cost of production is directly linked to the type of microorganism and the substrate employed. The strain used should have a high specific growth rate, using low cost substrates and a high conversion factor of substrate in PHA [144]. The selection of the proper raw material for biopolymer production has an additional impact on the ecological pressure of the entire process [148]. Renewable sources of polymeric materials offer an answer for sustainable development of economically and ecologically attractive technology [149, 150].

Sugarcane biomass represents an enormous reserve of renewable carbon source, which has the potential to be utilized as a feedstock for the production of biodegradable polymers. For instance, using sugarcane bagasse hydrolysate, Yu and Stahl [151] investigated the simultaneous detoxification and PHA production by the bacterium *Ralstonia eutropha* and accumulated PHA at a rate of 57 wt% of cell mass despite the large index of inhibitors. Silva et al. [152] studied the biopolymer production by *Burkholderia cepacia* IPT 048 and *B. sacchari* IPT 101 from sugarcane bagasse hydrolysate and obtained polymer contents and yields reached, respectively, 62% and 0.39 g/g with strain IPT 101 and 53% and 0.29 g/g with strain IPT 048.

4.3.9. Single cell protein

To create a balance between food versus fuel production from lignocellulosic residues, adequate land use, judicious usage of grain and corn/cane crop residues is essential [153]. Mathews et al. [153] presented a sugarcane 'feed+fuel' biorefinery model, which produces bioethanol and yeast biomass, a source of single-cell protein (SCP), that can be used as a high-protein animal feed supplement. The yeast SCP, which is synthesized as a part of the process of producing cellulosic bioethanol from sugarcane can be used as a supplement for grass in the feed of cattle grazing on pasture and thereby potentially release land for increased sugarcane production, with minimal land use change effects.

The production of SCP by growing microorganisms on organic wastes and its use in animal feed has a long history. Protein as an animal feed supplement has long been viewed as a potentially very significant development, with much discussion devoted to the topic of microbial SCP since the 1970s. The grounds for the intense interest in SCP is that feedstocks, in the form of agricultural and organic wastes are plentiful, and the rate of growth of microorganisms producing SCP is prodigious. Whereas a soybean crop is harvested after 1 season of growth, microorganisms double their cell mass within hours [154]. According to Tanaka et al. [155] the production of single-cell protein (SCP) from lignocellulosic materials needs four steps: (1) physical and chemical pretreatments; (2) cellulase production; (3) enzymatic hydrolysis; and (4) assimilation or fermentation of holocellulose. For each step the following topics need to be considerate: (1) effect and mode of action of each pretreatment; (2) optimization of culture media and operating conditions, and application of mutation, protoplast fusion and gene recombination; (3) elucidation of kinetics of cellulase reaction, and methods of immobilization, stabilization and recovery of cellulases; and (4) examples of SCP production by several types of cultivation and treatment of lignin. According to Zadrazil et al. [156] in order to convert a lignocellulosic material to obtain a more nutritive product, it is necessary to choose a microorganism or a microbial complex capable of synthesizing proteins with high nutritional value and, in the case of use of a substrate that has not been subjected to a previous hydrolysis step, able to degrade selectively the lignin present in the substrate. The lignocellulosic wastes can be fermented directly or they can be previously hydrolyzed chemically. Several reports have shown the application of substrates chemically pre-hydrolyzed for rapid protein enrichment by microbial fermentation. For example, Pessoa et al [157] hydrolyzed sugarcane bagasse using diluted sulphuric acid and the hydrolysate was fermented with *Candida tropicalis*. This process resulted in a 31.3% increase in protein content after 5 days of fermentation. However, for non-ruminant animals, which are not able to metabolize the natural fibers that comprise the bulk of lignocellulosic wastes, the bioconversion process must aim to transform these fibers into digestible components such as protein and sugars (mono- and disaccharides) as well as vitamins and minerals.

5. Conclusion and future recommendations

Sugarcane bagasse (SB) and straw (SS) constitute a sizeable fraction of agro-residues in many countries. Brazil is the largest producer of sugarcane residues in the world. Hemicellulose, in both raw materials, is an important fraction and could be a sustainable alternative for the production of second generation ethanol, industrial enzymes, food/feed and fine chemicals such as lactic acid, succinic acid, etc. It can be easily converted into simple sugars by thermochemical processes and the resultant sugar solution after conditioning and detoxification, can be converted into the aforementioned products by biotechnological routes. Alterations in thermochemical processes such as implication of counter-current, plug-flow, percolation and shrinking-bed reactors could be helpful to maximize the sugars recovery with minimum inhibitors generation. There are several promising detoxification strategies available which remove the inhibitors from hydrolysates. The detoxified sugar solution can be converted into valuable products including second generation ethanol by appropriate microorganisms under batteries of fermentation. Laboratories based research progress has clearly showed that it is quite possible to convert hemicellulose into commercially significant products with desired yields and productivities. However, it is necessary to build a robust process to be employed at industrial scale. Bio-products derived from hemicellulose of SB/SS have shown potential to replace chemically synthesize products. Owing to this, bio-industrial companies offer numerous opportunities to develop unique functionality and marketing benefits from the products derived from hemicellulose of SB/SS creating long term sustainability and green environment.

Acknowledgements

The authors acknowledge the funding sources Fapesp, Capes and CNPq. Editors would like to thank EEL/USP for providing necessary facilities and basic infrastructure. We are grateful

http://dx.doi.org/10.5772/53832

to BIOEN-FAPESP, CNPq and CAPES, Brazil for the financial assistance to our laboratory to carry out the research work on various aspects of lignocellulose biotechnology.

Author details

Larissa Canilha, Rita de Cássia Lacerda Brambilla Rodrigues,
Felipe Antônio Fernandes Antunes, Anuj Kumar Chandel*, Thais Suzane dos Santos Milessi,
Maria das Graças Almeida Felipe and Silvio Silvério da Silva*

*Address all correspondence to: silvio@debiq.eel.usp.br and anuj.kumar.chandel@gmail.com

Department of Biotechnology, School of Engineering of Lorena, University of São Paulo, Lorena, Brazil

References

[1] Chandel AK, Silva SS, Carvalho W, Singh OV. Sugarcane Bagasse and Leaves: Foreseeable Biomass of Biofuel and Bio-products. Journal of Chemical Technology and Biotechnology 2012; 87 11–20.

[2] Companhia Nacional de Abastecimento. CONAB: Acompanhamento da Safra Brasileira de Cana-de-açúcar. Primeiro Levantamento-Abril/12. http://www.conab.gov.br. (accessed 28 May 2012).

[3] Pandey A, Soccol CR, Nigam P, Soccol VT. Biotechnological Potential of Agro-industrial Residues I: Sugarcane Bagasse. Bioresource Technology 2000; 74 69-80.

[4] Krishnan C, Sousa LC, Jin M, Chang L, Dale BE, Balan V. Alkali Based AFEX Pretreatment for the Conversion of Sugarcane Bagasse and Cane Leaf Residues to Ethanol. Biotechnology and Bioengineering 2010; 107 441–450.

[5] Soccol CR, Vandenberghe LPS, Medeiros ABP, Karp SG, Buckeridge MS, Ramos LP, Pitarelo AP, Ferreira-Leitão V, Gottschalk LMF, Ferrara MA, Bon EPS, Moraes LMP, Araujo JA, Torres FAG. Bioethanol from Lignocelluloses: Status and Perspectives in Brazil. Bioresource Technology 2010; 101 4820–4825.

[6] Dias MOS et al. Production of Bioethanol and Other Bio-based Materials from Sugarcane Bagasse: Integration to Conventional Bioethanol Production Process. Chemical Engineering Research and Design 2009; 87 1206–1216.

[7] Ojeda K, Avila O, Suarez J, Kafaro V. Evaluation of Technological Alternatives for Process Integration of Sugarcane Bagasse for Sustainable Biofuels Production – Part 1. Chemical Engineering Research and Design 2010; 89 270–279.

[8] Saha BC. Hemicellulosic Bioconversion. Journal of Industrial Microbial Biotechnology 2003; 30 279-291.

[9] Carvalheiro F, Duarte LC, Gírio FM Hemicellulose Biorefineries: A Review on Biomass Pretreatments. Journal of Scientific & Industrial Research 2008; 67 849-864.

[10] Girio FM, Fonseca C, Carvalheiro F, Duarte LC, Marques S, Bogel-Lukasik R. Hemicelluloses for Fuel Ethanol: A Review. Bioresource Technology 2010; 101 4775-4800.

[11] Kuhad RC, Gupta R, Khasa YP, Singh A, Zhang Y.-H. P Bioethanol Production from Pentose Sugars: Current Status and Future Prospects. Renewable and Sustainable Energy Review 2011; 15 4950-4962.

[12] Chandel AK, Singh OV. Weedy Lignocellulosic Feedstock and Microbial Metabolic Engineering: Advancing of Biofuel. Applied Microbial and Biotechnology, 2011; 89 1289-1303, 2011.

[13] Ensinas A, Modesto M, Nebra SA, Serra L. Reduction of irreversibility generation in sugar and ethanol production from sugarcane. Energy 2009; 34 680–688. DOI:10.1016/j.energy.2008.06.001

[14] Neto MAT. Characterization of sugarcane trash and bagasse. In: Hassuani SJ, Leal MLRV, Macedo IC. Biomass Power Generation. Sugarcane Bagasse and Trash. Piracicaba: PNUD and CTC; 2005. p24.

[15] Singh P, Suman A, Tiwari P, Arya N, Gaur A, Shrivastava AK. Biological pretreatment of sugarcane trash for its conversion to fermentable sugars. World Journal of Microbiology and Biotechnology 2008; 24 667-673.

[16] Moriya RY. Use of microbial xylanases and laccases in the bleaching of organosolv pulps from sugarcane straw and study of the cellulosic derivatives obtained (Uso de xilanases e lacases de microrganismos no branqueamento de polpas organosolv de palha de cana-de-açúcar e estudo dos derivados celulósicos obtidos). PhD thesis. University of São Paulo, Engineering School of Lorena; 2007 [in Portuguese].

[17] Triana O, Leonard M, Saavedra F, Acan IC, Garcia OL, Abril A. Atlas of Sugarcane Bagasse. México: Geplacea and ICIDCA; 1990. p.139.

[18] Hatfield R, Fukushima RS. Can lignin be accurately measured? Crop Science 2005; 45 832-839.

[19] Paes LAD, Hassuani SJ. Potential trash and biomass of the sugarcane plantation, including trash recovery factors. In: Hassuani SJ, Leal MLRV, Macedo IC. Biomass Power Generation. Sugarcane Bagasse and Trash. Piracicaba: PNUD and CTC; 2005. p70.

[20] Filho JPR Characterization of sugarcane trash and bagasse. In: Hassuani SJ, Leal MLRV, Macedo IC. Biomass Power Generation. Sugarcane Bagasse and Trash. Piracicaba: PNUD and CTC; 2005. p78.

[21] Fengel D, Wegener G. Wood Chemistry, Ultrastructure, Reactions. Berlin: Walter de Gruyter; 1989, 613p.

[22] Pitarelo, A.P. Evaluation of susceptibility of sugarcane bagasse and straw on the bioconversion by steam-explosion and enzymatic hydrolysis (Avaliação da susceptibilidade do bagaço e da palha de cana-de-açúcar à bioconversão via pré-tratamento a vapor e hidrólise enzimática). Master thesis. Federal University of Paraná; 2007 [in Portuguese].

[23] Silva AS, Inoue H, Endo T, Yano S, Bon EPS. Milling pretreatment of sugarcane bagasse and straw for enzymatic hydrolysis and ethanol fermentation. Bioresource Technology 2010; 101 7402-7409.

[24] Canilha L, Santos VTO, Rocha GJM, Almeida e Silva JB, Giulietti M, Silva SS, Felipe MGA, Ferraz AL, Milagres AMF, Carvalho W. A study on the pretreatment of a sugarcane bagasse sample with dilute sulfuric acid. Journal of Industrial Microbiology and Biotechnology 2011; 38 1467-1475.

[25] Rocha GJM, Martin C, Soares IB, Souto Maior AM, Baudel HM, Abreu CAM. Dilute mixed-acid pretreatment of sugarcane bagasse for ethanol production. Biomass and Bioenergy 2011; 35 663-670.

[26] Singh A, Sharma P, Saran AK, Singh N, Bishnoi NR. Comparative study on ethanol production from pretreated sugarcane bagasse using immobilized Saccharomyces cerevisiae on various matrices. Renewable Energy 2012; 50 488-493.

[27] Martin C, Alriksson B, SJose A, Nilvebrant NO, Jonsson LJ. Dilute sulfuric acid pretreatment of agricultural and agro-industrial residues for ethanol production. Applied Biochemistry and Biotechnology 2007; 137-140 339-352. DOI 10.1007/s12010-007-9063-1

[28] Teixeira LC., Linden JC, Schroeder HA. Simultaneous saccharification and cofermentation of peracetic acid-pretreated biomass. Applied Biochemistry and Biotechnology 2000; 84-86 111-127. DOI 10.1385/ABAB:84-86:1-9:111

[29] Saad MBW, Oliveira LRM, Cândido RG, Quintana G, Rocha GJM, Gonçalves AR. Preliminary studies on fungal treatment of sugarcane straw for organosolv pulping. Enzyme and Microbial Technology 2008; 43 220-225.

[30] Luz SM, Gonçalves AR, Del'Arco Jr AP, Leão AL, Ferrão PMC, Rocha GJM. Thermal properties of polypropylene composites reinforced with different vegetable fibers. Advanced Materials Research 2010; 123-125 1199-1202.

[31] Rodrigues RCLB, Rocha GJM, Rodrigues Jr. DR, Filho HJI, Felipe MGA, Pessoa Jr. AP. Scale-up of diluted sulfuric acid hydrolysis for producing sugarcane bagasse hemicellulosic hydrolysate (SBHH). Bioresource Technology 2010; 101 1247-1253.

[32] Champagne P. Bioethanol from agricultural waste residues. Environmental Progress 2008; 27(1) 51-57

[33] Doherty WOS, Mousavioun P, Fellows CM. Value-adding to cellulosic ethanol: Lignin polymers. Industrial Crops and Products 2011; 33 259-276.

[34] Lavarack BP, Griffin GJ, Rodman D. The acid hydrolysis of sugarcane bagasse hemicellulose to produce xylose, arabinose, glucose and other products. Biomass and Bioenergy 2002; 23 367-380.

[35] Mosier N, Wyman C, Dale B, Elander R, Lee YY, Ladisch M. Features of promising technologies for treatment of lignocellulosic biomass. Bioresource Technology 2005; 96 673-686.

[36] Brodeur G, Yau E, Badal K, Collier J, Ramachandran KB, Ramakrishnan S. Chemical and physicochemical pretreatment of lignocellulosic biomass: A Review. Enzyme Research 2011;17p.

[37] Agbor VB, Cicek N, Sparling R, Berlin A, Levin D B. Biomass pretreatment: Fundamentals toward application. Biotechnology Advances 2011; 29 675-685.

[38] Sarkar N, Ghosh SK, Bannerjee S, Aikat K. Bioethanol production from agricultural wastes: An overview. Renewable Energy 2012; 37 19-27.

[39] Inoue H, Yano S, Endo T, Sakaki T, Sawayama S. Combining hot compressed water and ball milling pretreatments to improve the efficiency of the enzymatic hydrolysis of eucalyptus. Biotechnololy Biofuels 2008; 15 12p.

[40] Kumar P, Barret DM, Delwiche MJ, Stroeve P. Methods for pretreatment of lignocellulosic biomass for efficient hydrolysis and biofuel production. Industrial & Engineering Chemistry Research 2009. DOI: 10.1021/ie801542g

[41] Chornet E, Overend RP. Phenomenological kinetics and reaction engineering aspects of steam/aqueous treatments. In: ProcInt Workshop on Steam Explosion Techniques: Fundamentals and Industrial Applications. 1988. p.21-58.

[42] Kaar WE, Gutierrez CV, Kinoshita CM. Steam explosion of sugarcane bagasse as apretreatment for conversion to ethanol. Biomass and Bioenergy 1998; 14(3) 277- 287.

[43] Sun Y, Cheng J. Hydrolysis of lignocellulosic materials for ethanol production: A review. Bioresource Technology 2002; 83 1-11.

[44] Zheng Y, Pan Z, Zhang R. Overview of biomass pretreatment for cellulosic production. International Journal of Agricultural & Biological Engineering 2009; 2(3) 51-68.

[45] García-Cubero MT, Gonzalez-Benito G, Indacoechea I, Coca M, Bolado S. Effect of ozonolysis pretreatment on enzymatic digestibility of wheat and rye straw. Bioresource Technology 2009; 100 1608-1612.

[46] Vidal PF, Molinier J. Ozonolyzis of Lignin – Improvement of in vitro digestibility of poplar sawdust. Biomass 1988; 16 1-17.

[47] Vega A, Bao M, Lamas J. Application of factorial design to the modeling of organosolv delignification of Miscanthussinensis (Elephant grass) with phenol and dilute acid solutions. Bioresouresource Technology 1997; 61(1) 1-7.

[48] Taherzadeh MJ, Karimi K. Pretreatment of lignocellulosic wastes to improve ethanol and biogas production: A review. International Journal of Molecular Sciences 2008; 9(9) 1621-1651.

[49] Singh A, Mishra P. Microbial Pentose Utilization. Current Applications in Biotechnology Progress in Industrial Microbiology; 1995 33.

[50] Aguilar R, Ramírez JA, Garrote G, Vásquez M. Kinetic study of the acid hydrolysis of sugarcane bagasse. Journal of Food Engineering 2002; 55 304-318.

[51] Magee RJ, Kosaric N. Bioconversion of Hemicelluloses. Advances in Biomechanical Engineering and Biotechnology 1985; 32 60-93.

[52] Winkelhausen E, Kuzmanova S. Microbial Conversion of D-xylose to Xylitol. Journal of Fermentation and Bioengineering 1998; 86(1) 1-14.

[53] Palmqvist E, Hahn-Hagerdal B. Fermentation of Lignocellulosic Hydrolysates. I: Inhibition and Detoxication Bioresource Technology 2000; 74 17-24.

[54] Rodrigues RCLB, Felipe MGA, Almeida e Silva JB, Vitolo M, Gómez PV. The influence of pH, temperature and hydrolysate concentration on the removal of volatile and nonvolatile compounds from sugarcane bagasse hemicellulosic hydrolysate treated with activated charcoal before or after vacuum evaporation. Brazilian Journal of Chemical Engineering 2001; 18(3) 299-311.

[55] Lima LHA, Felipe MGA, Vitolo M, Torres FAG. Effect of acetic acid present in bagasse hydrolysate on the activities of xylose reductase and xylitol dehydrogenase in Candida guilliermondii. Applied Microbiology and Biotechnology 2004; 65 734-738.

[56] Anish R, Rao M. Bioethanol from Lignocellulosic Biomass Part III Hydrolysis and Fermentation. Handbook of Plant-Based Biofuels, 2009; 159-173.

[57] Mussatto SI, Roberto IC. Alternatives for detoxification of diluted-acid lignocellulosic hydrolyzates for use in fermentative processes: A review. Bioresource Technology 2004; 93 1-10.

[58] Grzenia DL, Schell DJ, Wickramasighe SR. Membrane extraction for detoxification of biomass hydrolysates. Bioresource Technology 2012; 111 248-254.

[59] Chandel AK, Silva SS, Singh OV. Detoxification of Lignocellulosic Hydrolysates for Improved Bioethanol Production, Biofuel Production-Recent Developments and Prospects. InTech; 2011. ISBN: 978-953-307-478-8

[60] Canilha L, Almeida e Silva JB, Solenzal AIN. Eucalyptus hydrolysate detoxification with active charcoal adsorption or ion-exchange resins for xylitol production. Process Biochemistry 2004; 39 1909-1912.

[61] Carvalho W, Batista MA, Canilha L, Santos JC, Converti A, Silva SS. Sugarcane bagasse hydrolysis with phosphoric and sulfuric acids and hydrolysate detoxification for xylitol production. Journal of Chemical Technology and Biotechnology 2004; 79 1308–1312.

[62] Nilvebrant N, Reimann A, Larsson S, Jnsson LJ. Detoxification of Lignocelluloses Hydrolysates with ion-Exchange Resins. Applied Biochemistry and Biotechnology 2001; 91-93 35-49.

[63] Palmqvist E, Hahn-Hägerdal B. Fermentation of Lignocellulosic Hydrolysates II: Inhibitors and Mechanisms of Inhibition. Bioresource Technology 2000; 74 25-33.

[64] Mussatto SI, Roberto IC. Hydrolysate detoxification with activated charcoal for xylitol production by Candida guilliermondii. Biotechnology Letters 2001; 23 1681–1684.

[65] Canilha L, Carvalho W, Giulietti M, Felipe MGA, Almeida e Silva JB. Clarification of wheat straw-derived medium with ion-exchange resins for xylitol crystallization. Journal of Chemical Technology and Biotechnology 2008; 83 715-721.

[66] Wilson JJ, Deschatelets L, Nishikawa NK. Comparative fermentability of enzymatic and acid hydrolysates of steam-pretreated aspen wood hemicellulose by Pichia stipitis CBS 5776. Applied Microbiology and Biotechnology 1989; 31(5-6) 592-596.

[67] Cantarella M, Cantarella L, Gallifuoco A, Spera A, Alfani F. Comparison of different detoxification methods for steam-exploded poplar wood as a substrate for the bioproduction of ethanol in SHF and SSF. Process Biochemistry 2004; 39 1533–1542.

[68] Chaud LCS, Silva DDV, Felipe MGA. Evaluation of fermentative performance of Candida guilliermondii in sugarcane bagasse hemicellulosic hydrolysate detoxified with activated charcoal or vegetal polymer. In: Mendez-Villas A. Microbes in Applied Research: Current Advances and Challenges. World Scientific Publishing Co. Pte. Ltd.; 2012. ISBN: 978-981-4405-03-4

[69] Hou-Rui Z, Xiang-Xiang Q, Silva SS, Sarrouh BF, Ai-Hua C, Yu-Heng Z, Ke J, Qiu X. Novel Isolates for Biological Detoxification of Lignocellulosic Hydrolysate. Applied Biochemistry and Biotechnology 2009; 152 199-212.

[70] Yang B, Wyman CE. Pretreatment: the key to unlocking low-cost cellulosic ethanol. Biofuels, Bioproducts and Biorefining 2008; 2 26-40.

[71] Spiriodon I, Popa VI. Hemicelluloses: Structure and Properties. In: Dimitriu S. Polysaccharides: Structural Diversity and Functional Versatility. New York: Marcel Dekker; 2005. p1204.

[72] Peng F, Peng P, Xu F, Sun RC. Fractional purification and bioconversion of hemicelluloses. Biotechnology Advances 2012; 30 879-903.

[73] Cipriani TR, Mellinger CG, De Souza LM, Baggio CH, Freitas CS, Marques MC et al. A polysaccharide from a tea (infusion) of Maytenus ilicifolia leaves with anti-ulcer protective effects. Journal of Natural Products 2006; 69 1018–1021.

[74] Kardosova A, Malovikova A, Patoprsty V, Nosalova G, Matakova T. Structural characterization and antitussive activity of a glucuronoxylan from Mahonia aquifolium (Pursh) Carbohydrate Polymer 2002; 47 27–33.

[75] Kulicke WM, Lettau AI, Thielking H. Correlation between immunological activity, molar mass, and molecular structure of different (1→3)-β-D-glucans. Carbohydrate Research 1997; 297 135–143.

[76] Kitamura S, Hori T, Kurita K, Takeo K, Hara C, Itoh W et al. An antitumor, branched (1→3)-β-D-glucan from a water extract of fruiting bodies of Cryptoporus volvatus. Carbohydrate Research 1994; 263 111–121.

[77] Stone AL, Melton DJ, Lewis MS. Structure–function relations of heparin-mimetic sulfated xylan oligosaccharides: Inhibition of human immunodeficiency virus-1 infectivity in vitro. Glycoconjugate Journal 1998; 15 697–712.

[78] Watson K, Gooderham NJ, Davies DS, Edwards RJ. Interaction of the transactivating protein HIV-1 tat with sulphated polysaccharides. Biochemical Pharmacology 1999; 57 775–783.

[79] Bisaria VS, Ghose TK. Biodegradation of cellulosic materials: substrates, microorganisms, enzyme and products. Enzyme and Microbial Technology 1981: 3 90-104.

[80] Emodi A. Xylitol: Its Properties and Food Applications. Food Technology 1978; 28-32.

[81] Tovani Benzaquen, Segmentos, Cárneos e Derivados. http://www.tovani.com.br/seg05.htm (accessed 08 August 2012).

[82] Munday JC. News about glyconutrition, general nutrition, and related health issues. Xylose - An essential nutrient. Sweet Nutrition News 2003; (4).

[83] Zhang YHP, Ding SY, Mielenz JR et al. Fractionating recalcitrant lignocellulose at modest reaction conditions. Biotechnology and Bioengineering 2007: 97 214-223.

[84] Hyvönen L, Koivistoinen P, Voirol F. Food Technological Evaluation of Xylitol. Advances in Food Research 1982; 28 373-403.

[85] Knill CJ, Kennedy JF. Cellulosic Biomass-Derived Products. In: Dimitriu S. Polysaccharides: Structural Diversity and Functional Versatility. New York: Marcel Dekker; 2005. p1204.

[86] Kamm B, Gruber PR, Kamm M. The Biofine Process - Production of Levulinic Acid, Furfural, and Formic Acid from Lignocellulosic Feedstocks. In: Hayes DJ, Fitzpatrick S, Hayes MHB, Ross JRH. (ed.) Biorefineries- Industrial Processes and Products: Status Quo and Future Directions. DOI: 10.1002/9783527619849.ch7

[87] Wittcoff HA, Reuben BG, Plotkin JS. Industrial Organic Chemicals. New Jersey: John Wiley & Sons, Inc.; 2004. p662.

[88] Rosatella AA, Simeonov SP, Frade RFM, Afonso CAM. 5-hydroxymethylfurfural (HMF) as a building block platform: Biological properties, synthesis and synthetic applications. Green Chemistry 2011; 13 754-793.

[89] Lewkowski J. Synthesis, chemistry and applications of 5-hydroxymethylfurfural and its derivatives. Arquive for Organic Chemistry (Arkivoc) 2001; 17-54.

[90] Tokay BA. Biomass Chemicals. Weinheim: Wiley-VCH Verlag GmbH & Co. KGaA; 2005. p1-7. DOI 10.1002/14356007.a04 099

[91] Yun J, Jin F, Kishita A, Tohji K, Enomoto H. Formica acid production from carbohydrates biomass by hydrothermal reaction. Journal of Physics: Conference series 2010; 215 1-4.

[92] Aminoff C, Vanninen E, Doty TE. The occurrence, manufacture and properties of xylitol. In: Counsell JN (ed) Xylitol. London: Applied Science Publishers; 1978.

[93] BAR A. Xylitol, In: Nabors L0, Gelardi RC (ed.) Alternative sweetener. Nova York: Marcel Dekker, p349-379; 1991.

[94] Zarif L, Greiner J, Pace S, Riess JG. Synthesis of perfluoroalkylated xylitol ethers and esters: New surfactants for biomedical uses. Journal of Medicinal Chemistry 1990; 33(4) 1262-9.

[95] Castillo E, Pezzotti F, Navarro A, Lopez-Múnguia A. Lipase-catalyzed synthesis of xylitol monoesters: solvent engineering approach. Journal of Biotechnology 2003; 102 251-259.

[96] Makinen KK. Latest dental studies on xylitol and mechanism of action of xylitol in caries limitation. In: Greenby TH (ed.) Progress in sweeteners, London: Elsevier Applied Science; 1992, p 331-362.

[97] Aguiar CL, Oetterer M, Menezes TJB. Caracterização e aplicações do xilitol na indústria alimentícia. Bol. SBCTA 1999; 33(2) 184-93.

[98] Michel ACS. Produção biotecnológica de xilitol e etanol a partir de hidrolisado de casca de soja. Master dissertation. Universidade Federal do Rio Grande do Sul; 2007.

[99] Cruz JM, Dominguez JM, Domingues H, Parajo JC. Xylitol production from barley bran hydrolysates by continuous fermentation with Debaromyces hansenii. Biotechnology Letters 2000; 22 1895–1898.

[100] Izumori K, Tuzaki K. Production of xylitol from D-xylulose by Mycobacterius smegmatis. Journal of Fermentation Technology 1988; 66 33–36.

[101] Saha BC, Bothast RJ. Microbial production of xylitol. In: Saha BC., Woodward J. (eds.) Fuels and Chemicals from Biomass. Washington DC: American Chemical Society; 1997, p307–09.

[102] Felipe MGA. et al. Environmental parameters affecting xylitol production from sugarcane bagasse hemicellulosic hydrolysate by Candida guilliermondii. Journal of Industrial Microbiology and Biotechnology 1997; 18 251-254.

[103] Guindea R, Csutak O, Stoica I, Tanase AM, Vassu T. Production of xylitol by yeasts. Romanian Biotechnological Letters 2010; 15(3).

[104] Santos JC. Processo fermentativo de obtenção de xilitol a partir de hidrolisado de bagaço de cana-de-açúcar em reator de leito fluidizado: avaliação das condições operacionais. PhD thesis. Faculdade de Engenharia Química de Lorena; 2005

[105] Prakash G, Varma AJ, Prabhune A, Shouche Y, Rao M., Microbial production of xylitol from D-xylose and sugarcane bagasse hemicellulose using newly isolated thermotolerant yeast Debaryomyces hansenii. Bioresource Technology 2011; 102 3304–3308.

[106] Pereira AFF, Silva TC, Caldana ML, Machado MAAM, Buzalaf MAR. Revisão de Literatura: Utilização do xilitol para a prevenção de otite média aguda. International Archives of Otorhinolaryngology 2009; 3(1) 87-92.

[107] Kimberley T, Farquharson B. Sweet deal: Nova Green to produce healthy sugar substitute xylitol. Media Advisory: Nova Green Inc. http://www.novagreen.ca/inthenews/08_05_12_NovaGreen_Xylitol.pdf, 2012 (accessed June 2012).

[108] Prakasham RS, Sreenivasrao R, Hobbs PJ. Current trends in biotechnological production of xylitol and future prospects. Current Trends in Biotechnology and Pharmacy 2009; 3(1) 8-36.

[109] Rao RS, Jyothi P, Rao V. Biotechnological production of xylitol from hemicelulose materials. In: Kuhad RC, Singh A (ed) Lignocellulose biotechnology, Future Prospects; 2007.

[110] Cortez DV. Permeabilização de Células de Candida guilliermondii empregando Processos Químicos e Físicos e seu potencial uso como biocatalisadores na Síntese de Xilitol. PhD thesis. Escolha de Engenharia de Lorena EEL-USP São Paulo; 2010.

[111] Mussato SI, Dragone G, Guimarães PMR, Silva JPA, Cameiro LM, Roberto IC, Vicente A, Domingues L, Teixeira JA. Technological trends, global market and challenges of bio-ethanol production. Biotechnology Advances 2010; 28 817-830.

[112] Pacheco TF. Produção de Etanol: Primeira ou Segunda Geração?. Embrapa Agroenergia. ISSN 2177-4420, 2011. http://www.embrapa.br/imprensa/artigos/2011/producao-de-etanol-primeira-ou-segunda-geracao/ (acceessed 10 August 2012).

[113] Cerqueira Leite RC, Leal MRLV, Cortez LAB, Griffin WM, Scandiffio MIG. Can Brazil replace 5% of the 2025 gasoline world demand with ethanol?. Energy 2009; 34 655–661.

[114] Olsson L, Hanh-Hägerdal B. Fermentation of lignocellulosic hydrolysates.Enzymes and Microbiology Technology 1996; 18 312-331.

[115] Jeffries TW. Engineering yeasts for xylose metabolism. Current Opinion in Biotechnology 2006; 17 320-326.

[116] Agbogbo FK, Coward-Kelly G. Cellulosic ethanol production using the naturally occurring xylose-fermenting yeast, Pichia stipitis. Biotechnology letters 2008; 30 1515-1524.

[117] Delgenes JP, Moletta R, Navarro JM. Effects of lignocellulose degradation products on ethanol fermentations of glucose and xylose by Saccharomyces cerevisiae, Zymomonas mobilis, Pichia stipitis, and Candida shehatae. Enzyme and Microbial Technology 1996; 19 220–225.

[118] Ha SJ, Galazkac JM, Kima SR, Choia JH, Yang X, Seoe JHN, Glassf L, Catec JHD, Jina YS. Engineered Saccharomyces cerevisiae capable of simultaneous cellobiose and xylose fermentation. Proceedings of the National Academy of Sciences (PNAS) 2011; 108 504–509.

[119] Canilha L, Carvalho W, Felipe MGA, Silva JBA, Giulietti M. Ethanol production from sugarcane bagasse hydrolysate using Pichia stipitis. Applied Biochemistry & Biotechnology 2010; 161 84–92.

[120] Hande A, Mahajan S, Prabhune A. Evaluation of ethanol production by a new isolate of yeast during fermentation in synthetic medium and sugarcane bagasse hemicellulosic hydrolysate. Annals of Microbiology 2012; DOI 10.1007/s13213-012-0445-4. Available from http://link.springer.com/article/10.1007/s13213-012-0445-4?null (accessed 10 August 2012).

[121] Chandel AK, Kapoor RK, Singh A, Kuhad RC. Detoxification of sugarcane bagasse hydrolysate improves ethanol production by Candida shehatae NCIM 3501. Bioresource Technology 2007; 98 1947-1950.

[122] Cheng KK, Cai BY, Zhang JA, Ling HZ, Zhou YJ, Ge JP, Xu JM. Sugarcane bagasse hemicellulose hydrolysate for ethanol production by acid recovery process, Biochemical Engineering Journal 2008; 38 105–109.

[123] Sindhu R, Kuttiraja M, Binod P, Janu KU, Sukumaran RK, Pandey A. Dilute acid pretreatment and enzymatic saccharification of sugarcane tops for bioethanol production. Bioresource Technology 2011; 102 10915-10921.

[124] Global Renewable Fuels Alliance. Global ethanol production to reach 88.7 billion liters in 2011. Available from http://www.globalrfa.org/pr_021111.php (accessed 21 June 2012).

[125] Licht's FO, F.O. Licht's World Ethanol & Biofuels Report. Available from http://www.agra-net.com/portal2/showservice.jsp?servicename=as072 (accessed 21 June 2012).

[126] Dürre, P. Fermentative butanol production. Annals of the New York Academy of Sciences 2008; 1125 353–362.

[127] Lee, S.Y., Park, J.H., Jang, S.H., Nielsen, L.K., Kim, J. and Jung, K.S. Fermentative Butanol Production by clostridia. Biotechnology and Bioengineering 2008; 101 209–228.

[128] Menon V, Prakash G, Rao M. Value Added Products from Hemicelluloses: Biotechnological Perspective. Global Journal Biochemistry 2010; 1 36–67.

[129] Atsumi S, Cann AF, Connor MR, Shen CR, Smith KM, Brynildsen MP, et al. Metabolic Engineering of Escherichia coli for 1-Butanol Production. Metabolic Engineering 2008; 10 305–311.

[130] Atsumi S, Hanai T, Liao JC. Non-Fermentative Pathways for Synthesis of Branched-Chain Higher Alcohols as Biofuels. Nature 2008; 451 86–89.

[131] Ezeji TC, Qureshi N, Karcher P, Blaschek HP. Butanol Production from Corn. In: S.D. Minter (ed.) Alcoholic fuels: fuels for today and tomorrow. New York: Taylor & Francis; 2006 p.99–122.

[132] Gapes JR. The Economics of Acetone-Butanol Fermentation: Theoretical and Market Considerations Journal of Molecular Microbiology and Biotechnology 2000; 2 27–32.

[133] Kosaric N, Magee RJ, Blaszczyk R. Redox Potential Measurement for Monitoring Glucose and Xylose Conversion by K. pneumonia. Chemical and Biochemical Engineering Quarterly 1992; 6 145–152.

[134] Tran AV, Chambers RP. The Dehydration of Fermentative 2,3-Butanediol into Methyl Ethyl Ketone. Biotechnology Bioengineering 1987; 29 343–351.

[135] Ji XJ, Nie ZK, Huang H, Ren LJ, Peng C, Ouyang PK. Elimination of Carbon Catabolite Repression in Klebsiella oxytoca for Efficient 2,3-Butanediol Production from Glucose-Xylose Mixtures. Applied Microbiology and Biotechnology 2011; 89 1119–1125.

[136] Willetts A. Butane 2,3-diol Production by Aeromonas hydrophila Grown on Starch. Biotechnology Letters 1984; 6 263–268.

[137] Groleau D, Laube VM, Martin SM. The Effect of Various Atmospheric Conditions on the 2,3-Butanediol Fermentation from Glucose by Bacillus polymyxa. Biotechnology Letters 1985; 7 53–58.

[138] Nakashimada Y, Marwoto B, Kashiwamura T, Kakizono T, Nishio N. Enhanced 2,3-Butanediol Production by Addition of Acetic Acid in Paenibacillus polymyxa. Journal of Bioscience and Bioengineering 2000; 90 661–664.

[139] Zhang L, Yang Y, Sun J, Shen Y, Wei D, Zhu J, Chu J. Microbial Production of 2,3-Butanediol by a Mutagenized Strain of Serratiamarcescens H30. Bioresource Technology 2010; 101(6) 1961-1967.

[140] Perego P, Converti A, Borghi AD, Canepa P. 2,3-Butanediol Production by Enterobacteraerogenes: Selection of the optimal Conditions and Application to Food Industry Residues. Bioprocess Engineering 2000; 23 613-620.

[141] Derraik JGB. The pollution of the marine environment by plastic debris: A review. Marine Pollution Bulletin 2002; 44 842-852.

[142] Reddy CSK, Ghai R, Rashmi, Kalia VC. Polyhydroxyalkanoates: An overview. Bioresource Technology 2003; 87 137-146.

[143] Thakor N, Trivedi U, Patel KC. Microbiological and biotechnological aspects of biodegradable plastics: Poly(hydroxyalkanoates). Indian Journal of Biotechnology 2006; 5 137-147.

[144] Castillo E, Pezzotti F, Navarro A, Lopez-Múnguia A. Lipase-catalyzed synthesis of xylitol monoesters: solvent engineering approach. Journal of Biotechnology 2003; 102 251-259.

[145] Ojumu TV, Yu J, Solomon BO. Production of Polyhydroxyalkanoates, a bacterial biodegradable polymer. African Journal of Biotechnology 2004; 3(1) 18-24.

[146] Suriyamongkol P, Weselake R, Narine S, Moloney M, Shah S. Biotechnological approaches for the production of polyhydroxyalkanoates in microorganisms and plants – A review. Biotechnology Advances 2007; 25 148-175.

[147] Lopes MSG, Rocha RCS, Zanotto SP, Gomez JGC, Silva LF. Screening of bacteria to produce polyhydroxyalkanoates from xylose. World Journal of Microbiology & Biotechnology 2009; 25 1751-1756.

[148] Koller M, Gasser I, Schmid F, Berg G. Linking ecology with economy: Insights into polyhydroxyalkanoate-producing microorganisms. Engineering in Life Sciences 2011; 11(3) 222-237.

[149] Mohanty AK, Misra M, Drzal LT, Selke SE, Harte BR, Hinrichsen G. Natural Fibers, Biopolymers, and Biocomposites: An Introduction. Natural Fibers, Biopolymers and Biocomposites, ISBN: 978-0849317415, Boca Raton: Taylor & Francis; 2005.

[150] Keenan TM, Nakas JP, Tanenbaum SW. Polyhydroxyalkanoate copolymers from forest biomass. Journal of Industrial Microbiology & Biotechnology 2006; 33 616-626.

[151] Yu J, Stahl H. Microbial utilization and biopolyester synthesis of bagasse hydrolysates. Bioresource Technology 2008; 99 8042-8048.

[152] Silva LF, Taciro MK, Ramos MEM, Carter JM, Pradella, JGC, Gomez JGC. Poly-3-hydroxybutyrate (P3HB) production by bacteria from xylose, glucose and sugarcane bagasse hydrolysate. Journal of Industrial Microbiology & Biotechnology 2004; 31 245-254.

[153] Mathews AJ, Tan H, Moore MJB, Bell G. A Conceptual Lignocellulosic 'Feed+Fuel' Biorefinery and its Application to the Linked Biofuel and Cattle Raising Industries in Brazil. Energy Policy. 2011; 39(9) 4932-4938.

[154] Kuhad RC, Singh A, Tripathi KK, Saxena RK, Eriksson KEL. Microorganisms as an Alternative Source of Protein. Topics in Food Science and Nutrition. 1997; 55 65–75.

http://dx.doi.org/10.5772/53832

[155] Lin Y, Tanaka S. Ethanol Fermentation from Biomass Resources: Current State and Prospects. Applied Microbiology Biotechnology 2006; 69 627-642.

[156] Zadrazil, F., Reiniger, P. Treatment of Lignocellulosics with White-rot Fungi. United Kingston: Elsevier; 1988, p117.

[157] Pessoa Jr A, Mancilha IM, Sato S. Cultivation of Candida tropicalis in sugarcane hemicellulosic hydrolysate for microbial protein production. Journal of Biotechnology 1996; 51(1) 83-88.

Chapter 3

Sugarcane and Woody Biomass Pretreatments for Ethanol Production

Ayla Sant'Ana da Silva,
Ricardo Sposina Sobral Teixeira,
Rondinele de Oliveira Moutta,
Viridiana Santana Ferreira-Leitão,
Rodrigo da Rocha Olivieri de Barros,
Maria Antonieta Ferrara and Elba Pinto da Silva Bon

Additional information is available at the end of the chapter

http://dx.doi.org/10.5772/53378

1. Introduction

Lignocellulosic biomass, which is chiefly composed of cellulose, hemicellulose, and lignin, has now been recognized and established as a source of energy, fuels, and chemicals. Amongst the several options for the use of lignocellulosic materials for energy generation, the production of ethanol has attracted particular attention worldwide, becoming the target of intense research and development (R&D) over the past 40 years. However, the technology required for industrial conversion of these materials into ethanol, in an economic manner, has not yet been fully developed. As an example, the necessary biomass pretreatment step has been under thorough investigation, as the production of sugar syrups with high concentrations and yields via enzymatic hydrolysis of biomass requires the pretreatment to be efficient to render the material accessible to the relevant enzyme pool.

A good choice for biomass pretreatment should be made by considering the following parameters: high yields for multiple crops, harvesting times, highly digestible pretreated solid, high biomass concentration, no significant sugar degradation, formation of a minimum level of toxic compounds, high yields of sugars after subsequent hydrolysis, fermentation compatibility, operation in reasonably sized and moderately priced reactors, lignin recovery, and minimum heat and power requirements [1].

Different pretreatment methods produce different effects on the biomass in terms of its structure and composition [2] (Figure 1). For example, the hydrothermal and acidic pretreatments conceptually remove mainly the biomass hemicellulose fraction and alkaline pretreatments remove lignin, whereas the product of a milling-based pretreatment retains its initial biomass composition. As such, the choice of pretreatment as well as its operational conditions determines the composition of the resulting biomass hexose and pentose syrups. Furthermore, cellulose crystallinity is not significantly reduced by pretreatments based on steam, or hydrothermal, or acidic procedures, whereas ionic liquid-based techniques can shift crystalline cellulose into amorphous cellulose, substantially increasing the enzymatic hydrolysis rates and yields. The activity profile of the enzyme blend and the enzyme load for an effective saccharification may also vary according to the pretreatment. Indeed, a low hemicellulase load can be used for a xylan-free biomass and a lower cellulase load will be needed for the hydrolysis of a low crystalline and highly amorphous pretreated biomass material.

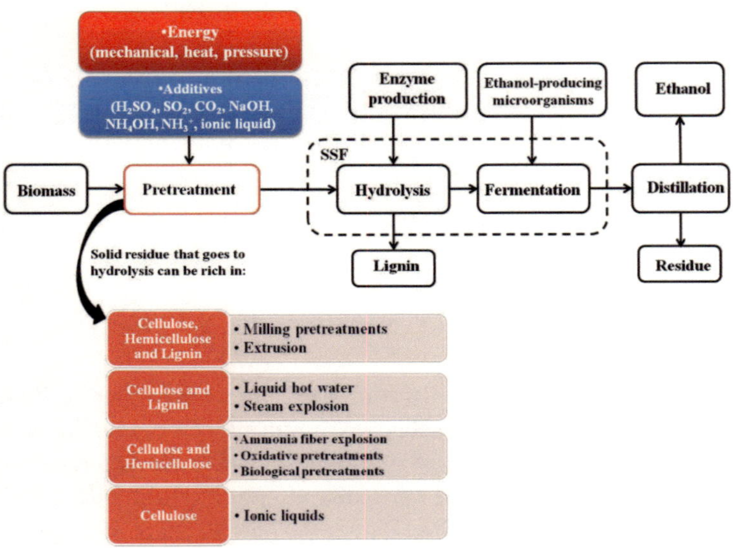

Figure 1. Flow diagram for biomass ethanol production showing different pretreatments options and the composition of the solid pretreated material. SSF: simultaneous saccharification and fermentation

As the pretreatment choice will also be affected by the type of biomass, the envisaged biorefinery model will need to consider the main types of biomass that will be used for the biorefinery operation so as to select an appropriate, and versatile pretreatment method [3]. To date, sugarcane and woody biomass, depending on the geographic location, are strong candidates as the main renewable resources to be fed into a biorefinery. However, due to major differences regarding their physical properties and chemical composition, the relevant pretreatments to be used in each case are expected to be selective and customized. Moreover, a necessary conditioning step for wood size reduction, prior to the pretreatment, may not be necessary for sugarcane bagasse, affecting the pretreatment energy consumption and costs.

Sugarcane is one of the major agricultural crops when considering ethanol production, especially in tropical countries. In Brazil, sugarcane occupies 8.4 million hectares, which corresponds to 2.4% of farmable lands in Brazil. The gross revenue of this sector is about US$ 20 billion (54% as ethanol, 44% as sugar, and 2% as bioelectricity) [4]. In addition, up to 50% of all vehicles in Brazil are flex fuel cars, which corresponds to approximately 15 million cars [5]. Given the above, Brazil is an important player in this scenario, and, consequently, sugarcane bagasse and straw are promising feed stocks for biomass ethanol. Brazil produced, in 2008, 415 million tons of sugar cane residues, 195 million tons of sugarcane bagasse, and 220 million tons of sugarcane straw, whereas the forecasted for the 2012 sugarcane production is 590 million tons, which would correspond to 178 million tons of bagasse, and 200 million tons of straw [6]. Although current R&D has been focused mainly on agricultural residues such as sugarcane residual biomass, woody biomass (hardwoods and softwoods) remains a very important feedstock for cellulosic ethanol production. It is estimated that 370 million dry tons of woody biomass can be sustainably produced annually in the United States. Woody biomass is also sustainably available in large quantities in various other regions of the world such as Scandinavia, New Zealand, Canada, Japan, and South America. Furthermore, short-rotation intensive culture or tree farming offers an almost unlimited opportunity for woody biomass production [7].

This chapter will address an overview of the pretreatments that have been studied for sugarcane and woody biomass aiming at ethanol production using diluted acid, hydrothermal processes, steam explosion, milling, extrusion, and ionic liquids. Advantages and disadvantages of each method will be presented and discussed. The chapter will also discuss the international scenario regarding the existing research and technological choices for the production of biomass ethanol.

2. Diluted acid pretreatment

The use of mineral acids for biomass processing has a historical record dating back to 1819, when concentrated acid was used for wood saccharification aiming at ethanol production [8]. Nevertheless, different technologies using mineral acids have been developed over the last two centuries for converting plant biomass into monosaccharides [9, 10]. The use of acid for biomass pretreatment is conducted with diluted sulfuric or hydrochloric acid (1 to 5%) at 150 °C and pressures up to 10 atm [11]. The efficiency of hemicellulose removal in acid pretreatments is approximately 90%, with sugar losses by degradation at around 1% [12].

The diluted acid pretreatment allows for the deconstruction of the lignocellulosic material structure and the release of sugar monomers, mostly derived from the hemicellulose. If acid pretreatment is carried out under mild conditions of acid concentration and temperature, the hemicellulose fraction can be extracted without significantly affecting the cellulose and lignin biomass content. Unlike cellulose, the hemicellulose is amorphous and branched, being more accessible to hydrolysis agents. This structure allows for the diffusion of acids, which accelerate the hydrolytic process. Therefore, in diluted acid pretreatment, the hemicellulose is preferably removed and hydrolyzed.

The process conditions are crucial in preventing undesirable reactions, which could promote a decrease in monosaccharide yields by the formation of sugar-derived toxic compounds. Temperatures lower than 150 °C reduce sugar degradation, but can result in the decrease of sugar extraction, while temperatures above 160 °C favor the unwanted hydrolysis of the cellulosic fraction, and the formation of toxic compounds [13, 14].

The mechanism of the acid hydrolysis reaction of lignocellulosic materials is described by the following steps (Figure 2) [15]:

1. The diffusion of protons through the wet lignocellulosic matrix;
2. The protonation of the ether–oxygen link between the sugar monomers;
3. The breakage of the ether bond and the generation of a carbocation as an intermediate;
4. The solvation of the carbocation with water;
5. The regeneration of protons and the cogeneration of sugar monomers, oligomers, or polymers, depending on the ether connection that is broken;
6. The distribution of products in the liquid phase (if permitted by their shape and size); and
7. The restart of the process from step 2.

With respect to the material to be treated, some intrinsic characteristics also have an influence during the pretreatment such as the sample phase, the structure and physical accessibility (in the case of heterogeneous hydrolysis), conformation effects, and, finally, the structure and substituents of the sugar ring [16].

The theoretical and fundamental relations among molecular structure, molecular conformation, and the inter-unities bonds of polysaccharides have been evaluated for numerous model experiments. The hydrolytic behavior of glycosidic bonds is also substantially influenced by the conformation of the sugar unit and the inductive effect in these molecules caused by certain substituents in the chain. The half-chair conformation occurring intermittently during the hydrolytic attack is caused by a small rotation of the substituents around the links between carbon atoms 2 and 3, and between carbon atoms 4 and 5, respectively. Generally, the hydrolysis is supported if the axial substituents change to an equatorial position. As the rate of hydrolysis increases with the number of axial groups, the β-anomers are hydrolyzed faster than the corresponding α-forms, with the exception of L-arabinose [16].

Other effects of conformation can accelerate hydrolysis; for example, reducing end bounds are easily hydrolyzed when compared to non-reducing end bounds in polysaccharide chains. C5 substituent's can also hinder hydrolysis reactions [16].

Furanosidic ring structures are hydrolyzed faster than the pyranosidic rings due to the difference in structural angular tension between furanosidic and pyranosidic rings. For example, in woods, α-D-galactofuranosides are hydrolyzed approximately 100 times faster than α-D-galactopyranosides [16].

The inductive effect describes the fact that different substituents on the ring promote changes in the electron density of the ring oxygen. Electrophilic substituents such as carbonyl and carboxyl groups reduce the protonation and inhibit the C–O fission, thus having a stabilizing effect on the glycosidic bond.

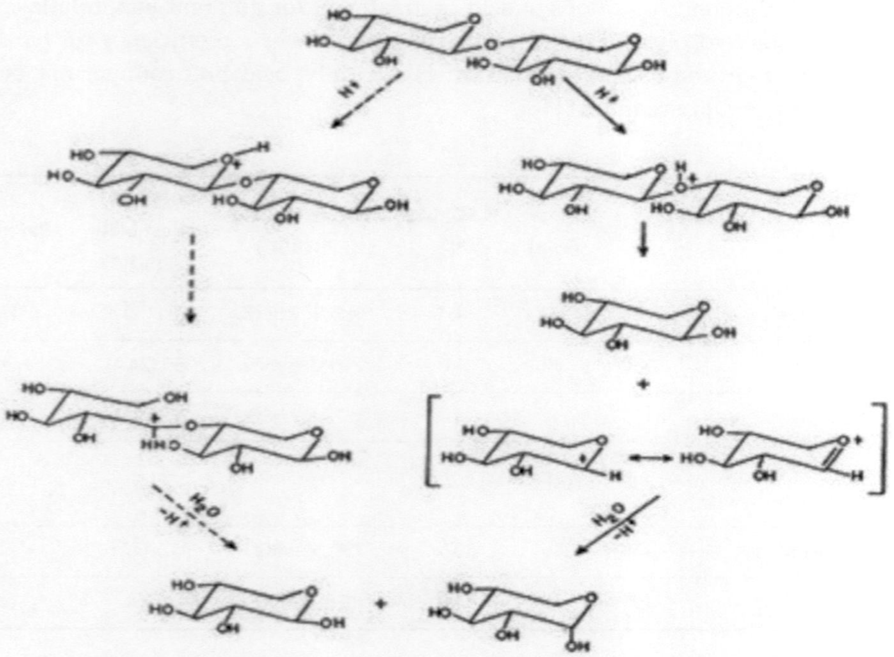

Figure 2. A simplified illustration for the mechanism of hemicellulose acid hydrolysis (adapted from [16]).

Figure 3 shows the inductive effect caused by the presence of glucuronic acid in the glycosidic chain. The carboxyl group induces different electron densities on the oxygen atom of the glycosidic bonds between A–B and B–C. The nucleophilicity is higher in the oxygen between B and C and this reduces the capacity for protonation. Thus, the bond is stabilized, while the glycosidic bond between A and B is activated by the same effect [16].

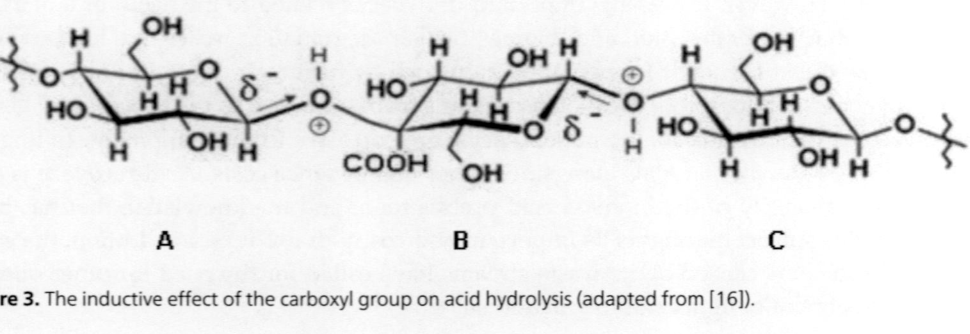

Figure 3. The inductive effect of the carboxyl group on acid hydrolysis (adapted from [16]).

The main problems associated with acid hydrolysis relate to the formation of toxic compounds from biomass degradation and from equipment corrosion. Such toxic products entail inhibition in cell metabolism when biomass hydrolyzates are used for bioconversion. Steps to remove these inhibitory compounds have been employed to improve the yields in bioconversion processes.

Table 1 presents different conditions of acid pretreatment for different lignocellulosic materials for enzymatic hydrolysis, as well as the cellulose conversion efficiency for hardwood, softwood, and sugarcane bagasse and straw. Historically, acid pretreatment has been the main choice for wood pretreatment [16].

Feedstock	T (°C)	Time (min)	H_2SO_4 (wt %)	Enzyme loading (FPU/BGU)	Enzymatic digestibility (%)	References
Athel pine[1]	165	8	1.4	15FPU/52.5BGU	60 (160 h)	[17]
Spruce[1]	180	30	5.0	15FPU/g solid	55 (24 h)	[18]
Eucalyptus camaldulensis[2]	165	8	1.4	15FPU/52.5BGU	38 (160 h)	[17]
Mixed wood (10% birch and 90% maple)[2]	230	0.12	1.17	---	95	[19]
Sugarcane bagasse	130	15	0.5	7FPU/3.5BGU	41.5 (72 h)	[20]
Sugarcane straw	195	10	1.0	15FPU/10BGU	72.4	[21]

[1]Softwood; [2]Hardwood

Table 1. Examples of sugarcane and woody biomass pretreated with diluted acid.

2.1. Advantages and disadvantages of acid pretreatment

Pretreatment with diluted sulfuric acid has been reported as one of the most widely used processes due to its high efficiency [14]. This pretreatment removes and hydrolyzes up to 90% of the hemicellulose fraction, rendering the cellulose fraction more accessible to hydrolytic enzymes. However, it presents important drawbacks related to the need for a neutralization step that generates salt and biomass sugar degradation with the formation of inhibitors for the subsequent fermentation step such as furfural from xylose degradation. The removal of inhibitors from the biomass sugar syrups adds cost to the process and generates a waste stream. Additionally, mineral acids are corrosive to the equipment, calling for the use of more sturdy materials alongside higher maintenance costs. Acid recovery is also costly. The availability of the biomass acid pretreatment and the knowledge that has been built up on this subject highlights its important and costly drawbacks. In addition, the environmental problems caused by its waste streams have called for the need for other options for the pretreatment of lignocellulosic materials.

3. Hydrothermal pretreatments

3.1. Liquid hot water (LHW) pretreatments

The liquid hot water (LHW) treatments are also called hot compressed water treatments, hydrothermolysis [22, 23], aqueous or steam/aqueous fractionation [24], uncatalyzed solvolysis [25, 26], and aquasolv [27]. LHW is based on the use of pressure to keep water in the liquid state at elevated temperatures (160–240 °C). This process changes the biomass native structure by the removal of its hemicellulose content alongside transformations of the lignin structure, which make the cellulose more accessible to the further enzymatic hydrolysis step [1, 28]. Differently from steam-explosion treatment, LHW does not use rapid decompression and does not employ catalysts or chemicals. Nevertheless, as with the acid treatment, LHW depolymerizes hemicelluloses to the liquid fraction. In this case, sugars are removed mostly as oligosaccharides, and the formation of the inhibitors furfural and 5-hydroxymethyfurfural (HMF) is at a slightly lower level [28], depending on the process conditions. To avoid the formation of inhibitors, the pH should be kept at between 4 and 7 during the pretreatment, because at this pH, hemicellulosic sugars are retained in oligomeric form, and monomer formation is minimized. The removal of hemicellulose also results in the formation of acetic acid in the liquid fraction.

LHW pretreatment, whose most important parameters are the biomass moisture content, the operation temperature, and the residence time [29], is usually done in a pressure tank reactor where two streams can be obtained after filtration of the biomass slurry: a solid, cellulose-enriched fraction and a liquid fraction rich in hemicellulose-derived sugars. The solid phase is therefore constituted by cellulose and lignin along with residual hemicellulose. There are three types of rector design for LHW pretreatment. For co-current reactors, the biomass liquid slurry passes through heat exchangers where it is heated to the appropriate temperature (140–180 °C) and kept for 10–15 minutes as the slurry passes through an insulated plug-flow snake-coil, followed by the slurry-cooling concomitant to heat recovery via the countercurrent heat exchange with the incoming slurry. Flow-through technologies pass hot water at 180–220 °C and approximately 350–400 psig. The resulting pretreated biomass has enhanced digestibility and a significant portion of the lignin is also removed. In countercurrent pretreatment, the biomass slurry is passed in one direction while water is passed in another in a jacketed pretreatment reactor. Temperatures, back pressures, and residence times are similar. In the flow-through pretreatment reactor, water or acid is passed over a stationary bed, and removes some of the biomass components including lignin. Although LHW can result in the partial depolymerization and solubilization of lignin, the re-condensation of lignin-derived, soluble compounds is also observed. Flow-through systems have been reported to be more efficient in terms of hemicellulose and lignin removal in comparison to batch systems for some types of biomass via the addition of external acid during the flow-through process [30].

There have been many studies on the use of LHW for the pretreatment of corn fiber [28, 30-33], wheat straw [34, 35], and sugarcane bagasse [36, 37]. Studies on woody biomass from *Eucalyptus* [38-40], and olive tree biomass [41] have also been reported.

Several works have reported about the optimal LHW pretreatment conditions in terms of temperature and residence time. For the pretreatment of corn stover the best conditions were reported to be 190 °C and 15 min, resulting in a 90% of cellulose conversion after enzymatic hydrolysis [33], while for wheat straw the optimum treatment temperature was found to be 188 °C during 40 min, which resulted in 79.8% of cellulose conversion and releasing of 43.6% of hemicellulose derived sugars to the liquid fraction. Nonetheless, when response variables were analyzed separately, the best conditions for the recovery of hemicellulose-derived sugars from wheat straw, at up to 71.2%, were found at 184 °C during 24 min, whereas the optimal conditions for a cellulose conversion of 90.6% were found to be 214 °C during 2.7 min [35].

For sugarcane bagasse, top-performing LHW runs are favored by high temperatures (≥ 220 °C) and a short residence time (≤ 2 min) associated with low solids concentrations (≤ 5%), reaching 87% of simultaneous saccharification and fermentation (SSF) conversion, and 81% of xylan recovery. However, it is reported that the use of LHW using a solid concentration of more than 1% can significantly decrease the ethanol fermentation rate due to inhibition [36].

The LHW pretreatment of *Eucalyptus* biomass was studied in two steps as follows: in the first step the pretreatment in which a temperature range from 180 to 200 °C was studied gave the highest total xylose recovery yield of 86.4% at 180 °C for 20 min. In the second step of the pretreatment, a temperature range from 180 to 240 °C was studied for intervals of time up to 60 min. The authors concluded that the efficiency of LHW for the cellulose conversion rate was more sensitive to temperature than residence time. The optimum reaction conditions for the second step of the pretreatment with minimal degradation of sugars were found to be 200 °C for 20 minutes, where the total sugar recovery from *Eucalyptus grandis* after 72 h of enzymatic saccharification reached 96.63%, which is superior to the yield from a single-step pretreatment with hot water or diluted acid [40].

LHW pretreatment of olive tree biomass resulted in a 72% glucose yield from cellulose hydrolysis after 72 h of saccharification using 2% of solids concentration during pretreatment, while for higher solids content the glucose yields were strongly affected reaching 70%, 60%, 57%, and 43% when using 5%, 10%, 20%, and 30% of solids, respectively [41]. A two-stage process which combines the LHW for hemicellulose removal and a treatment for delignification (e.g. ammonia pretreatment) has also been suggested for further improvement of enzymatic hydrolysis [42].

3.2. Steam pretreatment

The steam treatment is quite similar to LHW, with the major difference between the processes being related to the contact of the liquid phase with the biomass. For LHW, the biomass is in direct contact with the liquid phase at the bottom of the reactor, which prevents the use of high solids content, while in the steam pretreatment, the biomass is at the top of the reactor and not in direct contact with the liquid phase, in a similar manner to that of an autoclaving process. Using steam pretreatment means that it is possible to use a higher solids content of 50% or more, whereas for the LHW, in most cases, the solids content is lower than 10%. Sim-

http://dx.doi.org/10.5772/53378

ilarly to what is observed for LHW, the hemicellulose fraction is extracted by direct contact with water-saturated steam and due to the high temperatures and high solids concentration (around 50%), lignin and biomass polysaccharides can be extracted and degraded, releasing derived products such as furfural, HMF, and derived acids at high concentrations [36].

Sugarcane bagasse pretreatment by steam using high solids content (≥ 50%) at 200 °C for 10 min allowed a poor xylan recovery of 12%; nevertheless, an SSF yield of 79% was observed and the dissolved xylan content was found to be 89%. When the steam treatment was carried out at 220 °C for 2 min, the xylan recovery was increased to 48%, and the SSF yield and the dissolved xylan content were 85% and 88%, respectively, indicating the efficiency of high temperatures coupled with very short pretreatment times for high solids concentrations [36].

Aiming to improve the recovery of xylan sugars, most of the steam treatment studies report the use of SO_2 as the catalyst. When this procedure was used for sugarcane bagasse, it allowed for the recovery of 57% of hemicellulose-derived sugars and minimal amounts of sugar-degradation compounds were formed. The overall highest sugar yield achieved from the bagasse cellulose enzymatic hydrolysis was 87% [43]. There are several reports on the use of steam pretreatment associated with rapid decompression as a pretreatment for several biomass types. In this chapter, this type of treatment will be addressed as a steam-explosion treatment.

3.3. Advantages and disadvantages of LHW and steam pretreatments

LHW and steam pretreatments are attractive from a cost-savings perspective, as they do not require the addition of chemicals such as sulfuric acid, lime, ammonia, or other catalysts. Moreover, the reactors do not require high cost materials and maintenance due to their low-corrosion potential. Additionally, these treatments do not alter the biomass glucan content, as a glucose recovery rate of 97% was observed for sugarcane bagasse that was pretreated by both methods [36]. The main differences between the features of the two treatments relates to hemicellulose extraction, which is higher for the LHW, and the biomass load, which is higher for the steam pretreatment, with the obvious corresponding advantages and disadvantages. In contrast to steam pretreatment, LHW allows for a higher xylan recovery associated with the lower formation rate of inhibitors.

4. Steam-explosion pretreatment

Steam explosion is one of the most used methods for lignocellulosic biomass pretreatment. This process was initially developed in 1926 by the Masonite Corporation, Canada, for the production of fiberboard from wood [44]. From 1970–1980, the process was adapted to treat wood and agriculture residues aiming at improving the cellulose enzymatic hydrolysis and cattle feed production from lignocellulosic materials. A batch-type device was available from Iotech Corporation at pilot-plant scale in 1983 and a continuous device was available from Stake Technology, both from Canada, in the '80s.

In this process, the biomass is subjected to pressurized steam at temperatures ranging from 160 to 260 °C (corresponding to 0.69–4.83 MPa) for a few seconds to 20 minutes, followed by a rapid decompression of the reactor, by opening up an outlet, sharply reducing the temperature and interrupting the reactions. This approach combines chemical and mechanical forces in order to solubilize the hemicellulose fraction and render the cellulose more accessible to the enzymatic hydrolysis [1, 28, 45-47].

In the first step, called autohydrolysis, the water at high temperature causes the release of organic acids (mainly acetic acid) from the hemicellulose biomass moiety, followed by a chain reaction where the hemicellulose is partially solubilized and hydrolyzed [46]. The mechanical effects are caused by the rapid decompression, which results in the disruption of the cell wall fibers and, by extension, particle size reduction, and increased porosity [48]. In association with the partial hemicellulose hydrolysis, lignin is degraded to some extent, and a small portion is removed from the material [49].

At the end of the process, an insoluble solid fraction and a liquid fraction are obtained. The solid fraction contains primarily cellulose and the partially modified lignin. The liquid fraction—whose pH is in the range of 3.5 to 4.0 depending on how many of the acid chains are released [47]—contains soluble carbohydrates derived from hemicellulose in the form of oligomers and monomers in a proportion that is subject to the process conditions [11, 50]. A balance between the temperature and time of the pretreatment should be achieved in order to minimize the formation of phenolic compounds, furfural, and hydroxymethylfurfural from the degradation of lignin, five-carbon sugars (C5), and six-carbon sugars (C6), respectively. The formation of such compounds, as they may be inhibitory to the subsequent enzymatic hydrolysis and ethanol fermentation-production steps, should be minimized as much as possible.

Another approach for the steam-explosion process has been the use of impregnating agents such as SO_2, CO_2, and H_2SO_4, which can improve the effectiveness of the pretreatment, increasing the efficacy, and decreasing the residence time. In some materials, such as softwood, the intrinsic level of organic acids is not enough to promote the degradation of the hemicellulose backbone. Thus, the addition of a mineral acid can reduce the initial pH to below 2 and promote the hydrolysis. On the other hand, the use of acids as impregnating agents may require an additional step for pH adjustment. Moreover, such an approach may enhance partial carbohydrate and lignin degradation, resulting in an increase in the formation of toxic compounds, which will affect enzymatic hydrolysis, and fermentation [50]. Steam-explosion technology without additives has been successfully performed for ethanol production from hardwoods and for a wide range of agricultural residues [1].

The use of CO_2 was tested as an impregnating agent in the steam-explosion pretreatment of sugarcane bagasse and straw, and the pretreatment was assessed in terms of glucose yields after enzymatic hydrolysis [50]. For sugarcane bagasse, the highest glucose yield (86.6% of the theoretical level) was obtained after pretreatment at 205 °C for 15 min. For straw, the highest glucose yield (97.2% of the theoretical level) was obtained after pretreatment at 220 °C for 5 min. The reference pretreatment, using impregnation with SO_2 and performed at 190 °C for 5 min, resulted in an overall glucose yield of 79.7% and 91.9% for bagasse and

http://dx.doi.org/10.5772/53378

leaves, respectively. The production of toxic compounds from the dehydration of sugars (mainly furfural and hydroxymethylfurfural) was less than 1%.

Many patents [51-53] have been granted for steam-explosion, and these processes have been widely tested in pilot- and demonstration-scale ethanol plants, and are considered to be close to commercialization [47]. Iogen Corporation (Ottawa, Canada) was the first company to commercialize cellulosic ethanol; at full capacity, the demonstration plant was designed to process approximately 20–30 tons per day of wheat, barley, and oat straw, and to produce approximately 5000–6000 liters of cellulosic ethanol per day [54].

In Brazil, the pretreatment of lignocellulosic materials by steam explosion was initially developed at the Foundation of Industrial Technology in 1981. Studies which were performed on bench scale (a 1.6-L reactor) with sugarcane bagasse, *Eucalyptus*, and elephant grass, produced a wealth of data regarding the degree of hydrolysis and solubilization of the hemicellulose fraction, and the susceptibility of the treated material to enzymatic saccharification in the presence or absence of impregnating agents (SO_2 and H_2SO_4). These studies [55, 56] showed that the enzymatic saccharification enhancement due to steam explosion was more prominent for *Eucalyptus*, though steam exploded bagasse, and steam exploded grass were more accessible to attack by enzymes in comparison to treated *Eucalyptus*. In contrast to *Eucalyptus*, treatment conditions for achieving higher saccharification yields for bagasse and elephant grass were similar to those for optimal hemicellulose recovery in the process liquid stream. The use of the optimum time and temperature conditions for sugarcane bagasse (200 °C for 5–7 min) resulted in a 63% C-5 sugar recovery as soluble oligomers and an increase in glucose yield after enzymatic hydrolysis at standard conditions from 13% (untreated bagasse) to 56%. The use of 0.25% sulfuric acid as the impregnant agent promoted a sharp decrease in the optimal pretreatment time (1 min) and allowed a higher recovery of the hemicellulose fraction (90%) as monomers. Saccharification yields were at the same levels as those observed for the treatment without acid addition. For *Eucalyptus*, the best pretreatment conditions for C-5 sugar recovery (61%) were 200 °C for 3.5–5 min, while those for achieving higher enzymatic hydrolysis yields were 200 °C for 9 min, in which saccharification yields increased from almost zero (untreated wood) to 29.9%. Moreover, a study evaluated the effects of the explosion step in the steam-explosion process by comparing sugarcane bagasse pretreatment by steam explosion at 200 °C with that at the same temperature but without the explosion process [56]. Although both pretreatment processes showed almost identical chemical effects (hemicellulose autohydrolysis and cellulose and hemicellulose recovery), steam-exploded bagasse was 14.4% more accessible to attack by cellulases. Engineering studies were also carried out in order to expand the reactor scale to 0.2 and 2 m^3, both of which were tested and operated successfully [57]. These results allowed for the development of a joint project with the Sugar and Alcohol Plant, Iracema (Iracemápolis, SP), for the installation and operation of an unit with the capacity to produce treated bagasse enough to feed 150 cows/day. These studies have contributed to the implementation of the steam-explosion process of sugarcane bagasse for the production of cattle feed, currently operating in Brazil.

4.1. Advantages and disadvantages of steam-explosion pretreatment

The main advantages of steam explosion relate to the possibility of using coarse particles, thus avoiding a biomass-size conditioning step, the non-requirement for exogenous acid addition (except for softwoods, which have a low acetyl group content in the hemicellulosic portion), a high recovery of sugars, and the feasibility for industrial implementation. Moreover, the soluble stream rich in carbohydrates derived from hemicellulose in the form of oligomers and monomers may be easily removed and used as feedstock for the production of higher added-value products such as enzymes and xylitol [58]. Other attractive features include less hazardous process chemicals and conditions, the potential for significantly lower environmental impact, and lower capital investment [59]. The fact that the steam-explosion process does not require previous grinding of the raw biomass is an important feature, considering that the energy required to reduce the particle size before the pretreatment (pre-grinding) can represent up to one-third of the total energy required in the process [60].

The main drawbacks related to steam-explosion pretreatment are the enzyme and yeast inhibitors generated during the pretreatment, which include furfural and hydroxymethyl furfural; the formation of weak acids, mostly acetic, formic, and levulinic acids, the two latter acids being derived from furfural's and hydroxymethyl furfural's further degradation; and the wide range of phenolic compounds produced due to lignin breakdown. Several detoxification methods have been developed in order to reduce the inhibitory effect, which represent additional costs in the overall process. Other limitations of this method include the incomplete disruption of the lignin–carbohydrate matrix [11].

5. Mechanical pretreatments

Mechanical pretreatments of biomass aim primarily to increase the surface area by reducing the feedstock particle size, combined with defibrilization or reduction in the crystallinity degree. This approach facilitates the accessibility of enzymes to the substrate, increasing saccharification rates and yields. The most studied biomass mechanical pretreatment for biomass is the milling process, mainly the ball-milling and disk-milling pretreatments. Another mechanical treatment to be considered is extrusion, even though this process involves additional thermal and/or chemical pretreatments.

5.1. Milling

Different types of milling processes can be used to improve the enzymatic hydrolysis of lignocellulosic materials [61]. The main objective of milling pretreatment is to reduce particle size in order to increase the biomass-specific surface during biomass fibrillation and to reduce cellulose fiber organization, which is measured by a decrease in crystallinity. These effects can be produced by a combination of chipping, for final particle sizes of 10–30 mm, or grinding or milling, for final particle sizes of 0.2–2 mm [11]. It is important to emphasize that macroscopic particle size reduction does not lead to significant improvements of bio-

http://dx.doi.org/10.5772/53378

mass enzymatic saccharification, which is solely achieved by using a milling process that alters the biomass structure at a nanoscopic level. Ball milling and wet disk milling (WDM) are the most common biomass milling pretreatments.

5.1.1. Ball milling

The ball milling process uses mechanical shear stress and impaction to produce powdered material [62]. This process can be done in the wet or dry state. A combination of chemicals, such as acids, bases and organic solvents, can also be applied depending on the main treatment purpose. In general, the process uses a rotary drum and balls of different sizes made from different materials (tungsten, ceramic or stainless steel). The effect of ball milling on the biomass particle size, structure and crystalline degree depends on the rotation speed, operation time and ball size. Ball milling treatment can be considered a kind of ultra-fine grinding and fibrous materials can present between 10-20 μm in terms of particle size [63]. Highly crystalline cellulose has a strong interchain hydrogen-bonding network that confers a high resistance to enzymatic hydrolysis, whereas amorphous cellulose is readily digestible [64].

For biomass pretreatment after the material is fed into the rotary drum equipped with balls, drum rotation around a horizontal axis causes a reduction on the material particle size [65]. The ball milling process can drastically alter the complex heterogeneous network structure of wood cell walls, and with a long pretreatment time, cellulose crystallinity can be significantly reduced, which increases the ratio of amorphous cellulose, thus improving the saccharification yields [65]. However, it was found that nanofibrillation of woody biomass by ball milling in the wet state can improve the saccharification yield without a significant decrease in cellulose crystallinity [65]. They found the crystallinity index on the ball-milled biomass to be ca. 41% in comparison to 68% for the raw material. Thus, the decrease in particle size to a powder-like material and the increase of surface area seemed to be the main factors that promoted the hydrolysis of treated wood.

Results obtained with the pretreatment of sugarcane biomass have shown that ball milling treatment of bagasse for 60 min and sugarcane straw for 90 min results in glucose and xylose yields of 78.7% and 72.1% for bagasse and 77.6% and 56.8% for straw, respectively. In both cases, the enhancement in cellulose digestibility was related to the reduction of cellulose crystallinity to nearly an amorphous level [66]. In another study, the 20 min pretreatment of *Eucalyptus* using a planetary ball milling process was insufficient for improving enzymatic digestibility, even though the crystallinity index decreased from 59.7% to 7.6%. When a prolonged milling time of 120 min was used, the enzymatic digestibility of both glucan and xylan increased, while the degree of crystallinity of the material was almost the same as that milled for 20 min. Additionally, the digestibility of glucan and xylan and their total yield were 76.7%, 63.9% and 74%, respectively, even at a substrate concentration of 20% and an enzyme dosage of 4 FPU/g of substrate, indicating that ball milling is extremely efficient to enhance biomass reactivity to enzymes [39].

5.1.2. Advantages and disadvantages of ball milling

Mechanical disruption of cellulose by ball milling is a candidate method for a significant increase of cellulose-accessible surface area without the loss of low-molar mass components. After ball milling treatment, without the use of additive chemicals, the treated material keeps the same chemical composition of the untreated material and there is no generation of liquid fractions, gas or inhibitors. As such, ball milling is an environmentally friendly pretreatment method for lignocellulosic biomass. However, milling processes are known to be very energy intensive, depending on the material characteristics and the target particle size [36]. Taking into account the high energy requirements of milling and the continuous rise of energy prices, it is likely that this process is not economically feasible [67]. Moreover at an industrial scale, ball milling equipment requires high dimensions; nevertheless, in specific cases the balls can be replaced by bars for efficient milling depending on the amount of biomass to be used. Recently, a new milling pretreatment method for lignocellulosic biomass was described by using disk milling in a wet state, as described below [68].

5.1.3. Wet disk milling (WDM)

WDM is a recently introduced biomass pretreatment process able to produce milled biomass with low levels of inhibitors; it is considered to be feasible for industrial implementation. This technique has been shown to increase the degree of biomass fibrillation and the nano space between the microfibrils, thus promoting the accessibility of the cellulolytic enzyme pool to cellulose [69]. The disk mill is a type of crusher that can be used to grind, cut, shear, fiberize, pulverize, granulate or blend. In general, the suspended material is fed between opposing disks or plates that can be grooved, serrated or spiked. The force applied in the material will depend on the type of disks, the distance between the disks and its rotation speed. For biomass processing using WDM, a water suspension (1–5% of solids) of the lignocellulosic material is passed between two ceramic non-porous disks that are separated by a distance of 20–100 μm and that have a rotational speed of around 1800 rpm. This process can be repeated according to the required number of WDM cycles; very small particle sizes with high specific surface areas have been observed for a minimum of five cycles [66, 68].

A study on WDM pretreatment of sugarcane bagasse and straw showed that enzymatic hydrolysis yields increased with the number of WDM cycles; maximum sugar yields were obtained with 20 cycles, leading to glucose and xylose yields of 49.3% and 36.7% for sugarcane bagasse and 68.0% and 44.9% for sugarcane straw, respectively [66]. Hydrolysis yield data for 10 WDM cycles showed a glucose yield for bagasse of 31.5%, while a glucose yield of 56.1% was observed for straw, confirming that WDM is more efficient for straw.

As WDM is a recent procedure for biomass pretreatment, there is limited information on the pretreatment of several different types of biomass. However, reported data for the pretreatment of rice straw showed that after 10 cycles of WDM it was possible to achieve glucose and xylose yields of 78.5% and 41.5% respectively, with an energy consumption of 5.4 MJ/kg of rice straw. The authors evaluated energy consumption using 60 min ball milling for the pretreatment of rice straw. The process used 108 MJ/kg of rice straw, a value 20-fold higher than that for 10 cycles of WDM [69].

5.1.4. *Advantages and disadvantages of WDM*

Although WDM pretreatment presents lower energy consumption than that for ball milling, it requires large amounts of water due to low solids loading (1–5%), which is a drawback that may hinder its industrial application. WDM of rice straw has been reported to require almost the same energy (5.4 MJ/kg of biomass) as hydrothermal pretreatment, exemplifying the possibility of using milling for biomass pretreatment [69].

5.2. Extrusion process

Screw extruders were originally designed to extrude polymers and were also developed for food and feed processing [70-72]. An extruder can provide many functions, such as cooking, forming, kneading, degassing, dehydration, expansion, homogenization, mixing, sterilization, shaping, densification and shearing [73]. These functions can be performed in the same process, depending on the size of the extruder and the screw design. Since the 1990s, there has been an increase in the number of studies that use extrusion for biomass processing, such as for the extraction of compounds [74-76], densification [73, 77] and biomass pretreatment for enzymatic saccharification [78-87]. For lignocellulosic biomass pretreatment, extrusion processing can provide a unique continuous reactor working at higher throughput and solid levels. The extrusion equipment provides temperature control and efficient pulverization by applying a high shearing force. This process also allows the advantageous simultaneous combination of thermomechanical and chemical pretreatment.

The extruder consists of a barrel with a rotating screw (or screws) that squeezes and conveys the material continuously from the input to the output. The barrel is normally segmented, which allows temperature control (heating or cooling) along its length and feeding ports for additive injection. Different types of screw elements can be installed onto the shafts for screw configuration in accordance with the process requirements. The possibility to design the screw configuration using many combinations of elements renders the process very flexible. There are conveying, backward-conveying and kneading elements that can be threaded in different ways to provide mixing, shearing, elongation flow, and pressure build up. The use of kneading disks, which can be staggered at diverse angles (typically 30°, 45°, 60° and 90° of stagger) in forward or reverse directions, can impart a high shearing stress by forcing the material to pass through the small clearances between the disks and between the disks and the barrel surfaces [88, 89]. It is also possible to configure sealed regions where the pressure can be significantly higher than other zones.

There are several types of extruders that can be classified according to the number of shafts: single-screw, twin-screw or multiple-screw extruders. The single-screw extruder presents a unique screw rotating in a stationary barrel and is more applicable for distributive mixing without changes in physical properties of the material [89]. On the other hand, some types of twin-screw extruders can provide distributive and dispersive mixing; twin-screw extruders are normally applied to obtain changes in physical properties of materials, such as the reduction in particle size by high shearing forces [89]. The twin-screw extruder can be classified according to the rotation directions of the two screws: counter rotating (opposite directions) [90] or co-rotating (same directions) [88]. The counter-rotating design is used for a very high shear force; nevertheless, it can generate excessive wear and tear. The co-rotating twin screw can be operated at high screw speeds, resulting in high outputs, while maintaining the required shear force, mixing and conveying properties. Figure 4 shows an example of screw configuration and temperature control versatility of a twin-screw extruder used for biomass pretreatment [82].

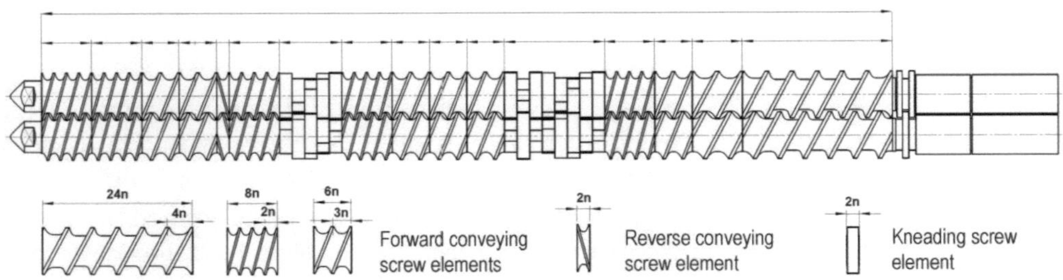

Figure 4. Example of a twin-screw configuration. The elements can be exchanged to be adapted to the type of biomass and the pretreatment conditions, such as retention time and shearing force.

Thus, the co-rotating twin-screw extruder is one of the most promising configurations for biomass processing due to its flexibility to carry out biomass deconstruction under relatively high stress, temperature and pressure. The twin-screw extruder is also easy to operate and economically suitable for large-scale production [80]. However, the use of kneading elements and sealed regions, with reverse elements, for biomass pretreatment at high rates and high solid concentration depends on the flow properties of the biomass. The extrusion of lignocellulosic biomass is difficult due to poor flow properties inside the extrusion barrel, leading to accumulation, burning and blocking of the die during the process [82]. This material can be mixed with water or additives to increase the viscosity and its flow capability, reducing the operational torque and transposing the aforementioned problems. The combination of chemical pretreatment inside the extruder, i.e., alkaline [91, 92], acid [93, 94] and ionic liquids [95], can also increase flow and reduce torque, allowing the use of extrusion as a continuous mixing reactor for biomass processing.

The extrusion process was reported for starch conversion [96] and for wood pulping [97]. A twin-screw extruder was used to fibrillate wood chips to produce individual fibers, which have higher aspect ratios than the wood flour particles usually used for wood–polymer composites [98]. Twin-screw extruders were also used as extractors of lipids [74] and essential oils [75]. The extraction of hemicelluloses via alkaline solubilization using a twin-screw extruder for pentose production from the hardwood *Populus tremuloides* was also reported [76]. The major advantage of using twin-screw for extraction of biomass compounds is that kneading and reverse screw elements can cause severe compression of the material and allow the simultaneous extraction and liquid/solid separation in a very efficient manner.

Significant improvements in sugar yield after enzymatic hydrolysis have been reported for biomass pretreatment based on extrusion; examples include corn stover [92], *Miscanthus*

[99], big bluestem, prairie cord grass, switchgrass, Indian grass [78, 83-87]. Nevertheless, there are no reports for the use of extrusion for pretreatment of sugarcane bagasse and straw aimed at cellulosic ethanol production.

The use of single-screw extrusion for the pretreatment of corn stover and soybean hulls at high solid loadings (75–80%) resulted in 54–61% sugar recovery [84, 100]. The pretreatment of Douglas fir wood (coniferous, *Pseudotsuga*) was performed by a counter twin-screw extruder [79] using cellulose affinity additives (ethylene glycol, glycerol and dimethyl sulfoxide) to effectively fibrillate the wood cell wall and lower the extrusion torque. However, it has been suggested that torque is more effective for fibrillation than temperature and the swelling effects of additives. The enzymatic conversion of extruded products into glucose was three to six times higher than that of untreated material [79]. Ethylene glycol was found to be the most effective additive for fibrillation, achieving a glucose yield of 62.4%. In another study, sawdust and wood chips were pretreated using a twin-screw extruder; this resulted in the recovery of 65% of glucose upon enzymatic hydrolysis, which was over 10-fold higher than that of untreated material [101].

Extrusion was also used in combination with hydrothermal and chemical pretreatment. A single-screw extruder was reported for the pretreatment of wheat straw in conjunction with NaOH, Na_2S and H_2O_2; nevertheless, the mass flow rates and concentration were limited to low values of 10–30 g wheat straw (dry basis)/min and 12–33% solids concentration [91]. The use of a twin-screw extruder in combination with the ammonia fiber explosion (AFEX) process for the pretreatment of milled dry corn stover showed that the extrusion process with ammonia compares well to batch AFEX pretreatment [92]. The pretreatment of milled (under 3 mm) Douglas fir (softwood) and *Eucalyptus* (hardwood) by LHW followed by a co-rotating twin-screw extruder was also reported [80]. The glucose yields obtained by extrusion after LHW were higher than those for the individual use of LHW for both types of biomass. Results for Douglas fir were fivefold higher, compensating for the limitations of LHW for this material as LHW is known to be less effective for softwood than for hardwood. The Douglas fir wood was also treated using a batch-type kneader with twin-screw elements [81]. The biomass was pulverized by ball milling for 20 min followed by kneading for 20 min. The maximum glucose yield was 54.20% (25.40 wt% based on initial wood weight). However, glucose yield was improved by heating the extruded biomass with water under pressure (135 °C and 0.25 MPa), revealing that only mechanical kneading with water showed limitation for enhancing the accessibility of cellulose to enzymes.

The thermomechanical extrusion of wheat bran and soybean hulls led to reduced sugars yields of 65–73% and 25–36%, respectively [102]. The combination of lower temperature and high residence time (low screw speed) or higher temperature and low residence time (high screw speed) led to higher sugar yields; these authors also tested the combination of chemicals (NaOH, urea and thiurea) with extrusion, with no significant improvements.

The combination of twin-screw extruder and diluted acid pretreatments was recently reported for rice straw [93] and rape straw [94]. When rice straw was pretreated with 3% sulfuric acid at 120 °C a low (32.9%) glucose yield was observed. However, the use of extrusion/acid pretreatment followed by a hot water extraction step enhanced the enzymatic hydrolysis yield from 32.9% to 60.9%. The hot water hemicellulose extraction step allowed the conversion of 83.7% of xylan to xylose and favored cellulose hydrolysis [93]. Rape straw extrusion pretreatment with 3.5% sulfuric acid at 165 °C obtained a glucose yield of 70.9% [94].

The high sugar recovery due to extrusion pretreatment is related to fibrillation, the increase in surface area [79-81, 86, 87] and pore size [103], which facilitate the access of enzymes to cellulose. Some authors have reported that the crystallinity, which confers resistance to enzymatic hydrolysis, was not significantly reduced in extruded biomass [79, 102] and therefore was not related to the increase in biomass digestibility. Moreover, an increase of 82% in the crystallinity of soybean hulls by thermomechanical pretreatment, using a twin-screw extruder was reported [82]; as there was no change in material composition, crystallization of the amorphous structure during thermomechanical extrusion was suggested. Some researchers have also noted the crystallinization of cellulose in the presence of moisture and heat, as has been observed for wood pretreated by steam explosion [104], cotton linter and wood treated in aqueous media after ball milling [105] and hemp cellulose treated by wet ball milling [106]. In accordance with the aforementioned, some researchers suggested that the opening of the cell wall structure at a microscopic scale is sufficient for enzymatic saccharification, regardless of the cellulose crystallinity index [79]. Furthermore, the combination of thermomechanical and/or chemical pretreatments can deconstruct the hemicelluloses chains and/or remove part of the hemicelluloses and lignin, facilitating biomass digestibility [82, 102, 107].

5.2.1. Advantages and disadvantages of extrusion pretreatment

The twin-screw extruder is highly efficiency for pulverization by applying high shearing forces and shows adaptability to different processes, such as chemical, high-pressure applications and explosion pretreatments (steam or other solvents) [79-81, 92, 108, 109]. The process is easy to operate and the extrusion process allows the continuous pretreatment of large amounts of biomass with high throughputs, which is advantageous in comparison to batch procedures for the industrial setting. Extrusion compares well to pretreatment technologies that have as drawbacks the batch processing mode, low solids loading or the use of large amounts of water, as already mentioned. Extrusion allows temperature control and does not require washing and conditioning steps, as required with diluted acid, alkali or ionic liquid pretreatments and does not produce effluent; thus there is no effluent disposal cost, no solids loss and no significant safety issues [86]. In comparison to other mechanical pretreatments, the extrusion process is normally less energy intensive than the milling pretreatment options. If extrusion is combined with chemical pretreatment, due to its effective ability of kneading and mixing, the process requires less chemical loadings and thus less residual effluents are formed; the combination of extrusion with chemical pretreatment can further reduce energy consumption as it is economically suitable for large-scale operation. Furthermore, extrusion does not produce fermentation inhibitors, such as furfural and hydroxyl methylfurfural; nevertheless, low concentrations of acetic acid have been reported [84-87]. However, the extrusion pretreatment of lignocellulosic biomass requires the use of additives to increase the flow ability inside the barrel and avoid the accumulation, burning

http://dx.doi.org/10.5772/53378

and blocking of the die during the process. Another disadvantage in comparison to other biomass pretreatments is the low yields of enzymatic hydrolysis achieved.

6. Ionic liquid pretreatment

Ionic liquids (ILs) can be defined as salts that melt below 100 °C and are composed exclusively of ions. The first report of a room temperature IL dates back to 1914 [110] and did not prompt any significant interest at that time. It was in the 1980s that these chemicals have come under intense worldwide attention due to the implications for their use as solvents [111, 112]. The fact that many ILs can be liquid at room temperature and, in general, present a negligible vapor pressure has justified the attention that this group of chemicals has received. They have also been suggested as candidates to substitute for low-boiling-point solvents, such as toluene, diethyl ether and methanol. In addition, ILs are versatile materials and often called designer solvents because their physical and chemical properties can be tuned to meet a specific purpose by preparing new ILs with different combinations of ions [113].

ILs have become increasingly trendy over the past few years in the biomass field due to the ability of some members of this class of chemicals to dissolve a wide variety of biomass types. ILs have been reported for the pretreatment of cellulose [114] and lignocellulosic materials, such as rice straw [115], sugarcane bagasse [116, 117], wheat straw [118], switchgrass [119], *Miscanthus* [120] and wood [121, 122, 123], among others. However, this concept is not new since in 1934 a patent claimed that certain organic salts were capable of dissolving cellulose and alter its reactivity [124]; nevertheless, at that time this publication did not generate any important reaction in the scientific community. In 2002, a research group from the University of Alabama investigated new compounds, now known as ILs, based on the concept of cellulose dissolution by a molten salt described by Graenacher in 1934. As result, they found that the IL 1-methyl-3-butyl imidazolium chloride ([Bmim][Cl]) could dissolve up to a 10% solution of cellulose by stirring cellulose with the IL while heating (100 °C). When heating was performed in a microwave oven, the dissolution achieved was up to 25% (wt%) [125]. Their pioneer work has now been cited over 1000 times and is considered a breakthrough that has set the basis for a novel concept for lignocellulosic biomass pretreatment.

Based on the concept of cellulose dissolution described by Swatloski and co-workers [124] and lately by another work that has shown that [Bmim][Cl] was also able to partially dissolve wood [126], many research groups have described processes of biomass pretreatment with ILs; most of these studies document the complete or partial dissolution of lignocellulose under heating conditions followed by precipitation with water as an antisolvent. The aim of this procedure is to recover a pretreated part of the biomass that is highly susceptible to enzymatic attack. After IL pretreatment, the biomass native structure is altered in the recovered material in such a manner that the reconstructed cellulose is essentially amorphous compared to highly crystalline untreated cellulose [127].

The mechanism for IL cellulose dissolution has been investigated by applying different analytical methods. In one study, nuclear magnetic ressonance (NMR) relaxation measurements on [Bmim][Cl] confirmed that chloride ions form hydrogen bonds with the cellulose hydroxyl group in a stoichiometric 1:1 ratio [128]. This interaction causes the break of intermolecular and intramolecular hydrogen bonding between cellulose fibrils, which ultimately leads to cellulose dissolution. Additionally, depending on the type of IL, an efficient extraction of lignin can be facilitated by the cellulose dissolution process, as more lignin can be exposed to the solvent [121, 126].

Different combinations of anion and cation compositions have been examined for biomass pretreatment, as the dissolution of biomass components is highly affected by the nature of the IL. In general, in order to dissolve cellulose, the anion of the IL must be a good hydrogen bond acceptor [123, 129]. The most promising anions have been shown to be chlorides, acetates, formates and phosphates. It has also been demonstrated that cations play a role in cellulose solubility as the imidazolium cation, whose electron-rich aromatic π system interacts with cellulose hydroxyl oxygen atoms via nonbonding and π electrons, prevents the cross-linking of cellulose molecules. In general, the most appropriate cations for cellulose dissolution are based on methylimidazolium and methylpyridinium cores, with allyl-, ethyl-, or butyl-side chains [130].

Considering the IL pretreatment of sugarcane bagasse, the IL 1-ethyl-3-methylimidazolium acetate ([Emim][Ac]) has been selected among six ILs studied as the best choice for pretreatment (120 min, 120 °C and 5% solid loading) as it was possible to reach a glucose yield of 98.2% after 48 h of enzymatic hydrolysis of 2.5% pretreated bagasse loading using commercial enzymes at a dosage of 15 FPU/g bagasse [116]. The authors suggested that the resulting pretreated biomass was highly digestible due to its amorphous-like structure, the high ability of [Emim][Ac] to extract lignin and the increased specific surface area (SSA) of 131.8 m^2/g compared to an SSA of 1.4 m^2/g measured for untreated bagasse. In another study, yields of 69.7% of reducing sugars were obtained for the enzymatic hydrolysis (30 FPU/g substrate) of 2% bagasse loading pretreated with [Emim][Ac] for 15 min at 145 °C, using a 14% solid loading during pretreatment [131]. Since high-solid loading during pretreatment was applied, [Emim][Ac] was ineffective in bagasse delignification, even though it was able to reduce the biomass crystallinity.

Some studies have combined other pretreatment strategies to IL pretreatment of sugarcane bagasse to reduce the pretreatment time and increase the efficiency. In an HCl-catalized pretreatment process in IL aqueous solutions, optimum conditions for the sugarcane bagasse pretreatment was obtained at 130 °C, 30 min, using a water:[Bmim][Cl]:HCl solution (%) of 20:78.8:1.2. Cellulose digestibility yields corresponding to 94.5% were obtained after 24 h saccharification of 2% glucan loading, using commercial enzymes at a dosage of 20 FPU/g glucan; the pretreatment for 120 min using solely [Bmim][Cl] resulted in 29.5% cellulose conversion [132]. Other reports of sugarcane bagasse pretreatment using [Bmim][Cl] have also reported low glucose yields of 38.6% (120 °C, 120 min) [116] and 62% (140 °C, 90 min) [117].

The reduction of cellulose crystallinity is usually reported as a main effect of IL pretreatment. However, in pretreatment of sugarcane bagasse catalyzed by acid, using HCl-[Bmim][Cl] [132] or H_2SO_4-1-butyl-3-methylimidazolium methylsulfate ([Bmim][$MeSO_4$]) systems [133], cellulose crystallinity remained unaltered even though a significant increase of cellulose digestibility was achieved. The authors suggested that digestibility increase was due to the highly effective and simultaneous removal of xylan and lignin that facilitated cellulose enzymatic saccharification efficiency. Table 2 presents the experimental conditions for the pretreatment of sugarcane bagasse with ILs.

IL	Pretreatment association	Temperature (°C)	Time (min)	Solid loading (%)	Enzyme dosage (FPU/g)	Substrate loading (%)	Glucose yield (%)	Reducing sugars yield (%)	Ref.
[Emim][Ac]	-	120	30/120	5	15	2.5	95.3/98.2	-	[116]
[Emim][Ac]	-	120	30	5	15	1.0	87.0	-	[134]
[Emim][Ac]	-	145	15	14	30	2.0	-	69.7	[131]
[Emim][Ac]	-	140	continuous	25	15	2.5	90.3	-	[95]
[Bmim][Cl]	-	140/150	90	5	15 [a]	1.0	62.0/100		[117]
[Bmim][Cl]	-	120	120	5	15	2.5	38.6	-	[116]
[Bmim][Cl]	-	130	120	10	20 [a]	2.0	29.5	-	[132]
[Bmim][Cl]	H_2SO_4/HCl	130	30	10	20 [a]	2.0	93.5/94.5	-	[132]
[Bmim][Cl]	NH_4OH-H_2O_2	100	60	3	20	2.0	-	90.0	[135]
[Amim][Cl]	-	120	120	5	15	2.5	43.3	-	[116]
[Amim][Cl]	NH_4OH-H_2O_2	100	60	3	20	2.0	-	91.4	[135]
[Bmim][$MeSO_4$]	-	125/150	120	10	60	1	79.0/100		[133]
[Bmim][$MeSO_4$]	H_2SO_4	100	120	10	60	1	74.0		[133]
[Mmim][DEP]	-	120	120	5	15	2.5	61.9	-	[116]

[a] Enzyme dosage per gram of cellulose

Table 2. Sugarcane bagasse IL pretreatment parameters and corresponding data for enzymatic saccharification and sugar yields

Many reports can be found for the pretreatment of wood biomass with ILs. Initial studies have focused on the use of ILs to dissolve lignocellulosic biomass aiming its fractionation [126, 136]. Moreover, the possibility to perform the derivatization of wood components *in situ* using the biomass IL solution was considered an interesting approach to reduce the number of steps to produce derivatives, such as acylated cellulose from raw materials [137]. More recently there have been reports on the enzymatic digestibility of recovered wood biomass after IL dissolution. Over 90% cellulose hydrolysis was obtained after *Pinus radiata* pretreatment with [Emim][Ac] at 120 °C for 180 min, using a 5% solid loading during pretreatment [138]. The authors demonstrated that the IL pretreatment induced compositional and structural changes in the wood, including extraction and deacetylation of the hemicellulose fraction and loss of lignin ether linkages. The cellulose crystallinity was altered, prompting the suggestion that cellulose I was transformed, to some extent, to cellulose II. However, in contrast to an earlier report for the pretreatment of maple wood flour with [Emim][Ac] in which up to 80% of delignification was achieved [121], no significant delignification of *P. radiata* was observed. The glucose saccharification yields obtained for the maple wood flour pretreated at 130 °C for 90 min reached 95%. In contrast to the high yields obtained after wood pretreatment with [Emim][Ac], the use of [Emim][Cl] was shown to be ineffective, as only 30% of total sugars were obtained after saccharification of pretreated eucalyptus at 150 °C for 60 min [139].

A comparison of the effects of newly synthesized ILs has also been performed for hardwood (barked mixed willow) and softwood (pine sapwood). The ILs 1-butyl-3-methylimidazolium hydrogen sulfate [Bmim][HSO_4] and 1-ethyl-3-methylimidazolium methyl sulfate [EMIM][$MeCO_2$] were mixed to 20% water and used for the pretreatment of both materials [120]. The pretreatment of the softwood sample with those ILs was ineffective as the maximum cellulose-to-glucose conversion achieved was 30%, while the pretreatments of hardwood samples with [Bmim][HSO_4] and [EMIM][$MeCO_2$] resulted in glucose yields of over 80% and 60%, respectively. Table 3 presents the experimental conditions for the pretreatment of woody biomass with ILs.

IL	Biomass	Temperature (°C)	Time (min)	Solid loading (%)	Enzyme dosage (FPU/g)	Substrate loading (%)	Glucose yield (%)	Reference
[Emim][Ac]	*Pinus radiata*	120/150	30	5	20	1.5	93/81	[138]
[Emim][Ac]	Maple wood flour	125	120	33	4.9	1.0	72	[140]
[Emim][Ac]	poplar	125	120	33	4.9	1.0	65	[140]
[Emim][Ac]	Maple wood flour	130	90	5	NI	NI	95	[121]
[Emim][Cl]	*Eucalyptus globulus*	150	60	5	180	5	30	[139]
[Emim][Cl]	*Nathofagus pumilo*	150	30	5	180	5	40	[139]
[Bmim][HSO_4]	Mixed willow	120	120	10	60 [a]	2	80	[120]
[Bmim][HSO_4]	Pine sapwood	120	120	10	60 [a]	2	30	[120]
[EMIM][$MeCO_2$]	Mixed willow	120	120	10	60 [a]	2	60	[120]
[EMIM][$MeCO_2$]	Pine sapwood	120	120	10	60 [a]	2	25	[120]

[a] Enzyme dosage per gram of cellulose; NI – Not informed

Table 3. Woody biomass IL pretreatment parameters and corresponding data for enzymatic saccharification and sugars yields

http://dx.doi.org/10.5772/53378

6.1. Advantages and disadvantages of IL pretreatment

ILs are able to disrupt the plant cell wall structure by the solubilization of its main components. This class of salts is also able to alter cellulose crystallinity and structure, rendering the amorphous cellulose prone to high rates and yields from enzymatic saccharification. Indeed, this combination of effects generates a pretreated material that can be easily hydrolyzed into monomeric sugars when compared to other pretreatment technologies, also rendering the enzymatic attack faster as the initial hydrolysis rate is greatly increased [116, 119]. In order to achieve high cellulose conversion yields (>80%) using other pretreatment processes, enzymatic saccharification times of 48–72 h are generally reported. However, in the case of IL pretreatment of bagasse and also some woody types, those yields can be obtained in less than 24 h with enzymatic hydrolysis. Nevertheless ILs are still too expensive to be used for biomass pretreatment at the industrial scale; however, the possibility of recovering the extracted lignin opens up the possibility for producing high-value products in addition to ethanol, which would favor the economics of a biorefinery based on IL biomass pretreatment. Indeed, modeling studies have shown that selling lignin can effectively lower the minimum selling price of ethanol to the point where lignin becomes the main revenue source of an IL-based biorefinery [141].

There are many challenges to be addressed before ILs can be considered as a real option for biomass pretreatment, including their high cost and the consequent requirement for ionic liquid recovery and recycling, and the high IL loading required for most IL pretreatment processes reported. It has been shown that the reduction in IL loading is more important than increasing the rate of IL recycling [141]. Aiming to tackle IL cost, two recent works have addressed this issue and were successful in reducing the IL requirement, demonstrating that it is possible to increasing biomass loading up to 33% [95, 140]. Moreover, a continuous pretreatment process using ILs by applying a twin-screw extruder as a mixing reactor has been developed [95]. Many works have also reported the use of recycled ILs up to 10 cycles without significant loss in pretreatment efficiency [121, 132, 140]; nevertheless the development of energy-efficient recycling methods for ILs for large-scale applications is still an open issue. It is also noteworthy that most studies have performed enzymatic saccharification of IL-treated biomass at low-biomass loadings (<5%). Data on saccharification yields obtained on high-biomass consistency hydrolysis assays (>15%) are also needed to truly evaluate the effectiveness of IL pretreatment on enhancing the enzymatic hydrolysis rate. ILs toxicity to enzymes and fermentative microorganisms must also be addressed as ILs trace residues may negatively affect the performance of enzymes [142] and inhibit fermentation [143]. Some research groups are now looking for new enzymes that are stable in ILs [144, 145].

Despite the current restrictions and the clear need for research and development to pave the way for the industrial use of ILs, ILs have great potential use within the engineering perspective of a biorefinery due to their uncommon and specific chemical features and their selectivity toward biomass processing.

7. Other pretreatment processes

According to the foregoing, this chapter has covered the most relevant pretreatment techniques for sugarcane and woody biomass as well as the new trends in this field. Below, we present other pretreatments, such as alkaline, ammonia fiber expansion and biological, which are also of relevance. Other important methods such as organosolv, ammonia percolation, and oxidative reactions using hydrogen peroxide or ozone will be dealt with elsewhere.

7.1. Alkaline pretreatment

This pretreatment is similar to the Kraft pulping process used in the pulp and paper industries. Nevertheless, sodium, potassium, calcium, and ammonium hydroxides have been employed for the pretreatment of lignocellulosic biomass, sodium hydroxide has been the most studied reagent [146-148]. However, calcium hydroxide is advantageous due to its low cost, higher safety besides its recovery as insoluble calcium carbonate through reaction with carbon dioxide [149]. Lime pretreatment has been used in studies carried out with several lignocellulosic materials, such as sugarcane bagasse [150], switchgrass [151], rice straw [152] and poplar wood [153].

The main effect of alkaline pretreatments is the biomass lignin removal thereby reducing the steric hindrance of hydrolytic enzymes and improving the reactivity of polysaccharides. It is believed that the mechanism involves saponification of intermolecular ester bonds between xylans and lignin, increasing the material porosity. The addition of air/oxygen to the reaction mixture dramatically improves delignification, especially in the case of materials with high lignin content. The removal of acetyl groups from hemicellulose by the alkalis also exposes the cellulose and enhanced its enzymatic hydrolysis [2]. The alkali pretreatment also causes partial hemicellulose removal, cellulose swelling and cellulose partial decrystallization [149].

In the alkaline process the biomass is soaked in the alkaline solution and mixed at a mild controlled temperature in a reaction time frame from hours to days. It causes less sugar degradation than the acidic pretreatments. The necessary neutralizing step, prior to the enzymatic hydrolysis, generates salts that can be partially incorporated to the biomass. Besides removing lignin the pretreated material washing also removes inhibitors, salts, furfural and phenolic acids.

7.2. Ammonia fiber expansion (AFEX) pretreatment

Another pretreatment that deserves attention is the ammonia fiber expansion (AFEX), which is a physicochemical process very similar to steam explosion, in which lignocellulosic biomass is exposed to liquid ammonia at high temperature and pressure for a period of time, with a subsequent quick reduction of the pressure [154]. In a typical AFEX process, the dosage of liquid ammonia is 1-2 kg of ammonia/kg of dry biomass and the temperature and residence time are around 170 °C and 30 min, respectively [2].

The AFEX technology has been used for the pretreatment of several lignocellulosic materials including wood, switchgrass, sugarcane bagasse and corn stover [154-158]. Over 90% hydrolysis of cellulose and hemicellulose was obtained after AFEX pretreatment of bermudagrass (approximately 5% lignin) and bagasse (15% lignin) [157]. Although hardwood pretreatment, like poplar, requires harsher AFEX conditions to obtain equivalent sugar yields upon enzymatic hydrolysis, poplar (*Populus nigra x Populus maximowiczii* hybrid) AFEX-pretreated at 180 °C, 2:1 ammonia to biomass loading, 30 minutes residence time by using various combinations of enzymes (commercial cellulases and xylanases) achieved high glucan and xylan conversion (93 and 65%, respectively) [159].

This process presents some disadvantages, such as the use of ammonia solvent itself, that should be recycled and handled with caution to make the process environmentally feasible, and also from an economic point of view the ammonia consumption needs to be minimized [47]. However, there are some advantages in this pretreatment, like the feasibly solvent recover and the hydrolysate from AFEX is compatible with fermentation microorganisms without the need for conditioning [160].

7.3. Biological pretreatment

Biological pretreatment employs various types of rot fungi, being the white-rot fungi the most effective for biological pretreatment of lignocellulosic biomass. The aim of biological pretreatment processes are the lignin degradation by microorganisms, through the action of lignin degrading enzymes such as peroxidases and laccases [2]. The most investigated fungus for lignin degradation is *Phanerochete chrysosporium* [161].

The biological pretreatment of sugarcane straw was evaluated by screening eight microorganisms, including bacteria and fungi, for an incubation time of 30 days. The fungus *Aspergillus terreus* was found as the most effective strain, resulting in 92% reduction in the lignin content [162]. The pretreatment of sugarcane straw was also evaluated using the fungus *Cariporiopsis subvermispora* with the objective to reduce cooking times and chemicals load for the organosolv pulping. The pretreatment was effective regarding the decomposition of lignin, however high cellulose losses were pointed as negative side effects [163]. Another study evaluated the pretreatment of sugarcane bagasse with the white-rot fungus *Pleurotus sajorcaju* PS 2001 using a 45 days incubation time, in order to modify its lignin content. However, in this case, the aim of the study was to provide a more digestible substrate for the production of cellulases by the fungus *Penicillium echinulatum* [164].

The pretreatment of the Japanese red pine *Pinus densiflora* was studied using three white-rot fungi. The fungus *Stereum hirsutum* was able to selectively degrade lignin resulting in a less recalcitrant biomass after eight weeks of pretreatment. As consequence, the sugar yields obtained after the hydrolysis of the pretreated red pine with commercial enzymes was 21% higher when compared to non pretreated control samples [165].

The main advantages of such processes are the low capital cost, low energy, no requirement for chemicals, fewer hydrolysis and fermentation inhibitors produced during pretreatment and mild environmental conditions [166]. However, the biological processes require a very long residence time, when compared to other pretreatment techniques and result in very low reaction rates. Additionally, most microorganisms consume part of the substrate as a nutrient for its growth during the pretreatment, which affects negatively the sugar yield at the end of the process [3]. In addition, the consumption of lignin also reduces the biomass energy utilization. At present, the use of biological pretreatments may represent a competitive option only if associated with other pretreatment techniques, in order to reduce the energy requirement of the total pretreatment process [167]. In future, if less recalcitrant genetically modified plant materials are available, biological pretreatments may represent an important alternative.

8. Conclusion

Sugarcane and woody biomass, which are abundant and readily available, are frontrunner materials as lignocellulosic feedstock for the production of biomass ethanol despite its differences in regard to structure and chemical composition, which relates to different responses for the same type of pretreatment.

In general, the biomass lignin content, which is an important parameter for enzymatic saccharification, is higher in woody biomass than in agricultural residues, such as sugarcane biomass. This is particularly true for softwood, which responds poorly to several pretreatment techniques, as shown throughout this chapter. This fact corroborates the need for the development of tailor-made pretreatments based on the biomass type, so that a suitable choice can benefit the subsequent bio-based conversion steps for enzymatic hydrolysis and ethanol fermentation.

The choice of pretreatment should also take into account the foreseen utilization of the main biomass molecular components (cellulose, hemicelluloses and lignin) for the ethanol production process or within the framework of the biorefinery concept. Considering the use of an ethanologenic microorganism able to ferment C6 and C5 sugars, it would be desirable to apply pretreatments such as milling or extrusion, avoiding the formation of a separated hemicelluloses stream, as observed for acidic pretreatment. However, even for the case of hydrothermal or steam pretreatments, the operational conditions can be fitted to minimize the removal of hemicellulose. Considering now a biorrefinery concept which broadens the biomass derived products, the C6 sugars could still be fermented into ethanol, while the C5 stream could be used for the production, via biotechnological routes, of a wide range of chemicals with higher added value. In that cause, the best suited pretreatments would be the acid pretreatment, which releases mostly C5 sugars, steam-based and LHW processes, which separates an oligosaccharides-rich stream. In both cases, lignin can be used as a valuable solid fuel or as a source of aromatic structures for the chemical industry.

Regarding innovative and promising biomass pretreatment technologies, the use of ILs stands out. These versatile class of chemicals can be tailored to suit the selective extraction and recovery of the biomass components, such as the recovery of a cellulose-hemicellulose rich material in an amorphous form which is prone to enzymatic hydrolysis with high yields and rates. Additionally, the possibility of recovering the extracted lignin broadens and increases the efficiency for the use of biomass.

Table 4 lists the pretreatment options presented in this chapter and its general effects in the biomass composition and structure. All pretreatments cause an increase in the surface area, which responds for the increased enzymatic digestibility of the treated materials. However, the substantial decrease in cellulose crystallinity is only observed for the treatments using ball milling and IL. This effect is of paramount importance for the increased rates and yields of cellulose enzymatic hydrolysis. The acid, LHW and steam explosion pretreatments are more effective on hemicelluloses and on the modification of the lignin structure, which also cause a higher formation of inhibitors in comparison to milling, extrusion and IL pretreatments.

Pretreatment	Increase of SSA	Reduction of CrI	Removal of hemicellulose	Removal of lignin	Modification of lignin	Formation of toxic compounds
Acid	++	-	+++	++	+++	+++
Alkali	-	-	+	+++	++	++
LHW	++	-	+++	+	++	++
Steam explosion	++	-	+++	+	+++	++
Ball milling	++	+++	-	-	-	-
WDM	+++	+	-	-	-	-
Extrusion	++	+	-	-	-	-
Ionic liquid	+++	+++	+	++	+	nd

+++ expressive effect; ++ moderate effect; + low effect; - no effect; nd: not determined
SSA: Specific surface area
CrI: Crystallinity index
LHW: Liquid hot water
WDM: Wet-disk milling

Table 4. General effects of different pretreatments on the composition and structure of the biomass.

Table 5 presents sixteen biomass ethanol plants (pilot, demonstration and commercial scale) which are operating or under construction. It is also presented, for each case, the feedstock and the biomass pretreatment that is used in these facilities. At the current scenario the majority of the units have implemented processes that generate a hemicelluloses rich stream: three units use diluted acid, three units use LHW and three units use steam-explosion pretreatment. Two units describe its process as a thermal-mechanical pretreatment which could also generate of a hemicelluloses rich stream. One unit applies a mild alkaline pretreatment that precludes lignin separation and the remaining four units have not disclosed the choice of pretreatment. A variety of feedstocks, such as pine wood chips, wood wastes, forest residues, garden waste, wheat, barley and oat straw, corn cob, corn stover, corn straw as well as perennial energy grasses, are used with different pretreatment types.

As the pretreatment step accounts for a substantial part of the biomass ethanol production cost, it is expected that the research in this field will continue to seek for improvements of existing methods or for the development of new and more advanced options.

Pretreatment	Company	Location	Biomass	Scale	Capacity m^3y^{-1}	Status	Ref.
Dilute acid hydrolysis[1]	SEKAB	Örnsköldsvik, Sweden	pine wood chips	D	5715	OP	[168]
Diluted acid	Abengoa Bioenergy	Salamanca, Spain	corn cob, corn stover and wheat straw	D	5080	OP	[169]
Diluted acid	Abengoa Bioenergy	Hugoton, USA	corn stover, wheat straw, and switchgrass	C	95000	UC	[169]
Hydrothermal	Inbicon	Fredericia, Denmark	wheat straw	P	1397	OP	170
Hydrothermal	Inbicon	Kalundborg, Denmark	wheat straw	D	5400	OP	[170]
Hydrothermal	Chemtex/ Proesa	Crescentino, Italy	perennial grass (*Arundo donax* - giant red) and wheat straw	C	50000	UC, Start by end 2012	[171]
Steam-explosion	IOGEN	Ottawa, Canada	wheat, barley and oat straw	D	1800	OP	[54]
Steam-explosion and wet oxidation[2]	BioGasol	Aakirkeby, Denmark	straw, garden waste, energy crops and grass	D	5080	UC	[172]
Steam-explosion and wet oxidation2	BioGasol	Boardman, USA	wheat straw, wood chips and corn stover	D	10233	UC	[172]
Low acid impregnation and thermal-mechanical	Blue Sugars Corp	Upton, USA	wood wastes	D	3600	OP	[173]
Thermal-mechanical	Süd-Chemie AG - Clariant	Straubing, Germany	wheat straw	D	1270	UC	[174]
Mild alkaline process	Dupont	Vonore, USA	corn cobs, corn stover and switchgrass	D	947.5	OP	[175]
Not informed	BP Biofuels	Highlands County, USA	perennial grass	C	136440	UC	[176]
Not informed	POET-DSM	Emmetsburg, USA	corn cobs, leaves, husk, and stalk	C	95000	UC, Start by end 2013	[177]
Not informed	Dupont	Nevada, USA	corn stover	C	102330	UC	[175]
Not informed	Procethol 2G, Futurol	Pomacle, France	wood wastes, agricultural and forest residue, garden waste and perennial grass	P	180	-	[178]

All pretreatments were followed by separated enzymatic hydrolysis and fermentation (SHF) or simultaneous saccharification and fermentation (SSF).

The data presented in this table was based on the official information provided in each company website.

Scale was defined as follow: Pilot – P; D – Demonstration; Commercial – C.

Operational status was defined as follow: Under construction – UC; Operational - O

[1] Two reactors in series, the hemicellulose is hydrolyzed in the first reactor and the cellulose is decomposed in the second reactor at >200 °C.

[2] Combination of steam-explosion and wet oxidation, applying both the addition of oxygen and a pressure release at high temperature (170-200° C)

Table 5. Pilot, demonstration and commercial scale biomass ethanol plants.

Author details

Ayla Sant'Ana da Silva[1], Ricardo Sposina Sobral Teixeira[1], Rondinele de Oliveira Moutta[1], Viridiana Santana Ferreira-Leitão[1,2], Rodrigo da Rocha Olivieri de Barros[1], Maria Antonieta Ferrara[3] and Elba Pinto da Silva Bon[1]

1 Departamento de Bioquímica, Instituto de Química, Universidade Federal do Rio de Janeiro, Centro de Tecnologia, Av. Athos da Silveira Ramos, Ilha do Fundão, Rio de Janeiro, RJ, Brazil

2 Instituto Nacional de Tecnologia (INT), Ministério da Ciência, Tecnologia e Inovação, Rio de Janeiro, Brazil

3 Instituto de Tecnologia em Fármacos FarManguinhos/ FIOCRUZ, Rua Sizenando Nabuco, Manguinhos, Rio de Janeiro, Brazil

References

[1] Alvira P, Tomás-Pejó E, Ballesteros M, Negro MJ. Pretreatments technologies for an efficient bioethanol production process based on enzymatic hydrolysis: A review. Bioresource Technology 2010;101 4851-4861.

[2] Kumar P, Barrett DM, Delwiche MJ, Stroeve P. Methods for pretreatment of lignocellulosic biomass for efficient hydrolysis and biofuel production. Industrial & Engineering Chemistry Research 2009;48(8) 3713-3729.

[3] Sousa LC, Chundawat SPS, Balan V, Dale BE. Cradle-to-crave assessment of existing lignocellulose pretreatment technologies. Current Opinion in Biotechnology 2009;20(3) 339-347.

[4] Sousa ELL, Macedo IC. Ethanol and bioelectricity: sugarcane in the future of the energy matrix. São Paulo: UNICA; 2010.

[5] Freitas LC, Kaneko S. Ethanol demand under the flex-fuel technology regime in Brazil. Energy Economics 2011;33 1146-1154.

[6] Soccol CR, Vandenberghe LPS, Medeiros ABP, Karp SG, Buckeridge M, Ramos LP, Pitarelo AP, Ferreira-Leitão V, Gottschalk LMF, Ferrara MA, Bon EPS, Moraes LMP, Araújo JA, Torres FAG. Boethanol from lignocelluloses: status and perspectives in Brazil. Bioresource Technology 2010;101 4820-4825.

[7] Zhu JY, Pan XJ. Woody biomass pretreatment for cellulosic ethanol production: technology and energy consumption evaluation. Bioresource Technology 2010;101 4992-5002.

[8] Braconnot H. Memoir on the conversion of wood particles in rubber, in sugar, and in a special natural acid, by means of sulfuric acid: conversion of the same woody substance in ulmin by potash. Annals of Chemistry and Physics 1819;12(2) 172-195.

[9] Lindsey JB, Tollens B. Ueber Holz-Sulfitflüssigkeit und Lignin. Annals Chemistry 1892;267 341-357.

[10] Freudemberg K. The kinetics of long chain disintegration applied to cellulose and starch. Transactions of the Faraday Society 1936; 32 74-75.

[11] Sun Y, Cheng J. Hydrolysis of lignocellulosic materials for ethanol production: a review. Bioresource Technology 2002;83(1) 1-11.

[12] Moutta RO, Chandel Anuj K, Rodrigues RCLB, Silva MB, Rocha GJM, Silva SS. Statistical optimization of sugarcane leaves hydrolysis into simple sugars by dilute sulfuric acid catalyzed process. Sugar Tech 2012;14(1) 53-60.

[13] Paiva LMC. Cultivo de *Candida utilis* em hidrolisados ácidos de bagaço de cana. Master thesis. Federal University of Rio de Janeiro; 1980.

[14] Sun Y, Cheng J. Dilute acid pretreatment of rye straw and bermuda grass for ethanol production. Bioresource Technology 2005;96(14) 1599-1606, 2005.

[15] Herrera A, Téllez-Luist SJ, Ramírez JA, Vásques M. Production of xylose from sorghum straw using hydrochloric acid. Journal of Cereal Science 2002;37(3) 267-274.

[16] Fengel D, Wegener G. Wood Chemistry, Ultrastructure, Reactions. Berlin: Walter De Gruyter Inc; 1989.

[17] Zheng Y, Pan Z, Zhang R, Wang D, Labavitch J, Jenkins BM. Dilute acid pretreatment and enzymatic hydrolysis of saline biomass for sugar production. July 9-12, 2006, Oregon Conventional Center, Portland, Oregon. An ASABE Meeting, paper. nº. 067003. American Society of Agricultural and Biological Engineers; 2006.

[18] Shuai, L. Yang, Q. Zhu, J.Y. Lu, F.C. Weimer, P.J. Ralph, J. Pan, X.J. Comparative study of SPORL and dilute-acid pretreatments of spruce for cellulosic ethanol production. Bioresource Technology 2010;101(9) 3106-3114.

[19] Grethlein HE, Allen DC, Converse AO. A comparative study of the enzymatic hydrolysis of acid-pretreated white pine and mixed hardwood. Biotechnology and Bioengineering 1984;26(12) 1498-1505.

[20] Rueda SMG, Andrade RR, Santander CMG, Costa AC, Maciel Filho R. Pretreatment of sugar cane bagasse with phosphoric and sulfuric diluted acid for fermentable sugars production by enzymatic hydrolysis. 2nd International Conference on Industrial Biotechnology 2010; ed. 1, UNICAMP, 2010 p321-326.

[21] Oliveira FMV. Avaliação de diferentes pré-tratamentos e deslignificação alcalina na sacarificação da celulose de palha de cana. Master thesis. Federal University of São Paulo (USP); 2010.

[22] Bobleter O, Concin R. Degradation of poplar lignin by hydrothermal treatment. Cellulose Chemistry and Technology 1979;13 583-593.

[23] Bobleter O, Binder H, Concin R, Burtscher E. The conversion of biomass to fuel raw material by hydrothermal pretreatment. In: Palz W, Chartier P, Hall DO (Eds). Energy from Biomass. London: Applied Science Publishers; 1981. p554-562.

[24] Bouchard J, Nguyen TS, Chornet E, Overend RP. Analytical methodology for biomass pretreatment. Part 2: characterization of the filtrates and cumulative product distribution as a function of treatment severity. Bioresource Technology 1991;36 121-131.

[25] Mok WSL, Antal Jr, MJ. Uncatalyzed solvolysis of whole biomass hemicellulose by hot compressed liquid water. Industrial Engineering Chemistry Research 1992;31 1157-1161.

[26] Mok WSL, Antal Jr MJ. Biomass fractionation by hot compressed liquid water. In: Bridgewater, A.V. (Ed.), Advances in Thermochemical Biomass Conversion, vol. 2. New York: Blackie Academic & Professional Publishers; 1994. p1572-1582.

[27] Allen SG, Kam LC, Zemann AJ, Antal Jr MJ. Fractionation of sugar cane with hot, compressed, liquid water. Industrial Engineering Chemistry Research 1996; 35 2709–2715.

[28] Mosier N, Wyman C, Dale B, Elander R, Lee YY, Holtzapple M, Ladisch M. Features of promising technologies for pretreatment of lignocellulosic biomass. Bioresource Technology 2005; 96 673–686.

[29] Negro MJ, Manzanares P, Ballesteros I, Oliva JM, Cabañas A, Ballesteros M. Hydrothermal pretreatment conditions to enhance ethanol production from poplar biomass. Applied Biochemistry and Biotechnology 2003;105(1-3) 87-100.

[30] Wyman CE, Dale BE, Elander RT, Holtzapple M, Ladisch MR, Lee YY. Coordinated development of leading biomass pretreatment technologies. Bioresource Technology 2005;96 1959-1966.

[31] Dien, BS, Li XL, Iten LB, Jordan DB, Nichols NN, O'Bryan PJ, Cotta MA. Enzymatic saccharification of hot-water pretreated corn fiber for production of monosaccharides. Enzyme and Microbial Technology 2006;39 1137-1144.

[32] Liu C, Wyman CE. Partial flow of compressed-hot water through corn stover to enhance hemicellulose sugar recovery and enzymatic digestibility of cellulose. Bioresource Technology 2005;96 1978-1985.

[33] Mosier N, Hendrickson R, Ho N, Sedlak M, Ladisch MR. Optimization of pH controlled liquid hot water pretreatment of corn stover. Bioresource Technology 2005; 96 1986-1993.

[34] Garrote G, Falqué E, Domínguez H, Parajó JC. Autohydrolysis of agricultural residues: study of reaction byproducts. Bioresource Technology 2007;98 1951-1957.

[35] Pérez JA, Ballesteros I, Ballesteros M, Sáez F, Negro MJ, Manzanares P. Optimizing liquid hot water pretreatment conditions to enhance sugar recovery from wheat straw for fuel–ethanol production. Fuel 2008;87 3640-3647.

[36] Laser M, Schulman D, Allen SG, Lichwa J, Antal MJ, Lynd LR. A comparison of liquid hot water and steam pretreatments of sugar cane bagasse for bioconversion to ethanol. Bioresource Technology 2002;81 33-44.

[37] Sasaki M, Adschiri T, Arai K. Fractionation of sugarcane bagasse by hydrothermal treatment. Bioresource Technology 2003;86 301-304.

[38] He J, Li Q. Forest Chemical Industry Handbooks, first ed. China, Beijing: Forestry Publishing House; 2001.

[39] Inoue H, Yano S, Endo T, Sakaki T, Sawayama S. Combining hot compressed water and ball milling pretreatments to improve the efficiency of the enzymatic hydrolysis of *Eucalyptus*. Biotechnology for Biofuels 2008;1(2) 1-9.

[40] Yu Q, Zhuang X, Yuan Z, Wang Q, Qi W, Wang W, Zhang Y, Xu J, Xu H. Two-step liquid hot-water pretreatment of Eucalyptus grandis to enhance sugar recovery and enzymatic digestibility of cellulose. Bioresource Technology 2010; 101(13) 4895-4899.

[41] Cara C, Ruiz E, Oliva JM, Saez F, Castro E. Conversion of olive tree biomass into fermentable sugars by dilute acid pretreatment and enzymatic saccharification. Bioresource Technology 2007; 99 1869-1876.

[42] Kim TH, Lee YY. Fractionation of corn stover by hot-water and aqueous ammonia treatment. Bioresource Technology 2006;97 224-232.

[43] Carrasco C, Baudel HM, Sendelius J, Modig T, Roslander C, Galbe M. SO_2-catalyzed steam pretreatment and fermentation of enzymatically hydrolyzed sugarcane bagasse. Enzyme Microbiology Technology 2010;46(2) 64-73.

[44] Taib RM, Ishak ZAM, Rozman HD, Glasser WG. Steam-exploded wood fibers as composite reinforcement. IN: Fakirov S, Bhattacharyya D. Handbook of Engineering Biopolymers: Homopolymers, Blends and Composites. Munich: Carl Hanser Verlag; 2007.

[45] Linde M, Galbe M, Zacchi G. Steam pretreatment of acid-sprayed and acid-soaked barley straw for production of ethanol. Applied Biochemistry Biotechnology 2006;130(1-3) 546-562.

[46] Ramos LP. The chemistry involved in the steam treatment of lignocellulosic materials. Química Nova 2003;26(6) 863-871.

[47] Galbe M, Zacchi G. Pretreatment: The key to efficient utilization of lignocellulosic materials. Biomass and Bioenergy 2012; doi:10.1016/j.biombioe.2012.03.026

[48] Pitarelo AP. Avaliação da susceptibilidade do bagaço e da palha de cana-de-açúcar à bioconversão via pré-tratamento a vapor e hidrólise enzimática. Master thesis. Federal University Paraná; 2007.

[49] Pan X, Xie D, Gilkes N, Gregg DJ, Saddler JN. Strategies to enhance the enzymatic hydrolysis of pretreated softwood with high residual lignin content. Applied Biochemistry and Biotechnology: Part A: Enzyme Engineering and Biotechnology 2005;124 1069-1079.

[50] Ferreira-Leitão V, Perrone CC, Rodrigues J, Franke APM., Macrelli S, Zacchi, G. An approach to the utilisation of CO_2 as impregnating agent in steam pretreatment of sugar cane bagasse and straw for ethanol production. Biotechnol Biofuels 2010;3 1-8.

[51] DeLong EA. Canadian patent 1 141 376, 1983.

[52] DeLong EA, DeLong EP, Ritchie GS, Rendall WA, US patent 4 908 098, 1990.

[53] Gibson WG, US patent 7 189 306 B2, 2007.

[54] Iogen Corporation. http://www.iogen.ca/ (accessed 17 July 2012).

[55] Kling SH, Carvalho Neto CC, Torres JCR, Magalhães DB, Ferrara MA. III Congresso Brasileiro de Energia: conference proceedings, October, 1984, COPPE/UFRJ, Rio de Janeiro, Brazil.

[56] Ferrara MA, Martins MRJ, Kling SH. Efeito da explosão no tratamento de bagaço de cana de açúcar por "steam explosion" a 200ºC. III Seminário de Hidrólise Enzimática de Biomassas: conference proceedings, December, Federal University of Maringá, Maringá, 1987.

[57] Kling SH, Carvalho Neto CC, Ferrara MA, Torres JCR, Magalhães DB, Ryu DDY. Enhancement of enzymatic hydrolysis of sugar cane bagasse by steam explosion pretreatment. Biotechnology and Bioengineering 1987;29(8) 1035-1039.

[58] Ferrara MA, Bon EPS, Araujo Neto, JS. Use of steam explosion liquor from sugar cane bagasse for lignin peroxidase production by *Phanerochaete chrysosporium*. Applied Biochemistry and Biotechnology 2002;98-100 289-300.

[59] Avellar BK, Glasser WG. Steam-assisted biomass fractionation. I. Process considerations and economic evaluation. Biomass Bioenergy 1998;14 205-218.

[60] Hamelinck CN, van Hooijdonk G, Faaij APC. Ethanol from lignocellulosic biomass: techno-economic performance in short-, middle- and long-term. Biomass Bioenergy 2005;28 384-410.

[61] Taherzadeh MJ, Karimi K. Pretreatment of lignocellulosic wastes to improve ethanol and biogas production: A review. International Journal of Molecular Science 2008; 9 1621-1651.

[62] Lin Z, Huang H, Zhang H, Zhang L, Yan L, Chen J. Ball milling pretreatment of cornstover for enhancing the efficiency of enzymatic hydrolysis. Applied Biochemistry and Biotechnology 2010;162 1872-1880.

[63] Silva GGD, Couturier M, Berrin Jean-Guy, Buléon A, Rouau X. Effects of grinding processes on enzymatic degradation of wheat straw. Bioresource Technology 2012; 103 192-200.

[64] Nishiyama Y, Langan P, Chanzy H. Crystal structure and hydrogen-bonding system in cellulose Iβ from synchrotron X-ray and neutron fiber diffraction. Journal of the American Chemical Society 2002;124 9074-9082.

[65] Endo T, Tanaka N, Sakai M, Teramoto Y, Lee SH. Enhancement mechanism of enzymatic saccharification of wood by mechanochemical treatment. The Third Biomass-Asia Workshop, 15-17 November 2006, Tokyo and Tsukuba; 2006.

[66] Silva AS, Inoue H, Endo T, Yano S, Bon EPS. Milling pretreatment of sugarcane bagasse and straw for enzymatic hydrolysis and ethanol fermentation. Bioresource Technology 2010;101 7402-7409.

[67] Hendriks ATWM, Zeeman G. Pretreatments to enhance the digestibility of lignocellulosic biomass. Bioresource Technology 2009;100 10-18.

[68] Endo T, Tanaka N, Yamasaki R, Teramoto Y, Lee SH. Wet mechanochemical treatment for enzymatic saccharification of wood. In: 15th Annual Meeting of the Cellulose Society of Japan. Kyoto 10-11 July 2008. p117-118.

[69] Hideno A, Inoue H, Tsukahara K, Fujimoto S, Minowa T, Inoue S, Endo T, Sawayama S. Wet disk milling pretreatment without sulfuric acid for enzymatic hydrolysis of rice straw. Bioresource Technology 2009;100 2706- 711.

[70] Mercier C, Feillet P. Modification of carbohydrate components by extrusion cooking of cereal products. Cereal Chemistry 1975;52 283-297.

[71] Riaz MN. Introduction to extruders and their principles. In: Extruders in food applications, M. N. Riaz, (Ed.), pp.1-23, CRC Press, ISBN 978-156-6767-79-8, Boca Raton, United States of America 2000.

[72] Lee SH, Teramoto Y, Tanaka N, Endo T. Improvement of enzymatic saccharification of woody biomass by nano-fibrillation using extruder, In: The 57th Annual Meeting of The Japan Wood Research Society; 2007.

[73] Gonzalez-Valadez M, Munoz-Hernandez G, Sanchez-Lopez R. Design and evaluation of an extruder to convert crop residues to animal feed. Biosystems Engineering 2008;100(1) 66-78.

[74] Isobe S, Zuber F, Uemura K, Nogushi A. A new twin-screw press design for oil extraction of dehulled sunflower seeds. Journal of the American Oil Chemists' Society 1992;69(9) 884-889.

[75] Bouzid N, Vilarem G, Gaset A. Extraction des huiles essentielles par des techniques non conventionnelles. *Proc. Conf. 'Valorisation Non-alimentaire des Grandes Productions Agricoles'*, Nantes, France, 18-19 May 1994, INRA edn (in press). In: N'Diaye S, Rigal L, Larocque P, Vidal PF. Extraction of hemicelluloses from poplar, *Populus tremu-*

loides, using an extruder-type twin-screw reactor: A feasibility study. Bioresource Technology 1996;57(1) 61-67.

[76] N'Diaye S, Rigal L, Larocque P, Vidal PF. Extraction of hemicelluloses from poplar, *Populus tremuloides*, using an extruder-type twin-screw reactor: A feasibility study. Bioresource Technology 1996;57(1) 61-67.

[77] Kaliyan N, Morey RV. Factors affecting strength and durability of densified biomass products. Biomass and Bioenergy 2009;33(3) 337-359.

[78] Muthukumarappan K, Julson JL. Pretreatment of biomass using a novel extrusion process. In: 15th European Biomass Conference & Exhibition from Research to Market Development, 2007, p583-586, Florence, Italy: ETA srl; 2007.

[79] Lee SH, Teramoto Y, Endo T. Enzymatic saccharification of woody biomass micro/nanofibrillated by continuous extrusion process I – Effect of additives with cellulose affinity. Bioresource Technology 2009a;100(1) 275-279.

[80] Lee SH, Inoue S, Teramoto Y, Endo T. Enzymatic saccharification of woody biomass micro/nanofibrillated by continuous extrusion process II: effect of hot-compressed water treatment. Bioresource Technology 2010a;101(24) 9645-9.

[81] Lee SH, Teramoto Y, Endo T. Enhancement of enzymatic accessibility by fibrillation of woody biomass using batch-type kneader with twin-screw elements. Bioresource Technology 2010b;101(2) 769-774.

[82] Yoo J, Alavi S, Vadlani P, Amanor-Boadu V. Thermomechanical extrusion pretreatment for conversion of soybean hulls to fermentable sugars. Bioresource Technology 2011;102(16) 7583-7590.

[83] Karunanithy C, Muthukumarappan K, Julson JL. Influence of high shear bioreactor parameters on carbohydrate release from different biomasses. ASABE Paper No. 084109. St. Joseph, Mich: ASABE, 2008a.

[84] Karunanithy C, Muthukumarappan K. Influence of extruder temperature and screw speed on pretreatment of corn stover while varying enzymes and their ratios. Applied Biochemistry Biotechnology 2010a;162 264-279.

[85] Karunanithy C, Muthukumarappan K. Effect of extruder parameters and moisture content of switchgrass, prairie cord grass on sugar recovery from enzymatic hydrolysis. Applied Biochemistry Biotechnology 2010b;162(6) 1785-1803.

[86] Karunanithy C, Muthukumarappan K. Influence of extruder and feedstock variables on torque requirement during pretreatment of different types of biomass e A response surface analysis. Biosystems Engineering 2011a;109 37–51.

[87] Karunanithy C, Muthukumarappan K. Optimization of alkali soaking and extrusion pretreatment of prairie cord grass for maximum sugar recovery by enzymatic hydrolysis. Biochemical Engineering Journal 2011b;54(2) 71-82

[88] Kalyon DM, Malik M. An integrated approach for numerical analysis of coupled flow and heat transfer in co-rotating twin screw extruders. International Polymer Processing 2007;22(3) 293-302.

[89] Senturk-Ozer S, Gevgilili H, Kalyon DM. Biomass pretreatment strategies via control of rheological behavior of biomass suspensions and reactive twin screw extrusion processing. Bioresource Technology. 2011;102(19) 9068-9075.

[90] Malik M, Kalyon DM. Three-dimensional finite element simulation of processing of generalized newtonian fluids in counter-rotating and tangential twin screw extruder and die combination. International Polymer Processing 2005;20(05) 398-409.

[91] Carr ME, Doane WM. Modification of wheat straw in a high shear mixer. Biotechnology and Bioengineering 1984;26(10) 1252-1257.

[92] Dale BE, Weaver J, Byers FM. Extrusion processing for ammonia fiber explosion (AFEX). Applied Biochemistry and Biotechnology 1999;77(1-3) 35-45.

[93] Chen WH, Xu YY, Hwang WS, Wang JB. Pretreatment of rice straw using an extrusion/extraction process at bench-scale for producing cellulosic ethanol. Bioresource Technology 2011;102(22) 10451-8.

[94] Choi CH, Oh KK. Application of a continuous twin screw-driven process for dilute acid pretreatment of rape straw. Bioresource Technology 2012;110 349-54.

[95] Silva AS, Teixeira RSS, Endo T, Bon EPS, Lee S-H. Use of a twin-screw extruder as a mixing tool to promote sugarcane bagasse pretreatment with ionic liquids at high solids contents. In: Proceedings of the 34th Symposium on biotechnology for fuels and chemicals, 30 April-3 May 2012, New Orleans, USA. p72. 2012.

[96] Colonna P, Buleon A. Transformations structurales de l'Amidon. In: Prat L, N'Diaye S, Rigal L, Gourdon C. Solid–liquid transport in a modified co-rotating twin-screw extruder-dynamic simulator and experimental validations. Chemical Engineering and Processing 2004;4(7) 881-886.

[97] de Choudens C, Angelier R. Les pâtes chimicothermomécaniques blanchies obtenues avec le procédé bi-vis. In: Prat L, N'Diaye S, Rigal L, Gourdon C. Solid–liquid transport in a modified co-rotating twin-screw extruder-dynamic simulator and experimental validations. Chemical Engineering and Processing 2004;4(7) 881-886.

[98] Hietala M, Niinimäki J, Oksman K. The use of twin-screw extrusion in processing of wood : The effect of processing parameter and pretreatment. Bioresource Technology 2011;6(4) 4615-4625.

[99] de Vrije T, de Haas GG, Tan GB, Keijsers ERP, Claassen PAM. Pretreatment of Miscanthus for hydrogen production by *Thermotoga elfii*. International Journal of Hydrogen Energy 2002;27(11-12) 1381-1390.

[100] Karuppuchamy V, Muthukumarappan K. Extrusion pretreatment and enzymatic hydrolysis of soybean hulls. ASABE Paper No. BIO-097989. ASABE. St. Joseph, MI, U.S., 2009.

[101] Litzen D, Dixon D, Gilcrease P, Winter R. Pretreatment of biomass for ethanol production. US Patent Application US2006/0141584 A1; 2006.

[102] Lamsal BP, Yoo J, Brijwani K, Alavi S. Extrusion as a thermo-mechanical pretreatment for lignocellulosic ethanol. Biomass Bioenergy 2010;34(12) 1703-1710.

[103] Jurisic V, Karunanithy C, Julson JL. Effect of extrusion pretreatment on enzymatic hydrolysis of *Miscanthus*. ASABE Paper no. 097178. St. Joseph: ASABE; 2009.

[104] Tanahashi M, Takada S, Aoki T, Goto T, Higuchi T, Hanai S. Characterization of explosion wood: 1. Structure and physical properties. Wood Research 1983;69 36-61.

[105] Bertran MS, Dale BE. Enzymatic hydrolysis and recrystallization behavior of initially amorphous cellulose. Biotechnology and Bioengineering 1985; 27 177-181.

[106] Ouajai S, Shanks RA. So6lvent and enzyme induced recrystallization of mechanically degraded hemp cellulose. Cellulose 2006;13(1) 31-44.

[107] Yoshida M, Liu Y, Uchida S, Kawarada K, Ukagami Y, Hinose H, Kaneko S, Fukuda K.. Effects of cellulose crystallinity, hemicellulose, and lignin on the enzymatic hydrolysis of miscanthus sinensis to monosaccharides. Bioscience Biotechnology, and Biochemistry 2008;72(3) 805-810.

[108] Kumagai H, Kumagai H, Yano T. Critical bubble radius for expansion in extrusion cooking. Journal of Food Engineering 1993;20(4) 325-338.

[109] Thymi S, Krokida MK, Pappa A, Maroulis ZB. Structural properties of extruded corn starch. Journal of Food Engineering 2005;68(4) 519-526.

[110] Walden P. Molecular weights and electrical conductivity of several fused salts. Bulletin of the Academy of Science of St. Petersburg 1914; 405-422.

[111] Fremantle M. Designer solvents—Ionic liquids may boost clean technology development. Chemical. Engeneering News 1998;76 32-37.

[112] Earle MJ, Seddon, KR. Ionic liquids. Green solvents for the future. Pure and Applied Chemistry 2000;7 1391-1398.

[113] Smiglak M, Metlen A, Rogers RD. The second evolution of ionic liquids: from solvents and separations to advanced materials -energetic examples from the ionic liquid cookbook. Accounts of Chemical Research 2007;40 1182-1192.

[114] Ha SH, Mai NL, Gwangmin A, Koo Y. Microwave-assisted pretreatment of cellulose in ionic liquid for accelerated enzymatic hydrolysis. Bioresource Technology 2011;102 1214-1219.

[115] Nguyen TD, Kim K, Han SJ, Cho HY, Kim JW, Park SM, Park JC, Sim SJ. Pretreatment of rice straw with ammonia and ionic liquid for lignocelluloses conversion to fermentable sugars. Bioresource Technology 2010;101 7432-7438.

[116] Silva AS, Lee S-H, Endo T, Bon EPS. Major improvement in the rate and yield of enzymatic saccharification of sugarcane bagasse via pretreatment with the ionic liquid 1-ethyl-3-methylimidazolium acetate ([Emim] [Ac]). Bioresource Technology 2011;102 10505-10509.

[117] Kimom KS, Alan EL, Sinclair DWO. Enhanced saccharification kinetics of sugarcane bagasse pretreated in 1-butyl-3-methylimidazolium chloride at high temperature and without complete dissolution. Bioresource Technology 2011; 102 9325-9329.

[118] Li Q, He YC, Xian M, Jun G, Xu X, Yang JM, Li LZ. Improving enzymatic hydrolysis of wheat straw using ionic liquid 1-ethyl-3-methyl imidazolium diethyl phosphate pretreatment. Bioresource Technology 2009;100 3570-3575.

[119] Li C, Knierim B, Manisseri C, Arora R, Scheller HV, Auer M, Vogel KP, Simmons BA, Singh S. Comparison of dilute acid and ionic liquid pretreatment of switchgrass: biomass recalcitrance, delignification and enzymatic saccharification. Bioresource Technology 2010;101 4900-4906.

[120] Brandt A, Ray MJ, To TQ, Leak DJ, Murphy RJ, Welton T. Ionic liquid pretreatment of lignocellulosic biomass with ionic liquid–water mixtures. Green Chemistry 2011;13 2489-2499.

[121] Lee SH, Doherty TV, Linhardt RJ, Dordick, JS. Ionic liquid-mediated selective extraction of lignin from wood leading to enhanced enzymatic cellulose hydrolysis. Biotechnology and Bioengineering 2009b; 102 1368-1376.

[122] Sun N, Rahman M, Qin Y, Maxim ML, Rodriguez H, Rogers, RD. Complete dissolution and partial delignification of wood in the ionic liquid 1-ethyl-3-methylimidazolium acetate. Green Chemistry 2009;11 646-655.

[123] Brandt A, Hallett JP, Leak DJ, Murphy RJ, Welton T. The effect of the ionic liquid anion in the pretreatment of wood chips. Green Chemistry 2010;12 672-679.

[124] Graenacher C. Cellulose solutions. US Patent 1 943 175, 1934.

[125] Swatloski RP, Spear SK, Holbrey JD, Rogers RD. Dissolution of cellulose with ionic liquids. Journal American Chemical Society 2002;124 4974-4975.

[126] Fort DA, Remsing RC, Swatloski RP, Moyna P, Moyna G, Rogers RD. Can Ionic Liquids Dissolve Wood? Processing and analysis of lignocellulosic materials with 1-n-butyl-3-methylimidazolium chloride. Green Chemistry 2007; 9 63-69.

[127] Dadi AP, Varansi S, Schall CA. Enhancement of cellulose saccharification kinetics using an ionic liquid pretreatment step. Biotechnology and Bioengineering 2006;95 904-910.

[128] Moulthrop JS, Swatloski RP, Moyna G, Rogers RD. High-resolution ^{13}C NMR studies of cellulose and cellulose oligomers in ionic liquid solutions. Chemical Communications 2005;1557-1559.

[129] Anderson JL, Ding J, Welton T, Armstrong DW. Characterizing ionic liquids on the basis of multiple solvation interactions. Journal of American Chemical Society 2002;124 14247-14254.

[130] Tadese H, Luque R. Advances on biomass pretreatment using ionic liquids: An overview. Energy and Environmental. Science 2011;4 3913-3929.

[131] Yoon LW, Ang TN, Ngoh GC, Chua ASM. Regression analysis on ionic liquid pretreatment of sugarcane bagasse and assessment of structural changes. Biomass and Bioenergy 2012;36 160-169.

[132] Zhang Z, O'Hara IM, Doherty WOS. Pretreatment of sugarcane bagasse by acid-catalysed process in aqueous ionic liquid solutions. Bioresource Technology 2012;120 149-156.

[133] Diedericks D, Rensburg E, Garcia-Prado MP and Gorgens JF. Enhancing the enzymatic digestibiity of sugarcane bagasse through the application of an ionic liquid in combination with an acid catalyst. Biotechnology Progress 2012;28 76-84.

[134] Qiu Z, Aita GM, Walker MS. Effect of ionic liquid pretreatment on the chemical composition, structure and enzymatic hydrolysis of energy cane bagasse. Bioresource Technology 2012;117 251-256.

[135] Zhu Z, Zhu M, Wu Z. Pretreatment of sugarcane bagasse with NH(4)OH-H(2)O(2) and ionic liquid for efficient hydrolysis and bioethanol production. Bioresource Technology 2012;119 199-207.

[136] Kilpeläinen I, Xie H, King A, Granstrom M, Heikkinen S, Argyropoulos DS. Dissolution of wood in ionic liquids. Journal of Agricultural and Food Chemistry 2007;55 9142-9148.

[137] Wu J, Zhang H J, Zhang J, He Q, Ren MG. Homogeneous acetylation of cellulose in a new ionic liquid. Biomacromolecules 2004;5 266-268.

[138] Torr KM, Love KT, Çetinkol OP, Donaldson LA, George A, Holmes BM, Simmons BA. The impact of ionic liquid pretreatment on the chemistry and enzymatic digestibility of *Pinus radiate* compression wood. Green Chemistry 2012;14 778-787.

[139] Pezoa R, Cortinez V, Hyvärinen S, Reunanen M, Hemming J, Lienqueo ME, Salazar O, Carmona R, Garcia A, Murzin DY, Mikkola JP. Use of ionic liquids in the pretreatment of forest and agricultural residues for the production of bioethanol. Cellulose Chemistry and Technology 2010;44(4-6) 165-172.

[140] Wu H, Mora-Pale M, Miao J, Doherty TV, Linhardt RJ, Dordick JS. Facile pretreatment of lignocellulosic biomass at high loadings in room temperature ionic liquids. Biotechnology and Bioengineering 2011;108(12) 2865-2875.

[141] Klein-Marcuschamer D, Simmons BA, Blanch HW. Techno-economic analysis of a lignocellulosic ethanol biorefinery with ionic liquid pre-treatment. Biofuels, Bioproducts and Biorefinering 2011; 5 562-569.

[142] Tuner MB, Spear JG, Huddleston JG, Holbrey JD, Rogers RD. Ionic liquid salt-induced inactivation and unfolding of cellulase from Trichoderma reesei. Green Chemistry 2003;5 443-447.

[143] Ouellet M, Datta S, Dibble DC, Tamrakar PR, Benke PI, Li C, Singh S, Sale KL, Adams PD, Keasling JD,. Simmons BA, Holmes BM, Mukhopadhyay A. Impact of ionic liquid pretreated plant biomass on Saccharomyces cerevisiae growth and biofuel production. Green Chemistry 2011;13 2743-2749.

[144] Pottkämper J, Barthen P, Ilmberger N, Schwaneberg U, Schenk A, Schulte M, Ignatiev N, Streit WR. Applying metagenomics for the identification of bacterial cellulases that are stable in ionic liquids. Green chemistry 2009;11 957-965.

[145] Datta S, Holmes B, Park JI, Chen Z, Dibble DC, Hadi M, Blanch HW, Simmons BA, Sapra R. Ionic liquid tolerant hyperthermophilic cellulases for biomass pretreatment and hydrolysis. Green Chemistry 2010; 12 338-345.

[146] Soto ML, Dominguez H, Nunez MJ, Lema JM. Enzymatic saccharification of alkali-treated sunflower hulls. Bioresource Technology 1994; 49(1) 53-59.

[147] Zhao Y, Wang Y, Zhu JY, Ragauskas A, Deng Y. Enhanced enzymatic hydrolysis of spruce by alkaline pretreatment at low temperature. Biotechnology and Bioengineering 2008; 99(6) 1320-1328.

[148] Zhu J, Wan C, Li Y. Enhanced solid-state anaerobic digestion of corn stover by alkaline pretreatment. Bioresource Technology 2010; 101(19) 7523-7528.

[149] Brodeur G, Yau E, Badal K, Collier J, Ramachandran KB, Ramakrishnan S. Chemical and Physicochemical Pretreatment of Lignocellulosic Biomass: A Review. Enzyme Research 2011; 1-17.

[150] Rabelo SC, Filho RM, Costa AC. Lime pretreatment of sugarcane bagasse for bioethanol production. Applied Biochemistry and Biotechnology 2009;153(1–3) 139-150.

[151] Hu Z, Wang Y, Wen Z. Alkali (NaOH) pretreatment of switchgrass by radio frequency-based dielectric heating. Applied Biochemistry and Biotechnology 2008; 148 1-3 71-81.

[152] Park JY, Shiroma R, Al-Haqetal MI. A novel lime pretreatment for subsequent bioethanol production from rice straw calcium capturing by carbonation (CaCCO) process. Bioresource Technology 2010;101(17) 6805–6811.

[153] Chang VS, Nagwani M, Kim CH, Holtzapple MT. Oxidative lime pretreatment of high-lignin biomass. Applied Biochemistry and Biotechnology 2001;94 1-28.

[154] Mes-Hartree M, Dale BE, Craig WK. Comparison of steam and ammonia pretreatment for enzymatic hydrolysis of cellulose. Applied Microbiology Biotechnology 1988;29 462-468.

[155] McMillan JD. Pretreatment of lignocellulosic biomass. In Enzymatic Conversion of Biomass for Fuels Production; Himmel ME, Baker JO, Overend RP (Eds). American Chemical Society: Washington, DC, p292-324 1994.

[156] Alizadeh H, Teymouri F, Gilbert TI, Dale BE. Pretreatment of switchgrass by ammonia fiber explosion (AFEX). Applied Biochemical Biotechnology 2005; 121-123 1133-1141.

[157] Holtzapple MT, Jun JH, Ashok G, Patibandla SL, Dale BE. The ammonia freeze explosion (AFEX) process: A practical lignocellulose pretreatment. Applied Biochemical Biotechnology 1991;28-29 59-74.

[158] Teymouri, F, Perez LL, Alizadeh H, Dale BE. Ammonia fiber explosion treatment of corn stover. Applied Biochemical Biotechnology 2004;113-116 951-963.

[159] Balan V, Sousa LC, Chundawat SP, Marshall D, Sharma LN, Chambliss CK, Dale BE. Enzymatic digestibility and pretreatment degradation products of AFEX-treated hardwoods (*Populus nigra*) Biotechnology Progress 2009; 25(2) 365-375.

[160] Teymouri F, Laureano-Peres L, Alizadeh H, Dale BE. Optimization of the ammonia fiber explosion (AFEX) treatment parameters for enzymatic hydrolysis of corn stover. Bioresource Technology 2005; 96 (18) 2014-2018.

[161] Müller HW, Trösch W. Screening of white-rot fungi for biological pretreatment of wheat straw for biogas production. Applied Microbiology and Biotechnology 1986; 24(2) 180-185.

[162] Singh P, Suman A, Tiwari P, Arya N, Gaur A, Shrivastava AK. Biological pretreatment of sugarcane trash for its conversion to fermentable sugars. World Journal of Microbiology and Biotechnology 2008; 24(5) 667-673.

[163] Saad MBW, Oliveira LRM, Cândido RG, Quintana G, Rocha GJM, Gonçalves AR. Preliminary studies on fungal treatment of sugarcane straw for organosolv pulping. Enzyme and Microbial Technology 2008;43 220-225.

[164] Camassola M, Dillon AJP. Biological pretreatment of sugar cane bagasse for the production of cellulases and xylanases by *Penicillium echinulatum*. Industrial Crops and Products 2009;29(2-3) 642-647.

[165] Lee J-W, Gwak K-S, Park J-Y, Park M-J, Choi D-H, Kwon M, Choi I-G. Biological pretreatment of softwood *Pinus densiflora* by three white rot fungi. Journal of Microbiology 2007;45(6) 485-491.

[166] Alvira P, Tomás-Pejó E, Ballesteros M, Negro MJ. Pretreatments Technologies for an efficient bioethanol production process based on enzymatic hydrolysis: A review. Bioresource Technology 2010;101 4851-4861.

[167] Balan V, Sousa LC, Chundawat SPS, Vismeh R, Jones AD, Dale BE. Mushroom spent straw: a potential substrate for an ethanol based biorefinery. Journal of Industrial Microbiology and Biotechnology 2008;35 293-301.

[168] SEKAB. http://www.sekab.com/cellulose-ethanol/demo-plant (accessed 28 July 2012).

[169] Abengoa Bioenergy. http://www.abengoabioenergy.com/corp/web/en/2g_hugoton_project/general_information/index.html (accessed 28 July 2012).

[170] Inbicon. http://www.inbicon.com/Biomass%20Refinery/Pages/Inbicon_Biomass_Refinery_at_Kalundborg.aspx (accessed 28 July 2012).

[171] Chemtex/Proesa. http://www.chemtex.com/templates/renewables_PROESA.html or http://www.betarenewables.com/Crescentino.html (accessed 28 July 2012).

[172] BioGasol. http://www.biogasol.com/Projects-177.aspx (accessed 28 July 2012).

[173] Blue Sugars Corp. http://bluesugars.com/technology-demonstration.htm (accessed 28 July 2012).

[174] Süd-Chemie AG – Clariant. http://www.sud-chemie.com/scmcms/web/content.jsp?nodeIdPath=2991,7756,7757,7760,7767&lang=en (accessed 28 July 2012).

[175] Dupont. http://ddce.com/cellulosic/index.html (accessed 28 July 2012).

[176] BP Biofuels. http://www.bp.com/sectiongenericarticle.do?categoryId=9040296&contentId=7073450 (accessed 28 July 2012).

[177] POET-DSM. http://www.poetdsm.com/ or http://www.poet.com/emmetsburg (accessed 28 July 2012).

[178] Procethol 2G, Futurol. http://www.projet-futurol.com/index-uk.php (accessed 28 July 2012).

Section 2

Biomass Processing

Effectiveness of Lignin-Removal in Simultaneous Saccharification and Fermentation for Ethanol Production from Napiergrass, Rice Straw, Silvergrass, and Bamboo with Different Lignin-Contents

Masahide Yasuda, Keisuke Takeo,
Tomoko Matsumoto, Tsutomu Shiragami,
Kazuhiro Sugamoto, Yoh-ichi Matsushita and
Yasuyuki Ishii

Additional information is available at the end of the chapter

http://dx.doi.org/10.5772/54194

1. Introduction

Second-generation biofuels from lignocellulosic materials have gained much attention since the lignocelluloses are not in competition with food sources and animal feed and will provide a new sustainable energy sources alternative to petroleum-based fuels (Galbe and Zacchi, 2007). Bioethanol production from herbaceous lignocellulose such as corn stover (Ryu and Karim, 2011), rice straw (Ko *at al.*, 2009), sweet sorghum bagasse (Cardoba *et al.*, 2010), switchgrass (Keshwani and Cheng, 2009), bamboo (Sathitsuksanoh *at al.*, 2010), wheat straw (Talebnia *et al.*, 2010), alfalfa stems (González-García *at al.*, 2010), and silvergrass (Guo *et al.*, 2008) has been extensively developed through a variety of processes combining the biological saccharification and fermentation steps with the pre-treatment methods. In almost all processes, the pretreatments to remove the lignin components and to promote an enzymatic digestibility of cellulosic components are carried out by the use of energy and cost which are frequently higher than those of bio-fuels gained (Alvira *et al.*, 2010). If lignocelluloses with low lignin-content are selected, the operation to remove the lignin might be excluded from the bio-ethanol process.

Among the many kinds of lignocelluloses, therefore, we (Yasuda *et al.*, 2011; Yasuda *et al.*, 2012) and other groups (Li *et al.*, 2011; Brandon *et al.*, 2011; Zhang *et al.*, 2011; Huang *et al.*,

2011; Lin *et al.*, 2011a; Lin *et al.*, 2010b; Kai *et al.*, 2010; Anderson *et al.*, 2008) have been interested in napiergrass (*Pennisetum purpureum* Schumach) which is herbaceous lignocellulose with its low lignin- content. During our investigations on bioethanol production, it was found that the alkali-pretreatment of napiergrass enhances scarcely the ethanol yield whereas the alkali-pretreatment of silvergrass (*Miscanthus sinensis* Anderss) remarkably enhances the ethanol yield (Yasuda *et al.*, 2011). Here, we compared the effectiveness of lignin-removal between napiergrass and other lingocelluloses with different lignin-contents (rice straw, silvergrass, and bamboo) in order to evaluate the availability of non-pretreated napiergrass as the raw materials of bio-ethanol.

2. Materials and methods

2.1. Chemical components of herbaceous lignocellulose

The lignocellulosic materials were cut, dried, and powdered until the 70 % of the particles became in a range of 32-150 μm in length to promote the cellulase- saccharification and to reduce varying in components in each experiment. The lignin-contents in lignocelluloses were determined as follows. The powdered lignocelluloses (30.0 g) was washed with MeOH and treated with a 1% aqueous solution of NaOH (400 mL) at 95 ºC for 1 h (Silverstein, *et al.*, 2007; Yasuda *et al.*, 2011; Yasuda *et al.*, 2012). After centrifugation at 10,000 rpm for 10 min to separate the precipitates, the supernatant solution was neutralized to pH 5.0 by a dilute HCl solution to give the lignin as a dark brown precipitate. The lignin-contents of napiergrass, rice straw, silvergrass, and bamboo were determined to be 14.9, 18.2, 21.7, and 26.2 wt%, respectively.

The holocellulose (cellulose and hemicellulose) was isolated as a pale yellow precipitate by the above centrifugation. The saccharide components of holocellulose were determined according to the methods published by the National Renewable Energy Laboratory (NREL) as follows (Sluiter *et al.*, 2010). Sulfuric acid (72%) was added to holocellulose and then diluted with water until the concentration of sulfuric acid became 4%. This was heated at 121 ºC for 1 h in a grass autoclave (miniclave, Büchi AG, Switerland). HPLC analysis of the hydrolyzate showed that holocellulose mainly composed of glucose and xylose along with the small amounts of arabinose and galactose. The ash component in lignocelluloses was obtained by the burning of the lignocelluloses (2.0 g) in an electric furnace (KBF784N1, Koyo, Nara, Japan) for 2 h at 850 ºC. Chemical components of lignocelluloses are shown in Table 1.

2.2. Saccarification

As has been previously reported (Yasuda *et al.*, 2011; Yasuda *et al.*, 2012), a cellulase from *Acremonium cellulolyticum* (Acremozyme, Kyowa Kasei, Osaka, Japan) was selected by the comparison in activity with other cellulase such as Meycellase (Kyowa Kasei), a cellulase from *Trichoderma viride* (Wako Chemicals, Osaka, Japan) and a cellulase from *Aspergillus niger* (Fluka Japan, Tokyo). The cellulase activity of Acremozyme was determined by the method of the breakdown of filter paper (Yasuda *et al.*, 2012). At first, cellulase activity was defined as 10,000 units when two sheets of filter papers (1 cm×1 cm) degraded at pH 5.0 and 45 °C by the cellulase for 150 min. The filter papers were entirely degraded in 114 min by 10 mg of Acremozyme. Thus, cellulase activity of Acremozyme was determined to be 1320 units mg^{-1} according to the following equation: cellulase activity (units mg^{-1}) = 150×10,000/(a×b) where a and b denoted weight of cellulase in mg and period in min required for the degradation, respectively.

Lignocelloloses	Components/g [a]			
	Holocellulose (hexose : pentose) [b]	Lignin	Ash	Others
Napiergrass [c]	57.3 (37.5 : 26.5)	14.9	12.7	15.1
Rice straw	61.3 (39.7 : 28.4)	18.2	17.7	2.8
Silvergrass	41.0 (34.2 : 11.4)	21.7	4.0	33.3
Bamboo	66.5 (43.9 : 30.0)	26.2	1.4	5.9

a) The amounts of components derived from 100 g of lignocellulose.

b) The values in the parenthesis are the amounts (g) of hexose and pentose derived from 100 g of lignocelluloses.

c) Referred from Yasuda *et al.*, 2012.

Table 1. Components of herbaceous lignocellolosic materials

The saccharification of the powdered cellulosic materials (10.0 g) was performed with Acremozyme (1.0 g) in an acetate buffer (60 mL, pH 5.0) under vigorous shaking at 45 °C. At the given saccharification time, the portion was taken from the reaction mixture and centrifuged at 12,000 rpm. The supernatant solutions were subjected to analysis for saccharides. The amounts of the reducing saccharides obtained from the saccharification reactions at 30, 40, and 45 °C were almost the same.

2.3. Simultaneous Saccharification and Fermentation (SSF)

Saccharomyces cerevisiae NBRC 2044 was grown at 30 ºC for 24 h in a basal medium (initial pH 5.5) consisting of glucose (20.0 g L^{-1}), peptone (1.0 g L^{-1}, Difco), yeast extract (1.0 g L^{-1}), $NaHPO_4$ (1.0 g L^{-1}), and $MgSO_4$ (3.0 g L^{-1}). After incubation for 24 h, the cell suspension of *S. cerevisiae* was obtained. The grown culture of *S. cerevisiae* showed a cell density of 7.7×10^7 cells mL^{-1}.

The suspension of cellulosic materials (1.33 g) in an acetate buffer solution (5 mL, pH 5.0) was introduced into the test tube (100 mL) and was autoclaved at 121 ºC for 20 min. After cooling the autoclaved suspension of cellulosic materials, the cell suspension (0.16 mL) of *S. cerevisiae* and the Acremozyme cellulase (133 mg) in an acetate buffer solution (3 mL, pH 5.0) were added (Yasuda *et al.*, 2012). The glucan contents were determined to be 436, 475, 410, and 525 mg in non-treated cellulosic materials (1.33 g) of napiergrass, rice straw, silvergrass, and bamboo, respectively. In the case of alkali-treated cellulosic materials (1.33 g), 761 (napiergrass), 774 (rice straw), 999 (silvergrass), and 790 mg (bamboo) of the glucan contents were included. The reaction vessel was connected by tube to messcylinder set in a water-bath to collect the evolved CO_2 gas. The reaction progress was monitored by the volume of CO_2. Thus, the simultaneous saccharification and fermentation (SSF) process was performed by stirring vigorously the reaction mixture with a magnetic stirrer at 34 °C, which is the optimal temperature.

2.4. Analysis

Saccharides were analyzed on a high-performance liquid chromatography system (LC-20AD, Shimadzu, Kyoto, Japan) equipped with RI detector (RID-10A) using anion exchange column (NH2P-50 4E; Shodex Asahipak, 250 mm in length and 4.6 mm in ID, Yokohama, Japan). Acetonitrile-water (8:2 v/v) was flowed at 1.0 mL min^{-1} as mobile phase. As a method to supplement LC analysis of saccarides, the amount of the reducing sugars released by the saccharification process was analyzed by a modified Somogyi–Nelson method (Kim and Sakano, 1996) assuming the composition of sugars to be $C_6H_{12}O_6$. The amounts of pentose were analyzed by a modified orcinol method using 5-methylresorcinol (orcinol), $FeCl_3$ $5H_2O$, and conc HCl (Fernell and King, 1953). Ethanol was analyzed by gas-liquid chromatography using a Shimadzu gas chromatograph (model GC–2014) and a glass column of 5% Thermon 1000 on Sunpak-A (Shimadzu) with 2-propanol as an internal standard. Scanning electron microscope (SEM) images were taken on a Hitachi S–4100 (Tokyo, Japan).

3. Results and discussion

3.1. Napiergrass (*Pennisetum purpureum* Schumach)

Napiergrass is a herbaceous tropical species, native to the east Africa. There are wide variation of phenotypes in napiergrass, reflected by plant breeding due to the crossing of dwarf genotype and relative species such as pearl millet (*Pennisetum americanum*) (Ishii *et al.*, 2005a, Hanna and Sollenburger, 2007). Dwarf variety of late-heading type originated from Florida, USA, via Thailand (Mukhtar *et al.*, 2003) was assessed to be suitable for both grazing (Ishii *et al.*, 2005b) and cut-and-carry systems among several sites of southern Kyushu, Japan (Utamy *et al.*, 2011). Dwarf variety of napiergrass meets the requirement of lignocellulose for the biofuel production, because it has low lignin-content and a high herbage mass per year and per area (Rengsirikul *et al.*, 2011). Therefore, we have continued to use this dwarf type of napiergrass for the bio-ethanol (Yasuda *et al.*, 2011) and bio-hydrogen production (Shiragami *et al.*, 2012) in University of Miyazaki.

3.2. Alkali-pretreatment

The powdered lignocelluloses (30.0 g) were washed with MeOH to remove lipids and treated with a 1% aqueous solution of NaOH (400 mL) at 95 ºC for 1 h (Silverstein, *et al.*, 2007).

The resulting lignin-removed holocellulose was isolated by centrifugation of the solution at 10,000 rpm for 10 min. Lignin remained in the alkali solution. The precipitate was washed by dispersion in water to remove the contaminated lignin. After the pH-adjustment to 7.0, the washed holocellulose was collected by centrifugation and dried.

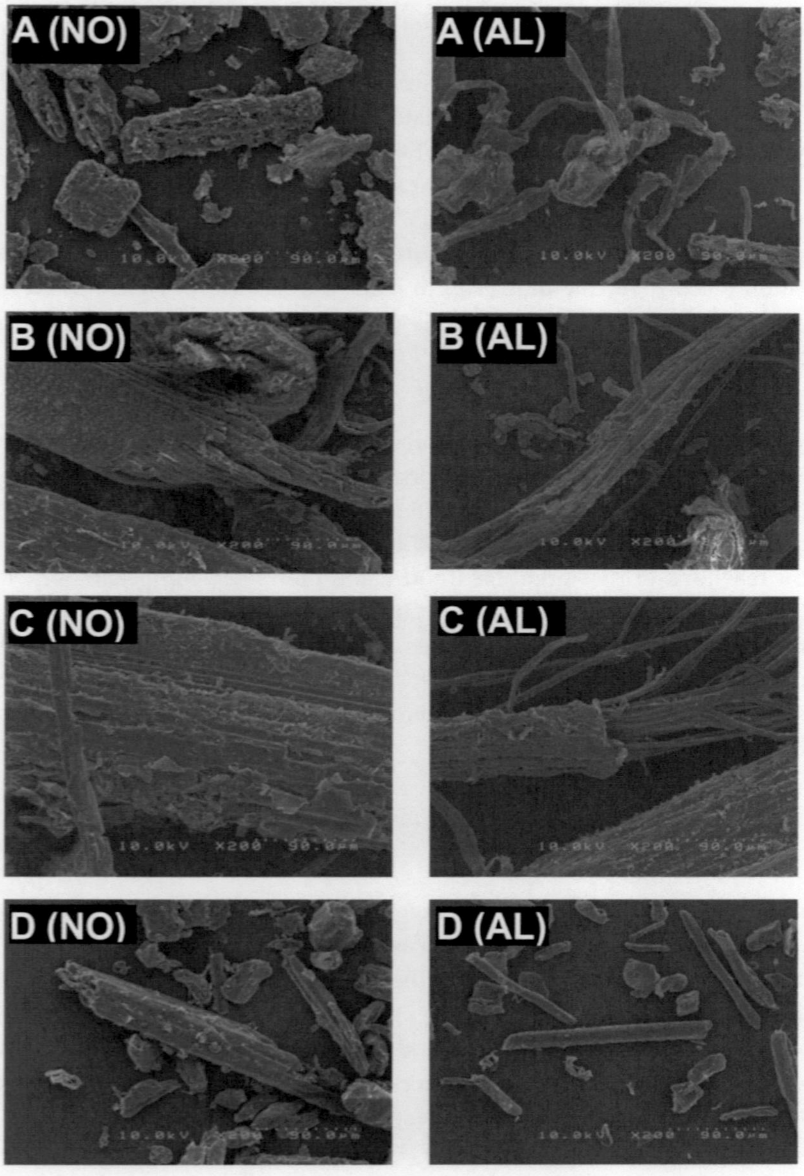

Figure 1. SEM images of non-treated (NO) and alkali-pretreated (AL) napiergrass (A), rice straw (B), silvergrass (C), and bamboo (D). The SEM images were taken under the magnification of 200.

Physical changes from non-pretreated lignocelluloses to alkali-pretreated lignocelluloses were studied using SEM images, as shown in Fig. 1. The fiber bundles observed in lignocelluloses were unloosened by the removal of lignin to change into the thin fibers in the alkali-pretreated lignocelluloses. It was expected that the accessibility of enzyme to the cellulose was increased by the alkali- pretreatment.

3.3. Lignin-removal effect on saccharification

The saccharification of alkali-pretreated lignocelluloses (holocellulose, 10.0 g) was performed with Acremozyme (1.0 g) in an acetate buffer (60 mL, pH 5.0) under vigorous shaking at 45 °C. The amounts of saccharides obtained from 1 g of alkali-pretreated napiergrass, rice straw, silvergrass, and bamboo were transformed to the amounts per 1.0 g of the alkali-untreated samples by multiplication with 0.573, 0.613, 0.410, and 0.665 g g^{-1} which were the contents of holocellulose. Table 2 summarizes the amounts of hexose and pentose after the saccharification reaction for the time (T_{SA}) to reach the maximum yields. In the cases of napiergrass and rice straw, the hexose yields (87.5 and 81.9 %) reached almost maximum yields whereas the pentose yields were still low. The largest amount of reducing saccharide was 451 mg obtained from 1.0 g of rice straw.

In order to examine the effectiveness of alkali-pretreatment, the saccharification of the non-pretreated lignocelluloses (10.0 g) was performed under conditions similar to the case of alkali-pretreated lignocelluloses. The largest amount of reducing saccharide was 307 mg g^{-1} obtained from non-pretreated napiergrass. Figure 2 shows the time-conversions of the saccharification reactions of non-pretreated and alkali-pretreated lignocelluloses. In all cases, the yields of saccharides from the alkali-pretreated lignocelluloses were higher than those from the non-pretreated lignocelluloses. The ratios (E_{SA}) of saccharide yields from the alkali-pretreated lignocelluloses to those from the non-pretreated lignocelluloses were used as a measure of the effectiveness of the lignin-removal on the saccharification process. The E_{SA} values are listed in Table 2.

3.4. Effectiveness of lignin-removal on Simultaneous Saccharification and Fermentation (SSF)

Ethanol was produced through a simultaneous saccharification and fermentation process (SSF) under optimal conditions as follows (Yasuda, *et al.*, 2012). Acremozyme (133 mg) in an acetate buffer solution (3.0 mL, pH 5.0) and the cell suspension (0.16 mL) of *S. cerevisiae* were added to the suspension of alkali-pretreated lignocelluloses (1.33 g) in an acetate buffer solution (5.0 mL, pH 5.0). The mixture was reacted at 35 °C under vigorous stirring until the CO_2 evolution ceased. The amounts of the products were transformed to the amounts per 1.0 g of the alkali-unpretreated lignocelluloses by the dividing by 1.33 and multiplication with 0.573 (napiergrass), 0.613 (rice straw), 0.410 (silvergrass), and 0.665 g g^{-1} (bamboo). Table 3 lists the amounts of ethanol and the recovered hexose and pentose which were determined by averaging the data of seven experiments. The maximum ethanol yield in SSF of alkali-pretreated lignocelluloses was 139 mg g^{-1} from rice straw.

http://dx.doi.org/10.5772/54194

Lignocelluloses	PT[a)]	T_{SA}/h[b)]	Product[c)]/mg g^{-1} (Yield/%)[d)]			E_{SA}
			Hexose	Pentose	Total	
Napiergrass	NO	120	215 (57.3)	91 (34.3)	307 (48.1)	1.36
	AL	120	328 (87.5)	90 (34.0)	419 (65.7)	
Rice straw	NO	120	192 (48.4)	51 (18.0)	244 (35.8)	1.85
	AL	120	325 (81.9)	125 (44.0)	451 (66.2)	
Silvergrass	NO	120	122 (35.7)	39 (34.2)	161 (35.3)	1.57
	AL	120	178 (52.0)	75 (65.8)	253 (55.5)	
Bamboo	NO	120	69 (15.7)	19 (6.3)	88 (11.9)	3.39
	AL	120	180 (41.0)	118 (39.3)	297 (40.2)	

a) Pretreatment (PT). NO: non-treatment, AL: lignin removal by alkali-pretreatment.

b) Saccharification time when the total yield of saccharides reached the maximum.

c) The amounts of products per 1 g of lignocellulosewhen the total yield of saccharides reached the maximum.

d) Yields were based on the amounts of hexose and pentose occurring in lignocelluloses.

Table 2. The lignin removal effects on saccharification processe

Lignocelluloses (EtOH/mg g^{-1})[a)]	PT[b)]	T_{SSF}/h[c)]	Product[d)]/mg g^{-1}			E_{SSF}
			Hexose	Pentose	EtOH (Yield/%)[e)]	
Napiergrass (192)	NO	24	18±5.2	99±1.6	102±3.5 (53.2)	1.18
	AL	96	38±5.3	125±5.0	121±4.6 (63.1)	
Rice straw (203)	NO	24	20±8.0	102±6.5	96±5.9 (47.3)	1.45
	AL	192	27±7.2	152±6.2	139±1.4 (68.5)	
Silvergrass (175)	NO	24	13±2.2	48±7.4	41±9.4 (23.5)	1.77
	AL	96	12±3.4	93±3.5	72±4.3 (41.2)	
Bamboo (224)	NO	24	6±5.1	18±5.8	34±1.7 (15.2)	2.28
	AL	96	22±4.3	111±1.5	78±5.6 (34.8)	

a) Theoretical amounts of ethanol obtained from glucan in lignocellulose (1 g).

b) Pretreatment (PT). NO: non-treatment, AL: lignin removal by alkali-pretreatment.

c) SSF time until the CO_2 evolution ceased.

d) The amounts of products per 1 g of lignocellulosewhen the SSF reaction reached the maximum. Data were determined by averaging the data of seven experiments.

e) Yield of ethanol based on the amounts of hexose occurring in lignocelluloses.

Table 3. The lignin removal effects on SSF processe

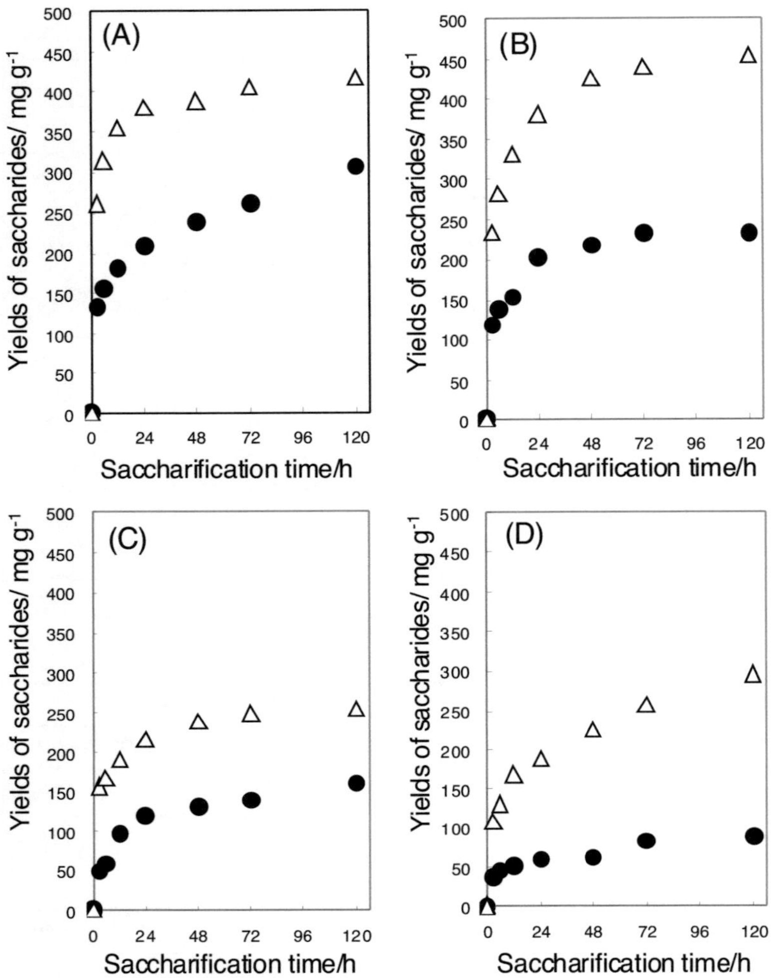

Figure 2. Time conversion of the saccharification of napiergrass (A), rice straw (B), silvergrass (C), and bamboo (D) for the non-pretreated lignocelluloses (●) and the alkali-pretreated lignocelluloses (△). The amounts of sugar from the alkali-pretreated lignocelluloses were transformed to the amounts per 1 g of the alkali-unpretreated samples by multiplication with 0.573 (napiergrass), 0.613 (rice straw), 0.410 (silvergrass), and 0.665 g g^{-1} (bamboo).

After the SSF, the pentose remained in the solution, although the hexose was consumed by the fermentation with *S. cerevisiae*. The amounts of pentose was compared between SSF and cellulase-saccharification processes under the optimized conditions. The amounts of pentose formed in SSF were larger than those in saccharification, except for the case of bamboo (Table 2 and 3). Therefore, the SSF process accelerated the hydrolysis of cellulosic components compared to the saccharification process. The consumption of saccharides by fermentation with *S. cerevisiae* might move the equilibrium to the product side in the hydrolysis of cellulosic components to saccharides with Acremozyme. In the case of bamboo, the ethanol yield was low, irrespective of higher content of hexose probably because of poor accessibility of the enzyme to holocellulosic components of bamboo (Yamashita *et al.*, 2010).

Also, the SSF process was applied to the non-pretreated lignocelluloses. The time- conversions of CO_2-evolution were compared between non-pretreated and the alkali-pretreated lignocelluloses, as shown in Fig. 3. The yields of ethanol from non- pretreated lignocelluloses were lower compared with the cases from alkali-pretreated lignocelluloses. Among the non-pretreated lignocelluloses, the largest amount of ethanol was 102 mg g^{-1} obtained from napiergrass. The enhanced effect of SSF yields by alkali-pretreatment was evaluated by the ratio (E_{SSF}) of ethanol yields from the alkali-pretreated lignocelluloses to those from non-pretreated lignocelluloses. The E_{SSF} values are listed in Table 3.

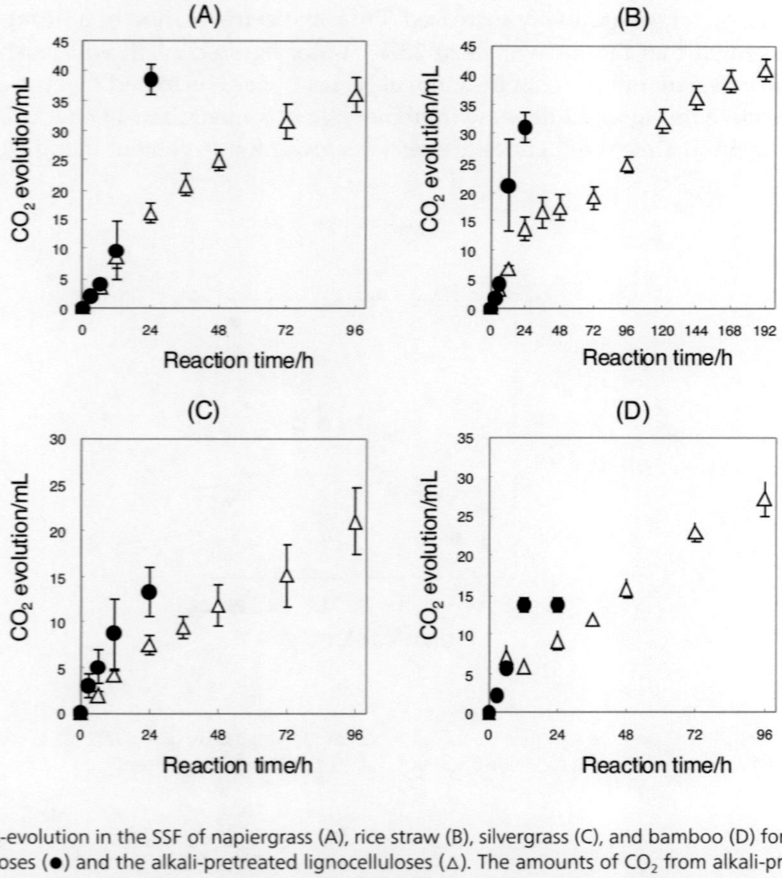

Figure 3. CO_2-evolution in the SSF of napiergrass (A), rice straw (B), silvergrass (C), and bamboo (D) for the non-treated lignocelluloses (●) and the alkali-pretreated lignocelluloses (△). The amounts of CO_2 from alkali-pretreated lignocelluloses was transformed to the amounts per 1 g of the alkali-unpretreated samples by multiplication with 0.573 (napiergrass), 0.613 (rice straw), 0.410 (silvergrass), and 0.665 g g^{-1} (bamboo).

It is noteworthy that the SSF of alkali-pretreated lignocelluloses was remarkably slowed down in all cases. In the fermentation by *S. cerevisiae* of the alkali-pretreated lignocelluloses, a nitrogen-source and a mineral were thought to be insufficient, since the aminoacids and the mineral were removed from lignocelluloses by alkali-pretreatment and the additional nutrients were not added in the SSF process (Alfenore *et al.*, 2003). Moreover, the fermentation process was affected by the inhibitory materials derived from the alkali-pretreatment since T_{SA} of both non-pretreated and the alkali-pretreated lignocelluloses were almost same (Alvira, 2010).

3.5. Availability of napiergrass as raw materials for ethanol production

In the cases of rice straw, silvergrass, and bamboo with relatively high lignin-contents (18.2–26.2 wt%), the lignin-removal was effective for both saccharification and SSF processes because of the larger E_{SA} (1.57–3.39) and E_{SSF} values (1.45–2.28). However, in the case of napiergrass with low lignin-content (14.9 wt%), the E_{SSF} value was small (1.18). Figure 4 shows the plots of the E_{SSF} values against the lignin-contents of lignocelluloses. As the lignin-contents increased, the E_{SSF} values gradually increased. From the extrapolation of a fitting line of the plots, it is assumed that the E_{SSF} values at 13.4 wt% of lignin-content will reach 1.0 which means no enhancement effect of lignin-removal. Thus, it was elucidated that the alkali-treatment was effective for lignocelluloses with higher lignin content than 13.4 wt%, but was not effective as the pretreatment of lignocelluloses with lower lignin content than 13.4 wt%.

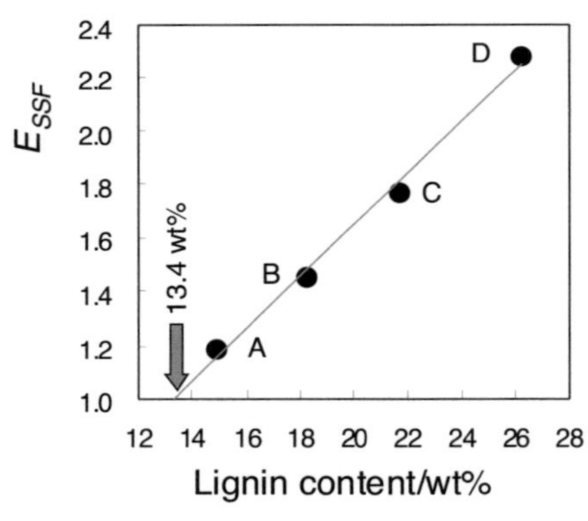

Figure 4. Dependence of E_{SSF} on the lignin contents in the SSF of napiergrass (A), rice straw (B), silvergrass (C), and bamboo (D). The plots showed that the E_{SSF} value became 1.0 at 13.4 wt% of lignin content.

4. Conclusion

In general, the alkali-pretreatment increases the accessibility of enzymes to the cellulose by the lignin-removal. Therefore alkali-pretreatment is effective for saccarification of the lignocellulose with higher lignin contents. In the case of napiegrass with low lignin- content, ethanol was produced in 102 mg g^{-1} and 121 mg g^{-1} from napiergrass through the SSF without and with alkali-pretreatment, respectively. Taking into consideration the low effectiveness of lignin-removal in ethanol yield, the retardation of fermentation rate, the loss of nutrients for the fermentation by *S. cerevisiae*, and the cost of lignin-removal, we concluded that ethanol production from napiergrass should be performed through the SSF process without the alkali-pretreatment. For example, Inoue and his coworkers (Hideno *et al.*, 2009) have recently proposed the enzymatic saccharification of rice straw treated by a wet disk milling method without chemical pretreatment. Even so, the development of a pretreatment method with low energy and low cost to enhance saccharification yields by the structural change of cellulosic components rather than lignin-removal are desired for economically viable bio-ethanol production. In our group, the development of more efficient pretreatment method other than alkali-pretreatment to produce effectively bioethanol from napiergrass is now in progress.

Moreover, the fermentation of the pentose remaining in SSF is important subject. We (Yasuda *et al.*, 2012) started the pentose fermentation using a recombinant *Escherichia coli* KO11. Pentose fermentation by *E. coli* KO11 produced additionally 31.4 mg g^{-1} of ethanol. Under the optimized conditions, the combination of the SSF and KO11 fermentation processes resulted in the production of 144 mg g^{-1} of ethanol from the non-pretreated napiergrass powder. The ethanol yield was 44.2% of the theoretical yield based on the hexose (375 mg) and pentose (265 mg) derived from 1 g of dry powdered napiergrass.

Acknowledgements

This study was done as a part of the project entitled "Research and Development of Catalytic Process for Efficient Conversion of Cellulosic Biomass into Biofuels and Chemicals" through Special Funds for Education and Research from the Ministry of Education, Culture, Sports, Science, and Technology of Japan.

Author details

Masahide Yasuda[1], Keisuke Takeo[1], Tomoko Matsumoto[2], Tsutomu Shiragami[1], Kazuhiro Sugamoto[1], Yoh-ichi Matsushita[1] and Yasuyuki Ishii[3]

1 Department of Applied Chemistry, Faculty of Engineering, University of Miyazaki, Gakuen-Kibanadai Nishi, Miyazaki, Japan

2 Center for Collaborative Research and Community Cooperation, University of Miyazaki, Gakuen-Kibanadai Nishi, Miyazaki, Japan

3 Department of Animal and Grassland Sciences, Faculty of Agriculture, University of Miyazaki, Gakuen-Kibanadai Nishi, Miyazaki, Japan

References

[1] Alfenore S, Molina-Jouve C, Guillouet SE, Uribelarrea J-L, Goma G., Benbadis L (2003). Improving ethanol production and viability of Saccharomyces cerevisiae by a vitamin feeding strategy during fed-batch process. *Applied Microbiology and Biotechnology* 60: 67-72.

[2] Alvira P, Tomás-Pejó E, Ballesteros M, Negro M J (2010). Pretreatment technologies for an efficient bioethanol production process based on enzymatic hydrolysis: A review. *Bioresour Technol* 101: 4851–4861.

[3] Anderson WF, Dien BS, Brandon SK, Peterson JD (2008). Assessment of bermudagrass and bunch grasses as feedstock for conversion to ethanol. *Applied Biochemistry Biotechnology* 145: 13-21.

[4] Brandon SK, Sharma LN, Hawkins GM, Anderson WF, Chambliss CK, Doran-Peterson J (2011). Ethanol and co-product generation from pressurized batch hot water pretreated T85 bermudagrass and Merkeron napiergrass using recombinant Escherichia coli as biocatalyst. *Biomass Bioenergy* 35: 3667-3673.

[5] Cardona CA, Quintero JA, Paz IC (2010). Production of bioethanol from sugarcane bagasse: Status and perspectives. *Bioresource Technol* 101: 4754–4766 .

[6] Fernell WR, King HK (1953). The simultaneous determination of pentose and hexose in mixtures of sugars. *Analyst* 78: 80–83 .

[7] Galbe M, Zacchi G. (2007). Pretreatment of lignocellulosic materials for efficient bioethanol production. *Advances Biochem Engineering Biotechnol* 108: 41-65.

[8] González-García S, Moreira MT, Feijoo G (2010). Environmental performance of lignocellulosic bioethanol production from alfalfa stems. *Biofuels Bioprod Bioref* 4: 118–131.

[9] Guo G-L, Chen W-H, Ehen W-H, Men L-C, Hwang W-S (2008). Characterization of dilute acid pretreatment of silvergrass for ethanol production. *Bioresour Technol* 99: 6046–6053.

[10] Hanna WW, Sollenberger LE (2007). Tropical and Subtropical Grasses. (In) Barnes, RF (Eds) Forages Volume II 6th Edition. Blackwell Pub, Iowa USA pp. 245-255.

[11] Hideno A, Inoue H, Tsukahara K, Fujimoto S, Minowa T, Inoue S, Endo T, Sawayama S (2009). Wet disk milling pretreatment without sulfuric acid for enzymatic hydrolysis of rice straw. *Bioresour Technol* 100: 2706-2711.

[12] Huang C-F, Jiang Y-F, Guo G-L, Hwang W-S (2011). Development of a yeast strain for xylitol production without hydrolysate detoxification as part of the integration of co-product generation within the lignocellulosic ethanol process. *Bioresour Technol,* 102: 3322-3329.

[13] Ishii Y, Yamaguchi N, Idota S (2005a). Dry matter production and *in vitro* dry matter digestibility of tillers among napiergrass (*Pennisetum purpureum* Schumach) varieties. *Grassl Sci* 51: 153–163.

[14] Ishii Y, Mukhtar M, Idota S, Fukuyama K (2005b). Rotational grazing system for beef cows on dwarf napiergrass pasture oversown with Italian ryegrass for 2 years after establishment. *Grassl Sci* 51: 209–220.

[15] Kai T, Tanimura T, Nozaki N, Suiko M, Ogawa K (2010). Bioconversion of soft celluosic resources into sugar and ethanol. *Seibutsu-kogaku Kaishi* 88: 66-72.

[16] Keshwani DR, Cheng JJ (2009). Switchgrass for bioethanol and other value-added applications: A review. *Bioresour Technol* 100: 1515–1523.

[17] Kim Y-K, Sakano Y (1996). Analyses of reducing sugars on a thin-layer chromatographic plate with modified Somogyi and Nelson reagents, and with copper bicinchoninate. *Biosci Biotechnol Biochem* 60: 594–597.

[18] Ko JK, Bak JS, Jung MW, Lee HJ, Choi I-G, Kim TH, Kim KH (2009). Ethanol production from rice straw using optimized aqueous-ammonia soaking pretreatment and simultaneous saccharification and fermentation processes. *Bioresour Technol* 100: 4374-4380.

[19] Li A-H, Lin C-W, Tran D-T (2011). Optimizing the response surface for producing ethanol from avicel by Brevibacillus strain AHPC8120. *J Taiwan Institute of Chemical Engineers* 42: 787-792.

[20] Lin C-W, Tran D-T, Lai C-Y, I Y-P, Wu C-H (2010b). Response surface optimization for ethanol production from Pennisetum Alopecoider by Klebsiella oxytoca THLC0409. *Biomass and Bioenergy* 34: 1922-1929.

[21] Lin C-W, Wu C-H, Tran D-T, Shih M-C, Li W-H, Wu C-F (2011b). Mixed culture fermentation from lignocellulosic materials using thermophilic lignocellulose- degrading anaerobes. *Process Biochemistry* 46: 489-493.

[22] Mukhtar M, Ishii Y, Tudsri S, Idota S, Sonoda T (2003). Dry matter productivity and overwintering ability in the dwarf and normal napiergrasses as affected by the planting density and cutting frequency. *Plant Prod Sci* 6: 65–73.

[23] Rengsirikul K, Ishii Y, Kangvansaichol K, Pripanapong P, Sripichitt P, Punsuvon V, Vaithanomsat P, Nakamanee G, Tudsri S (2011). Effects of inter-cutting interval on biomass yield, growth components and chemical composition of napiergrass (*Pennisetum purpureum* Schumach) cultivars as bioenergy crops in Thailand. *Grassl Sci* 57: 135–141.

[24] Ryu S, Karim MN (2011). A whole cell biocatalyst for cellulosic ethanol production from dilute acid-pretreated corn stover hydrolyzates. *Appl Microbiol Biotechnol* 91: 529–542.

[25] Sathitsuksanoh N, Zhu Z, Ho T-J, Bai M-D, Zhang Y-HP (2010). Bamboo saccharification through cellulose solvent-based biomass pretreatment followed by enzymatic hydrolysis at ultra-low cellulase loadings. *Bioresour Technol* 101: 4926–4929.

[26] Shiragami T, Tomo T, Tsumagari H, Ishii Y, Yasuda M (2012). Hydrogen evolution from napiergrass by the combination of biological treatment and a Pt-loaded TiO_2-photocatalytic reaction. *Catalyst* 2: 56-67.

[27] Silverstein RA, Chen T, Sharma-Shivappa RR, Boyette MD, Osborne J (2007). A comparison of chemical pretreatment methods for improving saccharification of cotton stalks. Rioresour Tech 98: 3000-3011.

[28] Sluiter A, Hames B, Ruiz R, Scarlata C, Sluiter J, Templaton D, Crocker D (2010). Determination of structural carbohydrates and lignin in biomass, Technical Report NREL/TP-510–42618, National Renewable Energy Laboratory, Golden, CO. http://www.nrel.gov/biomass/analytical_procedures.html.

[29] Talebnia F, Karakashev D, Angelidaki I (2010). Production of bioethanol from wheat straw: An overview on pretreatment, hydrolysis and fermentation. *Bioresour Technol* 101: 4744–4753.

[30] Utamy RF, Ishii Y, Idota S, Harada N, Fukuyama K (2011). Adaptability of dwarf napiergrass under cut-and-carry and grazing systems for smallholder beef farmers in southern Kyushu, Japan. *J. Warm Regional Society of Animal Science Japan* 54: 65-76.

[31] Yamashita Y. Shono M, Sasaki C, Nakamura Y (2010). Alkaline peroxide pretreatment for efficient enzymatic saccharification of bamboo. *Carbohydrate Polymers* 79: 914-920.

[32] Yasuda M, Miura A, Yuki R, Nakamura Y, Shiragami T, Ishii Y, Yokoi H (2011). The effect of TiO_2-photocatalytic pretreatment on the biological production of ethanol from lignocelluloses. *J Photochem Photobiol A: Chem* 220: 195-199.

[33] Yasuda M, Miura A, Shiragami T, Matsumoto J, Kamei I, Ishii Y, Ohta K (2012). Ethanol production from non-pretreated napiergrass through a simultaneous saccharification and fermentation process followed by a pentose fermentation with *Escherichia coli* KO11. *J Biol Bioeng* 114: 188-192.

[34] Zhang L, Yu CQ, Shimojo M, Shao T (2011). Effect of different rates of ethanol additive on fermentation quality of napiergrass *(Pennisetum purpureum)*. *Asian-Australasian J Animal Sci* 24: 636-642.

Chapter 5

The Effect of Washing Dilute Acid Pretreated Poplar Biomass on Ethanol Yields

Noaa Frederick, Ningning Zhang, Angele Djioleu,
Xumeng Ge, Jianfeng Xu and Danielle Julie Carrier

Additional information is available at the end of the chapter

http://dx.doi.org/10.5772/56129

1. Introduction

Ethanol sold in the US or Brazil is produced from feedstocks that contain starch or sucrose: corn starch in the US and sugar cane juice in Brazil. The use of these readily available fermentable sugar sources rouses societal discussions that are anchored on debates involving the use of food commodities for energy production (Wallington et al. 2012). From a sustainability perspective, conversion of cellulosic biomass to ethanol produces less greenhouse gases and particulate matter with a diameter less than 2.5 μm. Furthermore, the cost in dollars per liter in gas equivalent of using corn and corn stover as feedstock are 0.9 and 0.3, respectively (Hills et al. 2009). The production of fuels and biochemicals from cellulosic feedstock is desirable from both societal and environmental perspectives.

Although appealing, the deconstruction of cellulosic biomass into fermentable sugars is problematic. Cellulosic biomass conversion to industrial chemicals and fuels is performed via thermochemical, biochemical or a combination of these platforms. Unfortunately there is no clear technology winner and both conversion platforms have tradeoffs. The thermochemical platform is robust in terms of feedstock processing, but somewhat complicated in terms of the resulting product portfolio (Sharara et al. 2012). On the other hand, the biochemical platform can successfully yield industrial chemicals or fuels, but is delicate in terms of feedstock deconstruction into monomeric sugars (Lynd et al. 2008). This chapter is centered on biomass deconstruction using the biochemical platform.

In the biochemical platform, unfortunately, the deconstruction of plant cell wall into useable and fermentable carbohydrates remains challenging. Feedstock must be reduced in size, pretreated, and hydrolyzed with enzymes to produce a sugar stream that can be fermented into targeted products (Lynd et al. 2008). The cell wall is designed by nature as an elegant interwoven hemicellulose, cellulose and lignin tapestry that maintains its integrity, resulting in wood products that can sustain daily use for hundreds of years. To release the coveted carbohydrates from plant cell walls, the tapestry must be subjected to some form of pretreatment, ensuring the exposure of sugar polymers, which can subsequently be hydrolyzed.

There are a number of available pretreatment technologies (Tao et al. 2011). However, dilute acid pretreatment, though it contains many drawbacks, is most likely to be adopted at the deployment scale due to its relatively low cost and ease of use (Sannigrahi et al. 2011). Regrettably, dilute acid pretreatment results in the production of inhibitory compounds that inhibit downstream biochemical conversion processing steps. These inhibitory compounds are formed from the degradation of hemicellulose into furfural, acetic acid and formic acid; or lignin-derived phenolic compounds, oligomers and re-polymerized furans named humins (van Dam et al. 1986). Such compounds can inhibit enzymatic hydrolysis by at least 50% (Cantarella et al. 2004). In a sense, the dilute acid-based biochemical platform is caught in a chicken and egg situation: pretreatment is essential to loosen the sugar polymer tapestry, but pretreating biomass causes the formation of inhibitory products that hinder subsequent downstream processing steps. In other words, without pretreatment, the expensive processing enzymes cannot access the complex carbohydrates to release the coveted monomeric sugars, which will be fermented into fuels or bioproducts.

To circumvent the negative effects of dilute acid pretreatment, namely the production of inhibitory products, pretreated biomass is washed prior to enzymatic hydrolysis. Successive washes remove inhibitory products, resulting in biomass amenable to subsequent enzymatic hydrolysis. At the bench scale, inhibitory compounds are removed by washing with up to 30 volumes of water (Djioleu et al. 2012). At the pilot scale, inhibitory compounds are removed from pretreated biomass by washing with at least three volumes of water (Hodge et al. 2008). Washing pretreated biomass will be difficult to replicate at the deployment scale due to the daunting amount of water that will be required. Another approach consists of enhancing our understanding of which compounds critically impede enzymatic hydrolysis, and how to minimize their generation during pretreatment.

The conversion of cellulosic biomass into fuels and biochemicals can be conducted with a range of feedstocks. Cellulosic biomass can be sourced from various streams: forestry products and residues, agricultural byproducts, dedicated energy crops, food processing and municipal solid wastes. In particular, wood energy crops, such as hybrid poplars (*Populus deltoides*), are hardwoods that can find use as biorefinery feedstock. *P. deltoides* is being increasingly planted and managed in the United States as short-rotation plantations for timber, pulp and renewable energy (Studer et al. 2011). The use of *P. deltoides* as a feedstock and its response to various pretreatment technologies combined with enzymatic hydrolysis was reported by the Consortium for Applied Fundamentals and Innovation (CAFI), where the technologies were compared with identical characterized poplar feedstock (Kim et al. 2009). The series of papers were reported in one single 2009 issue of *Biotechnology Progress*. *P. deltoides* is an interesting feedstock that can be deconstructed into fermentable sugars. The production of a fermentable sugar stream was examined by our group (Martin et al. 2011; Djioleu et al. 2012), where high and low specific gravity poplar was pretreated in 1% (v/v) dilute acid in non-agitated batch reactors and hydrolyzed using Accelerase ® 1500 enzymes.

In this work, high specific gravity poplar was pretreated in 0.98% (w/v) dilute acid at 140 °C in a 1 L stirred reactor and the hydrolyzates were fermented with two ethanol producing strains. This work examined the side-by-side effect of washing and not washing the pretreated biomass on sugar yields and its effect on fermentation to ethanol.

2. Materials and methods

2.1. Biomass

High-density poplar was secured from University of Arkansas Pine Tree Branch Station. The material was identical to what was studied by Djioleu et al. (2012) and Martin et al. (2011). The biomass was transformed into chips, which were then ground to 10 mesh using a Wiley Mini Mill (Thomas Scientific, Swedesboro, NJ) as described by Torget et al. (1988). The moisture content was determined with an Ohaus MB45 Moisture Analyzer (Pine Brook, NJ). The poplar used in this study was reported to have a specific gravity of 0.48, as reported by Martin et al. (2011).

2.2. Pretreatment

Twenty-five grams of biomass were weighed and mixed with 250 ml of 0.98% (w/v) sulfuric acid (EMD, Gibstown, NJ), resulting in a solids concentration of 10%. The reaction mixture was placed in a 1 L Parr (Moline, IL) 4525 reaction vessel. The reaction temperature used in these experiments was 140 °C. Reaction time was set as the time when the reactor reached 140 °C. After 40 min, heating was halted and the reactor was cooled under a stream of cold tap water. Temperature decreased from 140 °C down to 100 °C in about four min. When the mixture inside the reactor reached a temperature lower than 60 °C, the contents were retrieved. On average, the cool down period lasted approximately 10 min. The mixture was filtered with a Buchner apparatus fitted with Whatman filter paper. The remainder of the reaction solids were removed from the vessel and likewise filtered through a Buchner apparatus. The volume of the hydrolyzate was recorded and saved for further testing. The mass of filtered solids was recorded and its moisture content determined, using Ohaus MB45 Moisture Analyzer. The filtered solids were either used as is (referred to throughout the work as non-washed) or washed (referred to throughout the work as washed) with three volumes of Millipore water as suggested by Hodge et al. (2008). The wash liquid was saved and kept at 4 °C for further testing.

2.3. Enzyme hydrolysis

The hydrolysis was essentially conducted as in Djioleu et al. (2012), but carried out in a 600 ml Parr reactor described by Martin et al. (2010). Forty grams of washed or non-washed pretreated biomass were placed in the Parr reactor with 20 ml of Accellerase ®1500 (Genencor), 200 ml of pH 4.9 sodium citrate buffer, and 180 ml of Millipore filtered water. The reactor was stirred continuously at a slow speed as reported by Martin et al. (2010) and maintained at 50 °C for 24 hours. The entire sample was collected at the end of the run and stored at 4 °C.

2.4. HPLC analysis

Aliquots from pretreatment hydrolyzates, wash waters and enzyme hydrolyzates were analyzed by high-pressure liquid chromatography (HPLC) for carbohydrates and inhibitory byproducts. Two instruments were used to conduct these analyses. Carbohydrates were analyzed with Waters 2695 Separations module (Milford, MA) equipped with Shodex (Waters, Milford, MA) precolumn (SP-G, 8 µm, 6 x 50 mm) and Shodex column (SP0810, 8 µm x 300 mm). Millipore filtered water (0.2 mL/min) was the mobile phase and the column was heated to 85 ˚C with an external heater. Carbohydrates were detected with a Waters 2414 Refractive Index Detector (Milford, MA) as described by Djioleu et al. (2012). Inhibitory byproducts were analyzed on a Waters 2695 Separations module equipped with a Bio-Rad (Hercules, CA) Aminex HPX-87H Ion Exclusion 7.8 mm X 30 mm column, heated to 55˚C. The mobile phase was 0.005 M H_2SO_4, flowing at 0.6 ml/min. Compounds were detected with a UV index using the Waters 2996 Photodiode Array detector. Furfural and hydroxymethylfurfural (HMF) were detected at 280 nm; whereas, formic acid and acetic acid were detected at 210 nm.

2.5. Fermentation

Fermentation was carried out in 50 ml shake flasks with two strains of yeast, self-flocculating SPSC01 and ATCC4126. The SPSC01 strain was provided by Dalian University of Technology, China (Bai et al. 2004). Preculture of both yeast strains was carried out in medium consisting of 30 g/L glucose, 5 g/L yeast extract and 5 g/L peptone. The overnight grown yeasts were harvested by centrifugation at 4,100 *g* for 30 min. The pellets of yeast cells were washed twice with de-ionized water, and then re-suspended in 50 mM sodium citrate buffer (pH 4.8) to reach a cell concentration of 2 to 4×10^9 /ml. The re-suspended yeast cells were inoculated into 10 ml of each hydrolysate to reach a yeast cell concentration of 8×10^7 /ml. Ethanol fermentations were performed at 30°C on a rotary shaker at 150 rpm for 8 hours. Glucose content of the samples was assayed using a glucose colorimetric assay kit (Cayman Chemical, MI). Produced ethanol was quantified by gas chromatography (GC) on the Shimadzu GC-2010 equipped with a flame ionization detector (FID) and a Stabilwax®-DA column (cross-bond polyethylene glycol, 0.25 mm ×0.25 µm ×30 m), as described early by Ge et al. (2011). Before injection into the GC, 50 µl of fermentation broth was diluted 10 times with de-ionized water and supplemented with 50 µl of 0.1 mg/ml n-butanol as an internal standard.

2.6. Statistical analysis

Experiments were conducted in duplicate (pretreatment and enzymatic saccharification) or triplicate (fermentation). Calculations of carbohydrate and degradation compounds, including HMF, furfural, formic acid, and acetic acid, were calculated using Microsoft Office Excel 2007. Analysis of the variance (ANOVA) was determined using JMP 9.0, LSMeans Differences Student's t, with α= 0.10.

3. Results and discussion

3.1. Pretreatment and enzymatic hydrolysis

Poplar biomass was pretreated at 140 °C for 40 min. This condition corresponded to a combined severity of 1.16 (Abatzoglou et al. 1992). The composition of the hydrolyzate was analyzed by HPLC and calculations were made to express the concentrations in terms of compounds obtained from 100 g of biomass. These pretreatment conditions resulted in the recovery of 12% and 41% of the possible glucose and xylose, respectively; these calculations were based on previously reported high specific gravity compositional analysis (Djioleu et al. 2012). Carbohydrate recoveries are presented in Table 1. Dilute acid pretreatment resulted in the release of xylose from hemicellulose as compared to that of glucose from cellulose, and results presented in Table 1 reflect this trend. Dilute acid hydrolyzates also contained furfural, acetic acid, formic acid and HMF. By determining HPLC concentrations, liquid volumes and initial feedstock masses, amounts of furfural, acetic acid, formic acid and HMF were calculated as 0.71, 1.56, 2.41 and 0.04 g per 100 g, respectively.

After pretreatment, the biomass was either washed with three volumes of water or used as is (non-washed), and the resulting wash waters were analyzed by HPLC. Table 1 presents the compositional analysis of the resulting wash waters; furfural, acetic acid, formic acid and HMF were 0.14, 0.31, 0.41 and 0.01 g per 100 g, respectively. Of the inhibitory compounds monitored, formic acid was generated in the highest concentration. In contrast to dilute acid hydrolyzates, wash waters contained similar proportions of glucose and xylose. Furfural, acetic acid, formic acid and HMF concentrations in the wash waters were at most 18% of those present in dilute acid hydrolyzates, indicating that inhibitory products could remain bound to the pretreated biomass.

The washed and non-washed pretreated pellets were subjected to enzymatic hydrolysis. The results are presented in Figure 1. Washing the pretreated pellet had a significant effect on glucose recovery, where glucose concentrations in the washed condition were 5.3 times higher than those from the non-washed samples. As expected, concentrations of furfural, acetic acid, formic acid and HMF were significantly higher in the enzymatic hydrolyzates of non-washed samples.

g/100 g	glucose	xylose	furfural	acetic acid	formic acid	HMF
Hydrolyzate	0.828 ± 0.030	4.420 ± 0.103	0.710 ± 0.028	1.560 ± 0.323	2.410 ± 0.231	0.037 ±0.003
Wash water	0.111 ± 0.077	0.103 ± 0.006	0.137 ± 0.023	0.311 ± 0.034	0.412 ± 0.126	0.007 ±0.002

Table 1. Composition of pretreatment hydrolyzate and wash water of high specific gravity poplar pretreated in dilute acid (0.98 % v/v) at 140 °C for 40 min.

3.2. Ethanol production from washed and non-washed hydrolyzates

The fermentability of the enzymatic hydrolysates was evaluated using two yeast strains, self-flocculating yeast SPSC01 and conventional *Saccharomyces cerevisiae* ATCC4126. Both yeast strains solely metabolize glucose and not xylose. A total of four hydrolysate samples, two from

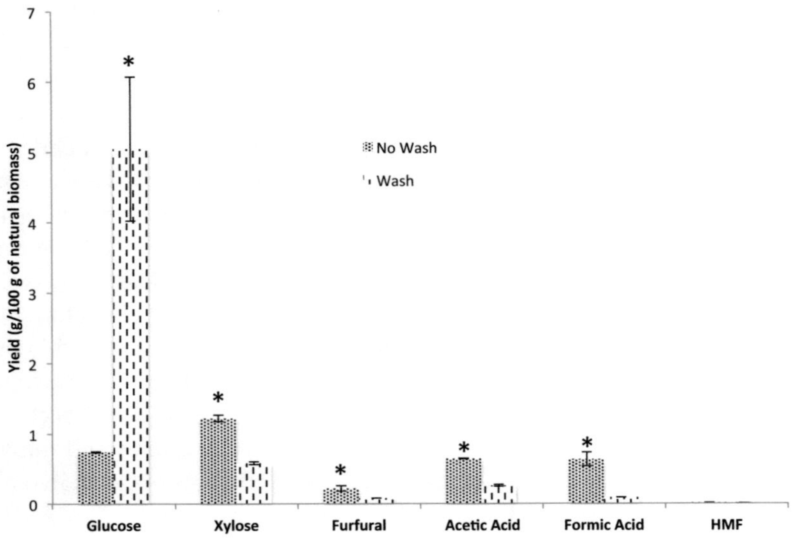

Figure 1. Carbohydrate, furan and aliphatic acid yields of washed and non-washed enzymatically hydrolysed dilute acid pretreated poplar.

washed pretreatments and two from non-washed pretreatments were directly used for the fermentation, and ethanol yields based on glucose ($Y_{E/G}$) were determined. Since the initial glucose concentrations in the hydrolyzates were low (less than 4.0 g/L) due to inefficient enzymatic saccharification (Table 2), all the fermentations were completed within 6 hours, as indicated by pre-experiments (data not shown).

Samples	Glucose content (g/l)	ATCC4126			SPSC01		
		Ethanol (g/l)	$Y_{E/G}$a (g/g)	g/100 g	Ethanol (g/l)	$Y_{E/G}$a (g/g)	g/100 g
Non-washed-A	0.20±0.00	0	0	0	0.08±0.01[b]	--[c]	--[c]
Non-washed-B	0.19±0.01	0	0	0	0.07±0.01[b]	--[c]	--[c]
Washed-A	2.32±0.06	0.41±0.03	0.18	0.90	0.48±0.05	0.21	1.05
Washed-B	3.76±0.11	1.08±0.09	0.29	5.75	1.46±0.15	0.39	7.74

[a]: $Y_{E/G}$ refers to ethanol yields based on the glucose contained in hydrolysates

[c]: Not accurately detected because of out of the detection limit of GC

[d]: Not calculated due to inaccurately determined ethanol concentration by GC

Table 2. Ethanol yields of the fermentation of four different enzymatic hydrolyzates with two yeast strains ATCC4126 and SPSC01. Of the four hydrolyzate samples, two were prepared from non-washed pretreated biomass and two from washed pretreated biomass. The pretreatments were conducted at 140 °C (A) and 160 °C (B), respectively.

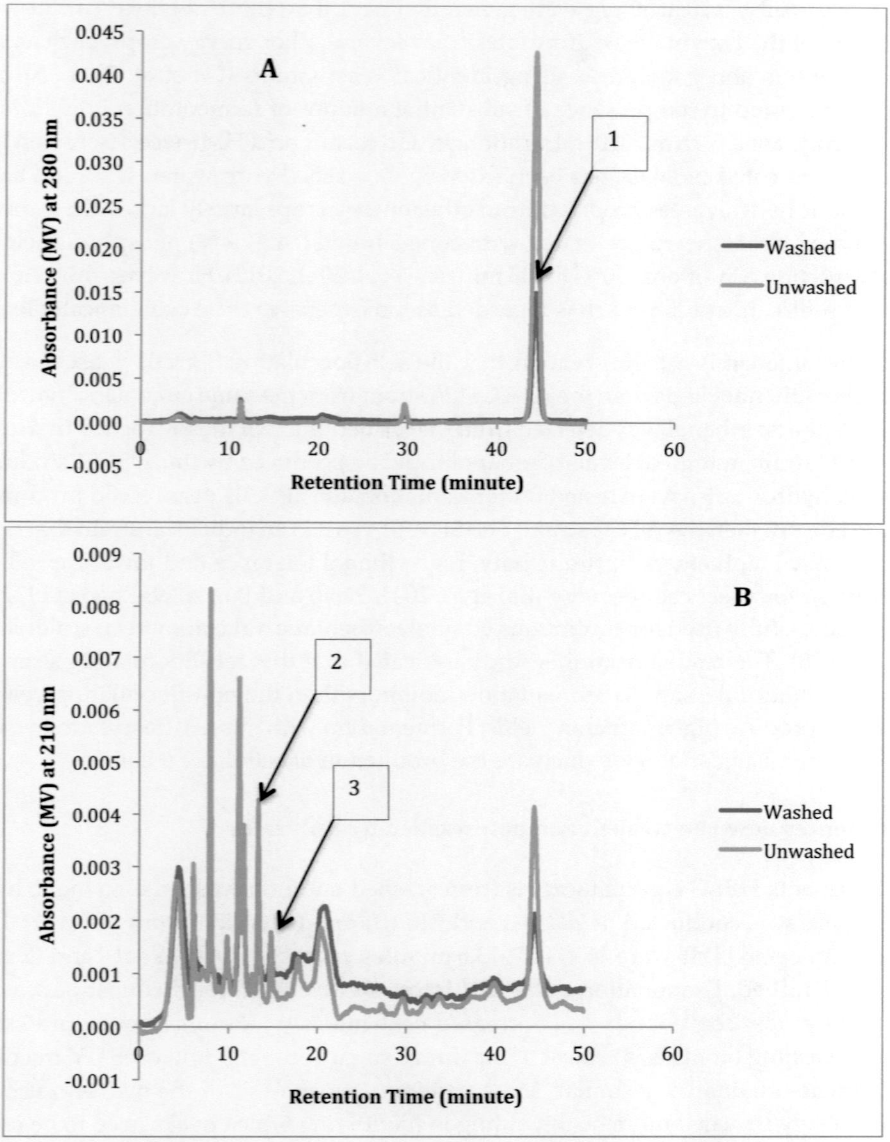

Figure 2. Chromatogram of enzymeatic hydrolyzate of washed and unwashed biomass after dilute acid pretreatment. Compounds detected at A) 280 nm and B) 210nm. Compounds are 1) Furfural, retention time = 44.35 min; 2) Formica-cid, retention time = 13.6 min; 3) Acetic acid, retention time = 14.7min

It is shown in Table 2 that trace amounts of ethanol were detected in the fermentation broth of the non-washed enzymatic hydrolysates. In contrast, significant amounts of ethanol were generated from the washed enzymatic hydrolyzates, in particular from the hydrolyzate sample that contained a glucose concentration of 3.76 g/L, producing 1.08 g/L and 1.46 g/L of ethanol by the ATCC4126 and SPSC01 strains, respectively. However, the ethanol yields $Y_{E/G}$ determined in this study, 0.18~0.39 g/g were generally lower than those obtained earlier with the fermentation of the enzymatic hydrolyzates from several other energy crops, such as switchgrass, miscanthus and gamagrass, using identical yeast strains (Ge et al. 2011, 2012). This should be attributed to the presence of substantial amount of fermentation inhibitors in the poplar hydrolyzates, such as furfural, acetic acid and formic acid HMF (see discussion below), though the pretreated biomass has been extensively washed with water. It should be noted that enzymatic hydrolyzates prepared from other energy crops largely lacked these inhibitors because these biomass were pretreated with concentrated (84%, w/v) phosphoric acid under moderate reaction conditions (50°C for 45 min) (Ge et al. 2011, 2012). However, this phosphoric acid-based pretreatment approach is regarded as too expensive to be economically feasible.

Of particular interest is the observation that the self-flocculating SPSC01 yeast always produced higher ethanol yields than the ATCC4126 strain from the same enzymatic hydrolyzates (Table 2). While no ethanol was detected from the fermentation of the unwashed hydrolyzates by the ATCC strain, marginal levels of ethanol could be produced by the SPSC01 strain. When the washed hydrolyzates were tested for fermentation, the SPSC01 yeast could produce up to 35% more ethanol then the ATCC strain. The SPSC01 yeast is an industrial strain that has been reported to have high ethanol productivity, high ethanol tolerance and lower capital investment required for yeast cell recovery (Bai et al. 2004; Zhao and Bai, 2009; Zhao et al. 2009). It has been successfully used for continuous ethanol fermentation at commercial scales in China (Bai et al. 2008). The results from this study indicated that this self-flocculating strain could also have a higher tolerance to fermentation inhibitors than the non-flocculating yeast, thus being able to produce higher ethanol yields. Fermentation with the self-flocculating yeast may represent a promising strategy to increase the production of cellulosic ethanol.

3.3. Differences between washed and non-washed hydrolyzates

Figure 2 presents HPLC chromatograms from washed and non-washed enzymatic hydrolyzates; analysis was conducted at 280 (A) and 210 (B) nm. Retention times of furfural, acetic acid and formic acid HMF were 44.4, 14.7, 13.6 minutes, respectively. Peaks at 9 and 12 minutes remain unidentified. Examination of the UV traces showed that, for the most part, washing did not remove any compounds, but decreased peak intensity. Results from Figure 1 demonstrate that washing biomass is critical to maximize sugar recovery; however UV traces at 280 and 210 nm are qualitatively similar. Mass spectrometry analysis of the hydrolyzates would have most likely revealed more peaks, aiding in identifying which peaks need to be removed and/or minimized prior to enzymatic hydrolysis and fermentation.

In related work, aliphatic acid and furans from wet distillers grain (Ximenes et al. 2010), corn stover (Hodge et al. 2008), wheat straw (Panagiotou and Olsson 2007) and poplar wood (Cantarella et al. 2004) wash waters were analyzed. Having detected and quantified compounds in wash waters, solutions were reconstituted and tested for their effect on saccharification cocktails. Cantarella et al. (2004) pretreated poplar in steam at a severity of 4.13 and tested the effect of washing the pretreated biomass. Cantarella et al. (2004) washed poplar-pretreated material with either 12.5 or 66.7 volumes of water to biomass ratio prior to enzymatic hydrolysis and fermentation steps. They reported that using washed biomass resulted

http://dx.doi.org/10.5772/56129

in the production of at least 20 g/L of ethanol, while the use of the non-washed control produced no ethanol.

Cantarella et al. (2004) showed that increasing formic acid concentrations of washed steam pretreated poplar from 3. 7 to 11.5 mg/ml decreased sugar recovery from enzymatic hydrolysis by 60%. Formic acid was also shown to inhibit enzymatic hydrolysis by Arora et al. (2012). They showed that adding 5 or 10 mg/ml formic acid to washed dilute acid pretreated poplar biomass resulted in recovery of 47% and 14%, respectively, of potential sugars. Using steam explosion-pretreated wheat straw as their system, Panagitou and Olsson (2007) reported the effects of adding 4 and 15 mg/ml of formic acid to their hydrolyzate; the higher concentration annihilated sugar recovery.

In work reported by Moreno et al. (2012), wheat straw was pretreated by steam explosion; the pretreated slurry was incubated with *Pycnoporus cinnabarinus* or *Trametes villosa* laccases prior to fermentation with *Kluyveromyces marxianus*. Biomass loadings of 5, 6 and 7% were tested. No differences in ethanol yields at 5% and 6% were observed; however, loadings at 7% resulted in an 86% reduction in ethanol yields compared to the control, which was not prior incubated with laccases. These results indicate that inhibitory byproducts are present in the pretreatment hydrolyzates. Incubation with *T. villosa* laccases removed almost 100 % of vanillin, syringaldehyde, p-coumaric acid and ferulic acid from pretreated hydrolyzates, enabling ethanol to glucose yields greater than 0.33 g/g.

Although this report is centered on the effects of aliphatic acids and furans on enzymatic hydrolysis and fermentation, it is important to note that other generated products may play key roles in inhibiting enzymatic hydrolysis and fermentation (Palmqvist and Hahn-Hägerdal 1999; Moreno et al 2012). Lignin derivatives can result in nonproductive binding of the saccharification cocktail with lignin derivatives (Berlin et al. 2006); and released sugars and their degradation compounds can deactivate or obstruct enzyme active sites (Kumar and Wyman 2008). It is critical to establish a better understanding of pretreatment chemistry in terms of generated degradation products. By understanding which compound plays a critical role in inhibiting enzymatic hydrolysis and fermentation, attempts can be made to minimize their generation, thereby improving processing yields. Pretreatments at 0.98% (w/v) dilute acid, 140 °C for 40 min resulted in the recovery of 12% and 41% of possible glucose and xylose, respectively. The authors recognize that these were low carbohydrate yields. Pretreatment were re-conducted at 0.98% (w/v) dilute acid, 160 °C for 40 min. Glucose recovery from non-washed and washed biomass was 0.92 and 19.85 g/100g, respectively, indicating that a 20 °C increase in temperature significantly augmented sugar recovery. Conversely, formic acid contents were 0.65 and 0.04 g/100 g non-washed and washed biomass, respectively; higher content was determined in non-washed biomass as for the 140 °C pretreatment conditions.

4. Conclusions

Dilute acid pretreatment processes resulted in the production of inhibitory byproducts, such as furfural, acetic acid, formic acid, and HMF that hindered both the enzymatic saccharification and fermentation steps. Washing the pretreated biomass with water did not entirely remove the inhibitory compounds, but significantly decreased their concentrations, which resulted in recovery of 5.3 times more glucose and substantially increased ethanol yields. The self-flocculating yeast strain SPSC01 showed higher tolerance to fermentation inhibitors than the non-flocculating ATCC4126 yeast, resulting in up to 35% increase in ethanol yield.

Acknowledgements

The authors would like to thank the University of Arkansas, Division of Agriculture, and the Department of Biological and Agricultural Engineering, for financial assistance. The authors would also like to acknowledge Department of Energy award 08GO88035 for pretreatment equipment and support; CRREES National Research Initiative award no. 2008-01499 for the HPLC instrument, and the Plant Powered Production (P3) Center through an NSF RII Arkansas ASSET Initiative (AR EPSCoR) for the support of undergraduate and graduate student stipend and equipment.

Author details

Noaa Frederick[1], Ningning Zhang[2], Angele Djioleu[1], Xumeng Ge[2], Jianfeng Xu[2] and Danielle Julie Carrier[1]

1 Department of Biological and Agricultural Engineering, University of Arkansas, Fayetteville, AR, USA

2 Arkansas Biosciences Institute and College of Agriculture and Technology, Arkansas State University, Jonesboro, AR, USA

References

[1] Abatzoglou, N, Chornet, E, & Belkacemi, K. (1992). Phenomenological kinetics of complex systems: The development of a generalized severity parameter and its application to lignocellulosics fractionation. *Chemical Engineering Science* , 153, 375-380.

[2] Arora A, Martin E, Pelkki M and Carrier DJ. (2013). "The effect of formic acid and furfural on the enzymatic hydrolysis of cellulose powder and dilute acid-pretreated poplar hydrolysates." *Sustainable Chemistry and Engineering* 1: 23-28.

[3] Bai, F. W, Chen, L. J, Anderson, W. A, & Moo-young, M. (2004). Parameter oscillations in a very high gravity medium continuous ethanol fermentation and their at-

tenuation on a multistage packed column bioreactor system. *Biotechnology and Bioengineering 88*, 558-566.

[4] Bai, F. W, Anderson, W. A, & Moo-young, M. (2008). Ethanol fermentation technologies from sugar and starch feedstocks. *Biotechnology Advances* , 26, 89-105.

[5] Berlin, A, Balakshin, M, Gilkes, N, Kadla, J, Maximenko, V, Kubo, S, & Saddler, J. (2006). Inhibition of cellulase, xylanase and β-glucosidase activities by softwood lignin preparations. *Journal of Biotechnology* , 125, 198-209.

[6] Cantarella, M, Cantarella, L, Gallifuoco, A, Spera, A, & Alfani, F. (2004). Subsequent enzymatic hydrolysis and SSF. *Biotechnology Progress* , 20, 200-206.

[7] Van Dam, H, Kieboom, A, & Van Bekkum, H. (1986). The conversion of fructose and glucose in acidic media: Formation of hydroxymethyl furfural. *Starch/Stärke* 3: S, 95-101.

[8] Djioleu, A, Arora, A, Martin, E, Smith, J. A, Pelkki, M, & Carrier, D. J. (2012). Sweetgum sugar recovery from high and low specific gravity poplar clones post dilute acid pretreatment/enzymatic hydrolysis. *Agricultural and Analytical Bacterial Chemistry* (in press).

[9] Ge, X. M, Burner, D. M, Xu, J, Phillips, G. C, & Sivakumar, G. (2011). Bioethanol production from dedicated energy crops and residues in Arkansas, USA. *Biotechnology Journal* , 6, 66-73.

[10] Ge, X. M, Green, S, Zhang, N, Sivakumar, G, & Xu, J. (2012). Eastern gamagrass as a promising cellulosic feedstock for bioethanol production. *Process Biochemistry* , 47, 335-339.

[11] Ge, X. M, Zhao, X. Q, & Bai, F. W. (2005). Online monitoring and characterization of flocculating yeast cell flocs during continuous ethanol fermentation. *Biotechnology and Bioengineering* , 90, 523-531.

[12] Hills, J, Polasky, S, Nelson, E, Tilman, D, Huo, H, Ludwig, L, Neumann, J, Zheng, H, & Bonta, D. (2009). Climate change and health costs of air emissions from biofuels and gasoline. *Proceeding of National Academy of Sciences* , 106, 2077-2082.

[13] Hodge, D, Karim, M, Schell, D, & Macmillan, J. (2008). Soluble and insoluble solids contributions to high-solids enzymatic hydrolysis of lignocellulose. *Bioresource Technology* , 99, 8940-8948.

[14] Kim, Y, Mosier, N, & Ladisch, M. (2009). Enzymatic digestion of liquid hot water pretreated hybrid poplar. *Biotechnology Progress* , 25, 340-348.

[15] Kumar, R, & Wyman, C. (2008). Effect of enzyme supplementation at moderate cellulose loadings on initial glucose and xylose release from corn stover solids pretreated by leading technologies. *Biotechnology and Bioengineering* , 102, 457-467.

[16] Lynd, L, Laser, M, Bransby, D, Dale, B, Davison, B, Hamilton, R, Himmel, M, Mcmillan, J, Sheehan, J, & Wyman, C. (2008). How biotech can transform biofuels. *Nature Biotechnology* , 26, 169-172.

[17] Martin, E, Bunnell, K, Lau, C, Pelkki, M, Patterson, D, Clausen, E, Smith, J, & Carrier, D. J. (2011). Hot water and dilute acid pretreatment of high and low specific gravity *Populus deltoids* clones. *Journal of Industrial Microbiology 38*, 355-361.

[18] Martin, E, Duke, J, Pelkki, M, Clausen, E, & Carrier, D. J. (2010). Sweetgum (*Liquidambar styraciflua* L.): Extraction of shikimic acid coupled to dilute acid pretreatment. *Applied Biochemistry and Biotechnology* , 162, 1660-1668.

[19] Moreno, A, Ibarra, D, Fernández, J, & Ballesteros, M. (2012). Different laccase detoxification strategies for ethanol production from lignocellulosic biomass by the thermotolerant yeast *Kluyveromyces marxianus* CECT 10874. *Bioresource Technology* , 106, 101-109.

[20] Palmqvist, E, Almeida, J, & Hahn-hägerdal, B. (1999). Influence of furfural on anaerobic glycolytic kinetics of *Saccharomyces cerevisiae* in batch culture. *Biotechnology and Bioengineering* , 62, 447-454.

[21] Panagiotou, G, & Olsson, L. (2007). Effect of compounds released during pretreatment of wheat straw on microbial growth and enzymatic hydrolysis rates. *Biotechnology and Bioengineering* , 96, 250-258.

[22] Sannigrahi, P, Kim, D, Jung, H, & Ragauskas, A. (2011). Pseudo-lignin and pretreatment chemistry. *Energy Environmental Sciences* , 4, 1306-1310.

[23] Sharara, M, Clausen, C, & Carrier, D. J. (2012). An overview of biorefinery technology" *in* Biorefinery Co-Products: Phytochemicals, Primary Metabolites and Value-Added Biomass Processing (Wiley Series in Renewable Resource). Bergeron C, Carrier DJ and Ramaswamy S. (*eds*). John Wiley & Sons. , 1-18.

[24] Studer, M, Demartini, J, Davis, M, Sykes, R, & Davison, B. Tuskan, Keller M, Tuskan G and Wyman C. (2011). Lignin content in natural Populus variants affects sugar release. *Proceeding of National Academy of Sciences* , 108, 6300-6305.

[25] Tao, L, Aden, A, Elander, R, Pallapolu, V, Lee, Y, Garlock, R, Balan, V, Dale, B, Kim, Y, Mosier, N, Ladisch, M, Falls, M, Holtzapple, M, Sierra, R, Shi, J, Ebrik, M, Redmond, T, Yang, B, Wyman, C, Hames, B, Thomas, S, & Warner, R. (2011). Process and technoeconomic analysis of leading pretreatment technologies for lignocellulosic ethanol production using switchgrass. *Bioresource Technology* , 102, 11105-11114.

[26] Torget, R, Himmel, M, Grohmann, K, & Wright, J. D. (1988). Initial design of a dilute sulfuric acid pretreatment process for aspen wood chips. *Applied Biochemistry and Biotechnology* , 17, 89-104.

http://dx.doi.org/10.5772/56129

[27] Wallington, T, Anderson, J, Mueller, S, Kolinski, E, Winkler, S, Ginder, J, & Nielsen, O. (2012). Corn ethanol production, food exports, and indirect land use change. *Environmental Science & Technology* 46:6379–6384.

[28] Ximenes, E, Kim, Y, Mosier, N, Dien, B, & Ladisch, M. (2010). Inhibition of cellulases by phenols. *Enzyme and Microbial Technology* , 46, 170-176.

[29] Zhao, X. Q, & Bai, F. W. (2009). Yeast flocculation: New story in fuel ethanol production. *Biotechnology Advances* , 27, 849-856.

[30] Zhao, X. Q, Xue, C, Ge, X. M, Yuan, W. J, Wang, J. Y, & Bai, F. W. (2009). Impact of zinc supplementation on the improvement of ethanol tolerance and yield of self-flocculating yeast in continuous ethanol fermentation. Journal of Biotechnology , 39, 55-60.

Chapter 6

Hydrolysis of Biomass Mediated by Cellulases for the Production of Sugars

Rosa Estela Quiroz-Castañeda and
Jorge Luis Folch-Mallol

Additional information is available at the end of the chapter

http://dx.doi.org/10.5772/53719

1. Introduction

Cellulose, the most abundant organic molecule on Earth is found mainly as a structural component of plant and algal cell walls, is also produced by some animals, such as tunicates, and several bacteria [1]. Natural cellulose is a crystalline and linear polymer of thousands of D-glucose residues linked by β-1,4-glycosidic bonds, considered the most abundant and renewable biomass resource and a formidable reserve of raw material.

It does not accumulate in the environment due to the existence of cellulolytic fungi and bacteria, which slowly degrade some of the components of plant cell walls. Both fungi and bacteria possess enzymes such as laccases, hemicellulases and cellulases, which efficiently degrade lignin, hemicellulose and cellulose, respectively [2-3].

In plant cell walls the cellulose microfibrils are encrusted in lignin and hemicellulose in a complex architecture that, together with the crystallinity of cellulose, makes untreated cellulosic biomass recalcitrant to hydrolysis to fermentable sugars [4]. However, a group of proteins with cellulose disrupting activity (expansins, expansin-like proteins, swollenins and loosenins) have the capacity of relaxing cell wall tension by disrupting the hydrogen bonds binding together cellulose fibrils and cellulose and other polysaccharides through a non-enzymatic process, improving subsequent sugar releasing [5-8].

An efficient degradation of this polysaccharide content into fermentable sugars could improve the production of biofuels. Rising energy consumption, depletion of fossil fuels and increased environmental concerns have shifted the focus of energy generation towards biofuel use [3].

INTECH
open science | open minds

In this chapter, we focus on cellulose degradation by cellulases in order to enhance sugars release from biomass. Cellulose structure, allomorphs and its hydrolysis by cellulolytic organisms such as fungi and bacteria, is also reviewed, as well as cellulases structure, CAZY classification, their synergistic activity and the recently cellulases identified by metagenomic analysis, an excellent tool in this search of better cellulolytic activity.

Another theme analyzed in this chapter is related to crystalline structure of cellulose, the main impediment to achieve full cellulose hydrolysis, and the role of proteins recently reported with cellulose disrupting activity that have improved saccharification processes. These proteins represent good candidates as an additive to enhance sugar production from plant biomass.

2. Structure and composition of the cell wall

It has been estimated that the net CO_2 fixation by land plants per year is approximately 56 X 10^9 tons and that the worldwide biomass production by land plants is 170–200 X 10^9 tons (Table 1). Of this amount, 70% is estimated to represent plant cell walls (revised in [9]).

Lignocellulose is a renewable organic material and is the major structural component of all cell plants. Lignocellulose plant biomass consists of three major components: cellulose (40–50 %), hemicellulose (20–40 %) and lignin (20–30 %) (Figure 1).

Production	Tons	Reference
Assimilated CO_2	56 X 10^9	[10]
Plant biomass	170-200 X 10^9	[11]
Cell walls	150-170 X10^9	[9]
Lignocellulose	200 X10^9	[12]
Cellulose	100 X 10^9	[13-14]
Wheat straw	540 X 10^9	[15]
Soybean straw	200 X 10^9	[16]
Sugar cane bagasse	54 X 10^9	[17]

Table 1. Worldwide annual production of biomass

Minor components are proteins, lipids, pectin, soluble sugars and minerals (Table 2) [9]. It has a thickness of ~0.1 a 10 μm contrasting with <0.01 μm of cell membrane formed by proteins and phospholipids [18].

Examples of such biomass are angiosperms (hardwoods), gymnosperms (softwoods) and graminaceous plants (grasses such as wheat, giant reed and *Miscanthus*).

http://dx.doi.org/10.5772/53719

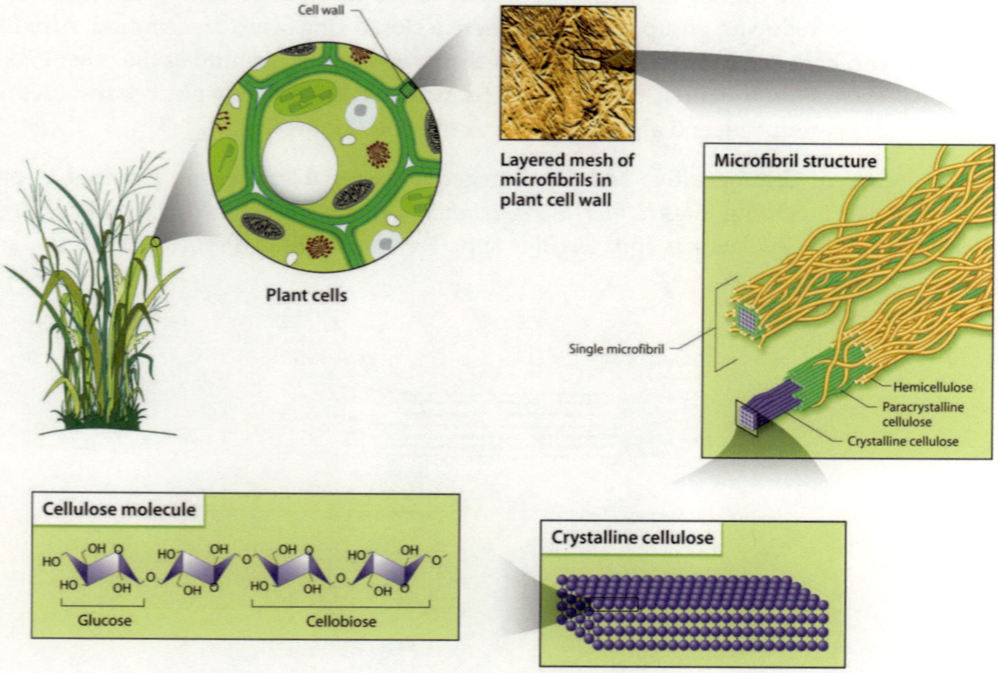

Figure 1. Structural organization of the plant cell wall. Cellulose is protected of degradation by hemicelluloses and lignin. Source: Office of Biological and Environmental Research of the U.S. Department of Energy Office of Science. science.energy.gov/ber/

Cell walls should play a wide array of disparate and sometimes opposing roles: the resistance to mechanical stress is necessary as well as the shape of the cell and protection against pathogens; at the same time, besides it must be reasonably flexible to withstand shear forces, and permeable enough to allow the passage of signalling molecules into the cell [19].

3. Cellulose structure

Cellulose is the main component in the plant cell walls, and is made of parallel unbranched D-glucopyranose units linked by β-1,4-glycosidic bonds that form crystalline and highly organized microfibrils through extensive inter and intramolecular hydrogen bonds and Van der Waals forces, amorphous cellulose correspond to regions where this bonds are broken and the ordered arrangement is lost (Figure 2).The cellulose chains aggregated into microfibrils are reported to consist of 24 to 36 chains based on scattering data and information about the cellulose synthase [20-21].

Consecutive glucose molecules along chains in crystalline cellulose are rotated by 180º, meaning that the disaccharide (cellobiose) is the repeating unit [22].

Two different ending groups are found in each cellulose chain edge. At one end of each of the chains, a non-reducing group is present where a closed ring structure is found. A reducing group with both an aliphatic structure and a carbonyl group is found at the other end of the chains. The cellulose chain is thus a polarized molecule and the new glucose residues are added at the non-reducing end allowing chain elongation (Figure 2) [23].

A wide variety of Gram-positive and Gram-negative bacterial species are reported to produce cellulose, including *Clostridium thermocellum, Streptomyces* spp., *Ruminococcus* spp., *Pseudomonas* spp., *Cellulomonas* spp., *Bacillus* spp., *Serratia, Proteus, Staphylococcus* spp., and *Bacillus subtilis* [24].

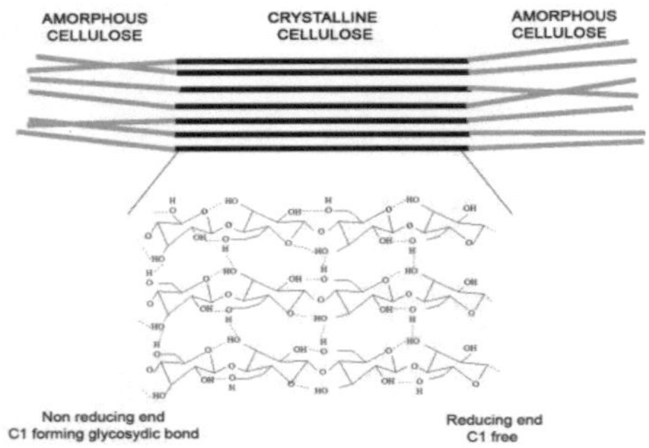

Figure 2. Crystalline and amorphous structure of cellulose. The crystalline structure is conserved by hydrogen bonds and Van der Waals forces, in amorphous structure exists twists and torsions that alter the ordered arrangement. Reducing and non-reducing are shown.

4. Cellulose crystallinity

In plants, cellulose is synthesized by CESA proteins (Cellulose Synthase) embedded in plasmatic membrane arranged in hexameric groups called rosettes particles [25].

Cellulose crystallites are thought to be imperfect, the traditional two-phase cellulose model describes cellulose chains as containing both crystalline (ordered) and amorphous (less ordered) regions. Crystalline structure of cellulose implies a structural arrangement in which all atoms are fixed in discrete position with respect to one another. An important feature of the crystalline array is that the component molecules of individual microfibrils are packed sufficiently tightly to prevent penetration not only by enzymes, but even by small molecules such as water. While its recalcitrance to enzymatic degradation may pose problems, one big advantage of cellulose is its homogeneity [1, 26-27].

Highly ordered, crystalline regions are interspersed with regions containing disorganized or amorphous cellulose, which constitute 5 to 20% of the microfibril. Many studies have shown that completely disordered or amorphous cellulose is hydrolysed at a much faster rate than partially crystalline cellulose; this fact supports the idea that the initial degree of crystallinity is important in determining the enzymatic digestibility of a cellulose sample. Crystallinity, is a measure of the weight fraction of the crystalline regions, is one of the most important measurable properties of cellulose that influences its enzymatic digestibility [19, 28-30].

A parameter termed the crystallinity index (CI) has been used to describe the relative amount of crystalline material in cellulose. Generally, in nature, crystallinity indexes range from 40% to 95%, the rest is amorphous cellulose [31]. The degree of polymerization, (DP) is the number of monomeric units in a polymer molecule, which in cellulose it ranges from 500 to 15,000 but varies depending the substrate (Table 2).

Substrate	Crystallinity index	Degree of polymerization	Ref.
Carboxymethyl cellulose (CMC)[a]	NA	100-2000	[32]
Cellodextrins [a]	NA	2-6	[32]
Avicel [b]	0.5-0.6	300	[13]
BC [b]	0.76-0.95	2000	[13]
PASC [b]	0-0.04	100	[13]
Cotton [b]	0.81-0.95	1000-3000	[13]
Filter paper [b]	0-0.45	750	[13]
Wood pulp [b]	0.5-0.7	500-1500	[13]
Fluka Avicel PH-101 [b]	0.56-0.91	200-240*	[26]
Fluka cellulose [b]	0.48-0-82	280*	[26]
Sigma α-cellulose [b]	0.64	2140-2420*	[33]

*According to manufacturer's data. [a], Soluble; [b], Insoluble.

Table 2. Some physical properties of cellulosic substrates

5. Cellulose allomorphs

The crystalline structure of cellulose has been studied since its discovery in the 19th century, its structure was first established by Carl von Nageli in 1858, and the result was later verified by X-ray crystallography [34-35].

In the past decades, many data on the polymorphism of cellulose were analysed, being the most reliable data published after 1984, when the results of NMR spectroscopic studies of cellulose were reported [36].

The repeating unit of the cellulose macromolecule includes six hydroxy groups and three oxygen atoms. Therefore, the presence of six hydrogen bond donors and nine hydrogen bond acceptors provides several possibilities for forming hydrogen bonds. Due to different arrangements of the pyranose rings and the possible conformational changes of the hydroxymethyl groups, cellulose chains can exhibit different crystal packings [37].

Four different crystalline allomorphs of cellulose have been identified by their characteristic X-ray diffraction patterns and solid-state ^{13}C nuclear magnetic resonance (NMR) spectra: celluloses I, II, III ($III_{I,}$ III_{II}) and IV (IV_I, IV_{II}). The most important allomorphs are cellulose I and II [22].

Some difference in symmetry and chain geometry have been found in unit cell dimensions of various allomorphs and some parameters have been established: a, interchain distance, b unit chain length and c, intersheet distance, as well as the angles α, β and γ which are the angles between b and c, a and c, and a and b, respectively, (Table 3) [38-40].

Allo morph	Unit cell parameters						Ref.
	Bond lengths (Å)			Angles (°)			
	a	*b*	*c*	*A*	*β*	*γ*	
I	6.717(7)	5.962(6)	10.400(6)	118.08(5)	114.80(5)	80.37(5)	(41)
I_β	7.784(8)	8.201(8)	10.380(10)	90	90	96.5	(42)
II	8.10(1)	9.03(1)	10.31(1)	90	90	117.10(5)	(43)
II	8.03(1) *8.03(1)(*	9.04(1) *9.02(1)*	10.35(1) *10.34(1)*	90 *90*	90 *90*	117.11(2) *117.11(2)*	(44)
III_I	4.450(4)	7.850(8)	10.310(10)	90	90	105.10(5)	(45)

I, Fresh water algae *Glaucocystis nostochinearum;* I_β, Tunicate *Halocynthia roretzi;* II, Ramie cellulose (mercerized); II, Regenerated cellulose (Fortisan); III_I, Marine algae *Cladophora.* All crystal structures have been determined at 293ºK, except allomorph II (Fortisan) that was also determined at 100ºK (italics).

Table 3. Unit cell parameters of different cellulose allomorphs obtained by X-ray diffractions.

Cellulose I is the most abundant form found in nature, is a mixture of two distinct crystalline forms: cellulose I_α, the predominant form isolated from bacteria (*Acetobacter xylinum*) and fresh water algae (*Glaucosystis nostochinearum*); and cellulose I_β is the major form in higher plants such as cotton and wood celluloses, ramie and animal celluloses, for example in the edible ascidian *Halocynthia roretzi* [4]. Cellulose from the marine algae *Claudophora* sp. and *Valonia ventricosa* is a mixture of both forms, predominating I_α [37]. Currently, cellulose I is receiving increased attention due to its potential use in bioenergy production.

Cellulose I_α has a triclinic one-chain unit cell where parallel cellulose chains stack through van der Waals interactions, with progressive shear parallel to the chain axis. *Cellulose* I_β has a monoclinic two-chain unit cell, which means parallel cellulose chains stacked with alternating shear (Figure 3) [46].

Cellulose II is the most crystalline thermodynamic stable form, it can also be obtained from cellulose I by two distinct routes: mercerization (alkali treatment) and regeneration (solubilization and subsequent recrystallization) [47]. Cellulose II, like cellulose I_β, has the monoclinic unit cell (space group $P2_1$). The different arrangement of the chains (parallel in cellulose I_β and antiparallel in cellulose II) is the most substantial difference between these two polymorphs. The cellulose is a highly rigid macromolecule due to the presence of a three-dimensional hydrogen bond network in addition to the C-O-C bonds between the glucopyranose rings. In the absence of such hydrogen bond networks the chains are much more flexible. These hydrogen bonds are responsible for both the poor solubility of cellulose and the difference in the reactivity of the hydroxy groups in esterification reactions (Figure 4) [37].

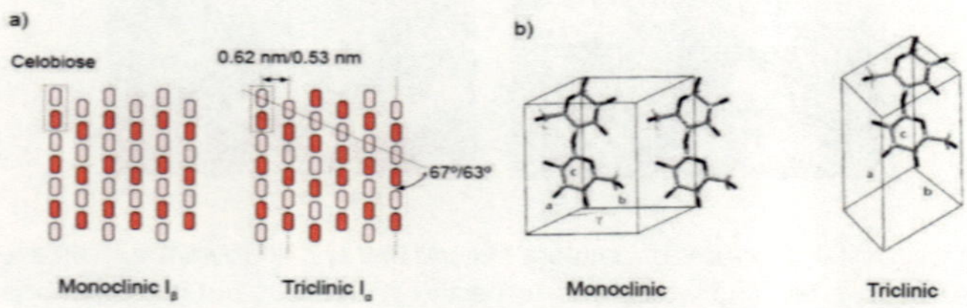

Figure 3. Differences between the monoclinic and triclinic forms of cellulose I. a) In the monoclinic form, cellobiose units stagger with a shift of a quarter of the c-axis period (0.26 nm), whereas the triclinic form exhibits a diagonal shift of the same amount. The angles shown depend on which crystallographic face is being viewed. A glucose unit is represented by rectangles (cellobiose, a dimer of glucose); image reproduced with publisher´s permission [23]. b) Mode of packing in the unit cell of cellulose I: mono and triclinic unit cell. Notice that the monoclinic angle γ is obtuse. Image reproduced with permission from PNAS Copyright (2012).

Cellulose III_I and III_{II} can be formed from cellulose I and II, respectively, by treatment with ammonia; in a reversible reaction. Besides producing the different allomorphs of cellulose, this chemical treatment can also alter other physical properties of cellulose, such as the degree of crystallinity and therefore enhanced cellulase accessibility and chemical reactivity. The degree of conversion of cellulose I to cellulose IIII depends on the reaction period and the temperature used in the final stage of the treatment [47-48].

In [45] solved the crystal structure of cellulose IIII by synchrotron X-ray and neutron fiber diffraction analyses, and showed that it has a lower packing density than cellulose I_α or I_β (Figure 4).

Cellulose IV can be most easily prepared by heating cellulose III, and therefore, two polymorphs of it also exist -celluloses IV_I and IV_{II} obtained respectively, from celluloses III_I and $III_{II.}$ In general, cellulose IV could be prepared by treatment in glycerol at 260 °C after transformation into cellulose II or III. Cellulose I cannot be transformed directly into cellulose IV [46, 49].

Fibrillation makes cellulose IV_I less suitable for crystallographic analysis: that is, it makes it more difficult to interpret cellulose IV_I as a crystal. For these reasons, it is unclear whether is a crystal with an orthogonal unit cell or a less crystalline form of cellulose I [49]. A thorough review of cellulose crystalline allomorphs can be found elsewhere [46-47].

Although considerable progress has been made in elucidating the crystal structures of cellulose in microfibrils, they are still not well understood, and a deeper understanding of cellulose structure is required [50-51].

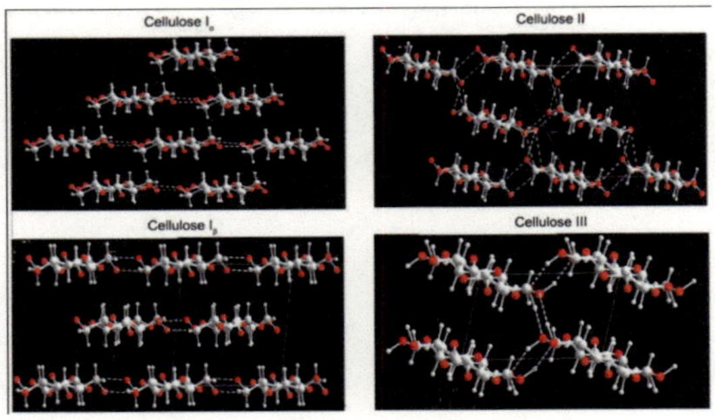

Figure 4. Projections of the crystal structures of cellulose I (α,and β) II and III down the chain axes directions. C, O, and H atoms are represented as gray, red, and white balls, respectively. Covalent and hydrogen bonds are represented as full and dashed sticks, respectively. H atoms involved in hydrogen bonding are explicitly represented for only cellulose IIII. Only the major components of hydrogen bonds are represented. Adapted with permission from [45]. Copyright (2012) American Chemical Society.

6. Cellulose-degrading microorganisms

Since cellulose is very difficult to degrade as a component of plant cell walls, only a few microorganisms specialized for plant cell wall degradation can hydrolyse cellulose. Among these, anaerobic and aerobic genera of Domain Bacteria and fungi of Domain Eukarya are included.

Generally speaking, two types of systems occur in regards to plant cell wall degradation by microorganisms. In one type, the organism produces a set of free enzymes that act synergistically to degrade plant cell walls. In the second type, the degradative enzymes are organized into an enzyme complex located in cellular surface called the cellulosome. This complex is very effective in degrading plant cell walls [52].

Anaerobic and aerobic bacteria have different strategies to degrade cellulolytic substrates; whereas anaerobic bacteria degrade cellulose using cellulosomes, aerobic bacteria secretes enzymes capable of degrading cellulose that freely diffuse to reach the substrate.

Anaerobic bacteria of the order *Clostridiales* (Phylum *Firmicutes*) are generally found in soils, decaying plant waste, the rumen of ruminant animals, compost, waste water, and wood processing plants; these bacteria have also been found in insects like termites (*Isopteran*), bookworm (*Lepidoptera*), and so, in a symbiotic relationship in their guts responsible for cellulosic feed digestion. Anaerobic hydrolysis represents 5% to 10% of global cellulose degradation [53-55].

http://dx.doi.org/10.5772/53719

Aerobic bacteria with cellulolytic activities of the order *Actinomycetales* (phylum *Actinobacteria*) have been found on soils, water, humus, agricultural waste (sugar cane) and decaying leaves, these bacteria excretes enzymes capable of degrading cellulose (cellulases) [52]. In aerobic bacteria *Pseudomonas fluorescens* subsp. cellulosa, *Streptomyces lividans* and *Cellulomonas fimi* cellulolytic systems of degradation have been reported [56-58].

Some anaerobic bacteria with cellulolytic activity are Butyrivibrio fibrisolvens, Fibrobacter succinogenes, Ruminococcus flavefaciens, Clostridium cellulovorans, C. cellulolyticum and C. thermocellum [59-61].

Due to the significant diversity in the physiology of cellulolytic bacteria, sometimes is difficult to classify bacteria as mentioned above, therefore, on this basis, they can be placed into three diverse physiological groups: (1) fermentative anaerobes, typically Gram-positive, (*Clostridium* and *Ruminococcus),* but with a few Gram-negative species (*Butyvibrio* and *Acetivibrio*) that are phylogenetically related to the *Clostridium* assemblage (*Fibrobacter*); (2) aerobic Gram-positive bacteria (*Cellulomonas* and *Thermobifida)* and (3) aerobic gliding bacteria, (*Cytophaga* and *Sporocytophaga)* [1, 53].

The ability to utilize lignocellulosic material is widely distributed among fungi, from chytridiomycetes to basidiomycetes. Among fungi, the most efficient at using wood as substrate are the basidiomycetes, considered the principal taxonomic group involved in the aerobic degradation of wood with all its components, they are the main organic material decomposition agents. These aerobic fungi produce extracellular enzymes allowing lignocellulose degradation (lacasses, hemicellulases, and cellulases), although some Ascomycetes are able to degrade cellulosic compounds as well. Unlike aerobic fungi, some of the Chytridiomycetes anaerobic fungi, have multienzymatic complexes similar to cellulosomes of bacteria [1, 3, 62-63] some members are anaerobic species living in the gastrointestinal tract of ruminants such as *Anaeromyces, Caecomyces, Neocallimastix, Orpinomyces* and *Piromyces.*

Examining the taxonomic composition of cellulolytic fungi inhabiting the decaying leaves and rotting woods of forest soils, zygomycetes are represented by a single genus, *Mucor,* while ascomycetes and basidiomycetes are represented by genera such as *Chaetomium, Trichoderma, Aspergillus, Penicillium, Fusarium, Coriolus, Phanerochaete, Schizophyllum, Volvariella, Pycnoporus* and *Bjerkandera.* Two of the most studied fungi, due to their industrial relevance, are *Trichoderma reesei* and *Phanerochaete chrysosporium*.

Nowadays, more than 14,000 fungi, which are active against cellulose and other insoluble fibres, are known [1, 24, 64-66]. A more detailed list of cellulose degrading bacteria and fungi is listed in Table 5.

7. Cellulose degradation mediated by cellulosome

Selective pressure of evolution is the force driving microorganisms to adapt a new environment, in anaerobic conditions is necessary a machinery for the extracellular degradation of substrates, such as the recalcitrant crystalline components of the plant cell wall. Due to this, the anaerobes tend to adopt different strategies for degrading plant components, being the cellulosomes the most remarkable feature.

Group		Fungi	Enzymes**	Substrate	Ref
Aerobic fungi (extracellular cellulolytic enzymes)	Ascomycetes	*T. reesei*	Cel, Xyl	Wheat straw	[68-69]
		T. harzianum	Cel	Cellulose	[70]
		A. niger	Cel	Sugar cane bagasse	[71]
		Schizophyllum commune	Cel, Xyl	Microcrystalline cellulose, rice xylan	[72]
	Basidiomycetes	*P. chrysosporium*	Cel, Xyl	Red oak, grape seed, barley bran, sorghum	[2, 63, 73]
		B. adusta Pycnoporus sanguineus	Cel, Xyl	Oak and cedar sawdust, rice husk, corn stubble, wheat straw and *Jatropha* seed husk	[74]
		Fomitopsis palustris	Cel	Microcrystalline cellulose	[75]
Anaerobic rumen fungi (cellulosomes)	Chytridiomycetes	*Anaeromyces mucrunatus*	Cel, Xyl	Orchard grass hay	[76]
		Caecomyces communis	Cel	Microcrystalline cellulose, alfalfa hay	[77]
		Neocallimastix frontalis	Cel, Xyl	wheat straw	[78]
		Orpinomyces sp.	Cel	Avicel	[79]
		N. patriciarum	Cel	CMC	[80]
Aerobic bacteria (cellulosomes)	Actinobacteria	*Acidothermus cellulolyticus*	Cel	Whatman paper No1, Microcyrtsalline cellulose, Azurine crosslinked hydroxyethylcellulose (AZCL-HEC)	[81]
		Actinospica robiniae			
		Actinosynnema mirum			
		Catenulispora acidiphila			
		Cellulomonas flavigena			
		Thermobispora bispora			
		Xylanimonas cellulosilytica			
Anaerobic bacteria (cellulosomes)	Firmicutes	Clostridium thermocellum*	Cel	Crystalline cellulose	[82]
		*Thermomonospora fusca**	Cel, Xyl	wheat straw, oat spelt xylan	[83]
		*Caldicellulosiruptor kristjanssonii**	Cel	Microcrystalline cellulose	[84]
		*Anaerocellum thermophilum**	Cel, Xyl	Microcrystalline cellulose, xylan	[85-86]

*Termophylic bacteria. **Cel: cellulases; Xyl: xylanases.

Table 4. Fungi and bacteria with cellulolytic activity.

http://dx.doi.org/10.5772/53719

The occurrence of a cellulosome was first observed in the thermophilic bacterium, *C. thermocellum* and has now been described in a number of mesophilic anaerobic bacteria and with some anaerobic fungi particularly *Piromyces* sp.[52, 67].

Cellulosomes are large extracellular enzyme complexes capable of degrading cellulose, hemicelluloses, and pectin; they may be the largest extracellular enzyme complexes found in nature, although the individual cellulosomes size range from 0.65 MDa to 2.5 MDa, some polycellulosomes have been reported to be as large as 100 MDa, [87].

The cellulosome structure is characterized by two components: (a) the non-enzymatic scaffolding proteins with enzyme binding sites called cohesins, (b) enzymes with dockerins proteind interacting with cohesins in the scaffolding protein (Figure 5).

Depending of the bacterial species, the scaffolding protein varies in the number of cohesins and cellulose binding modules (CBM) that binds the cellulosome tightly to the substrate and concentrates the enzymes to a particular site of the substrate. Recently, a more complex cellulosome structure with multiple interacting scaffolding proteins that allows the binding of as much as enzymes has been revealed [52].

The cohesin-dockerin interconnect the different scaffoldin components, whereby the specificities among the individual cohesin-dockerin complexes dictate the overall supramolecular architecture of the participating components [88].

In short, the enzymatic cellulosome system may exceed the potential of non-cellulosomal degradative system due to its structural organization, efficient binding to the substrate, the variety of hydrolytic enzymes acting synergistically [52]. Cellulosomes have not been identified in bacteria (or eukarya) that grow above 65 ºC, and have not been identified in the Archaea [89].

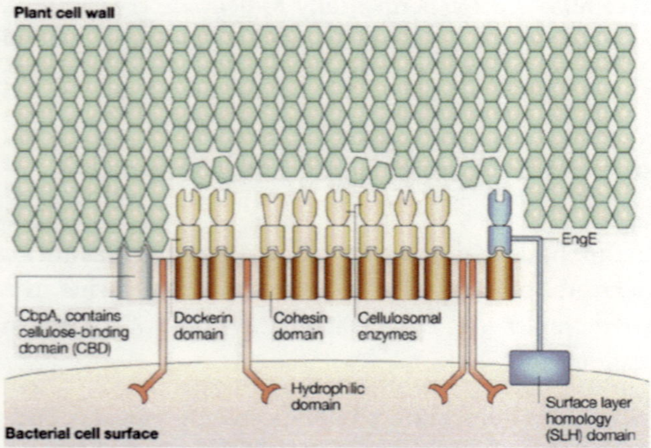

Figure 5. Cellulosome structure. A dockerin is appended to catalytic (enzyme) and noncatalytic carbohydrate-binding modules (CBMs). Dockerins bind the cohesins of a noncatalytic scaffoldin, providing a mechanism for cellulosome assembly. Image reproduced with publisher´s permission [90].

8. Cellulose degradation mediated by non-cellulosomal enzymes

Aerobic cellulolytic bacteria and fungi use a system for cellulose degradation consisting of sets of soluble cellulases. Cellulases are inducible enzymes by cellulosic substrates, which are synthesized by a large diversity of microorganisms including both fungi and bacteria during their growth on cellulosic materials.These microorganisms can be aerobic, anaerobic, mesophilic or thermophilic. Among them, the genera of *Clostridium, Cellulomonas, Thermomonospora, Trichoderma,* and *Aspergillus* are the most extensively studied cellulases producers [91].

The aerobic cellulase mechanism evolved in terrestrial microorganisms that colonise solid substrates and therefore secrete cellulases to enable degradation of the substrate. Because of the recalcitrance of plant cell walls some cellulolytic microorganisms secrete up to 50% of their total protein during growth on biomass or cellulose [53, 90].

Cellulases are composed of independently folding and structurally and functionally discrete units called domains, making cellulases modular enzymes. Structurally fungal cellulases are simpler as compared to bacterial cellulosomes [32, 88, 92].

Fungal cellulases have two independent domains: a catalytic domain (CD) and a cellulose-binding domain (CBD), which is joined by a short poly linker region to the catalytic domain at the N-terminal. The CBD is comprised of approximately 35 amino acids, and the linker region is rich in serine and threonine [93].

It is clear that the role of the CBD is to bind the enzyme to the cellulose so that the CD keep closer to the substrate and it also gives the CD time to move the chain into its active site before the enzyme diffuses away from the cellulose. It is still not clear whether the CBD also can modify cellulose or otherwise assist cellulose hydrolysis by the catalytic domain [94].

The mixture of free cellulases act synergistically to degrade crystalline cellulose increasing the specific activity up to fifteen fold higher than that of any individual cellulase [95].

9. Cellulolytic organisms from extreme environments

Novel enzymes with application in industry require improved features to tolerate extreme conditions of temperature, pH and salinity. Some microorganisms live in these environments, so called extremophiles and are considered a source of enzymes with potential biotechnological applications.

Extreme environments host a number of cellulolytic microorganisms, such as the Gram–negative Antarctic bacterium *Pseudoalteromonas haloplanktis,* collected from seawater, which secretes a psychrophilic cellulase, Cel5G, this cold adapted enzyme displays a high specific activity at low and moderate temperatures and a rather high thermosensitivity induced by a decrease of the intramolecular interactions [96-97].

Extremely thermophilic cellulose-degrading microorganisms are of particular and biotechnological interest owing to the presence of highly thermostable enzymes. A deeper analysis of these organisms is reported in [89, 98-99].

The group of thermophilic cellulolytic prokaryotes includes two aerobic species, *Rhodothermus marinus* and *Acidothermus cellulolyticus*, and numerous anaerobes of the genera *Caldicellulosiruptor*, *Clostridium*, *Spirochaeta*, *Fervidobacterium* and *Thermotoga* (reviewed by (100)).

All members of the genus *Caldicellulosiruptor* are extremely thermophilic, cellulolytic, and non-spore-forming anaerobes with Gram-positive type cell wall, capable of fermenting different types of carbohydrates and have been isolated mostly from neutral or slightly alkaline geothermal springs in New Zealand, Iceland and California [100].

Recently, thermostable cellulases have also been reported in the thermophilic *Geobacillus* sp. R7 that produces a cellulase with a high hydrolytic potential when grown on pretreated agricultural residues (corn stover and prairie cord grass). In fact, it was demonstrated that *Geobacillus* sp. R7 can ferment the lignocellulosic substrates to ethanol in a single step, improving bioethanol production with important potential for cost reductions. Cellulases genes were also identified in several *Sulfolobales* strains, however, their physiological function is not well understood [101-102].

Another thermophlic bacterium *Anaerocellum thermophilum* degrade lignocellulosic biomass untreated as well as crystalline cellulose and xylan [86].

While cellulases are widespread in Fungi and Bacteria, only one archaeal cellulase, an endoglucanase from *Pyrococcus furiosus*, has been reported. This enzyme exhibits a significant hydrolyzing activity toward crystalline cellulose even tough it lacks a CBD, the role of this intracellular enzyme in Archaea is unclear, given that Archaea are apparently unable to grow on cellulose [103-104].

In the alkali tolerant fungus *Penicillium citrinum* an alkali tolerant and thermostable cellulases were found which may have potential effectiveness as additives to laundry detergents [105].

In this search to improve cellulases activity, hybrids of hyperthermostable glycoside hydrolases have been constructed as reported by [106], for example, using the structural compatibility of two hyperthermostable family 1 glycoside hydrolases, *P. furiosus* CelB and *Sulfolobus solfataricus* LacS a library of hybrids using DNA family shuffling was created.

This study demonstrates that extremely thermostable enzymes with limited homology and different mechanisms of stabilization can be efficiently shuffled to form stable hybrids with improved catalytic features.

Alkaliphilic, thermophylic and halophilic microbial species have the potential to yield valuable new products for biotechnological industry. Alkaliphilic polymer-degrading enzymes such as proteases, lipases and cellulases are most frequently isolated from *Bacillus* or related species. Cellulases and lipases are important not only as components of washing detergents, but they are also applied in the paper and pulp, pharmaceutical, food, leather, chemical or waste treatment industries [107-108].

10. Cellulases structure

Proteins with hydrolytic activity such as cellulases and hemicellulases comprises a complex molecular architecture of discrete modules (a catalytic domain (CD) and one or more CBDs), which are joined by unstructured linker sequences [109].

The *catalytic domain* spans more than 70% of protein sequence. A sequence analysis of these domains in different cellulases shows a significant variability between them, in fact, active site of the enzyme has distinct three dimensional arrangements: in tunnel shape for a processive exo degradation or in a cleft shape for an endo degradation. This domain is N-glycosylated and is responsible of the cleavage of the glycosidic bond, which occurs through an acid hydrolysis mechanism, using a donor of protons and a nucleophyle or base such as glutamic and aspartic acid [1, 110-111].

The *cellulose binding domain* facilitates hydrolysis by keeping the catalytic domain nearby the substrate, therefore the presence of CBD is important for cellulases starting and processivity [112]. The CBDs, which is usually O-glycosylated, contain from 30 to about 200 amino acids, and exist as a single, double, or triple domain in a protein. Their location in the protein can be both, C or N terminal and occasionally is centrally positioned.

The CBDs bring the enzyme into a closer and prolonged association with the substrate, increasing the rate of catalysis, this domain was found to function more efficiently in substrate degradation, and removing the CBM from the enzyme or from the scaffolding in cellulosomes dramatically decrease its enzymatic activity (revised in [109]).

In the union of CBD and cellulose, some non polar residues left exposed, mostly tyrosines and tryptophans, showing the flat face of their aromatic ring towards the pyranose ring, this interaction is stabilized by polar residues that form hydrogen bonds [61].

Besides cellulases, CBDs have also been found in other polysaccharides degrading enzymes: hemicellulases, endomannanases and acetilxylanesterases [113].

The *linker peptide* is a sequence of amino acids connecting the cellulose binding domain and the catalytic domain. This linker contains from 6 to 59 amino acids and functions as a flexible hinge that allows the independent function of each domain [114]. The sequence of the linker varies between enzymes, however, the composition is typically rich in proline, treonine and serine, like in the sequence PTPTPTPTT$(PT)_7$ of the endoglucanase of *C. fimi* and NPSGGNPPGGNPPGTTTTRRPATTTGSSPG of the cellobiohydrolase CBHI of *T. reesei*.

Treonine and serine residues of the peptide linker are highly O-glycosylated to be protected from proteolysis; if the linker is completely absent or is too short then both domains, CBD and CD, obstruct each other and the affinity reduces. Based on the similarities of the linker between cellulases it has been suggested that it could be acting as a flexible hinge facilitating independent function of the domains (Figure 6) [115-116].

11. Mechanisms of cellulose biodegradation

Once the cellulase has recognized a free chain end, it threads the chain into the tunnel to form a catalytically active complex (CAC). Because cellulose decrystallization in water is free-energetically unfavourable, the tunnels or clefts of cellulase CDs contain hydrophobic and polar residues that form favourable contacts with the chain.

Several studies have mutated hydrophobic residues in the CD tunnels of cellulases and chitinases (structurally similar to cellulases), and have demonstrated that hydrophobic residues need to be present in the CD tunnels for digestion of crystalline cellulose to occur [117].

Additionally, in [27] have shown that removal of hydrophobic residues in cellulase and chitinase tunnels can increase processivity rates on more accessible polymers.

Once a cellulase forms a CAC with a cellodextrin chain, the hydrolysis reaction occurs usually via a retaining or inverting mechanism, depending on the directionality of the enzyme. After the reaction occurs, the product must be expelled and another CAC formed by threading another cellobiose unit into the CD (Figure 6) [117].

In most cases, the hydrolysis of the glycosidic bond is catalysed by two amino acid residues of the enzyme: a general acid (proton donor) and a nucleophile/base [111]. Depending on the spatial position of these catalytic residues, hydrolysis occurs via overall retention or overall inversion of the anomeric carbon. Recently, a completely unrelated mechanism has been demonstrated for two families of glycosidases utilizing NAD^+ as a cofactor [118-119].

The *retaining glycoside hydrolase mechanism* leads to a net retention of the configuration at the anomeric carbon (C1] of the substrate after cleavage, since the hydrolysis of a glycosidic bond creates a product with the same configuration at the anomeric carbon as the substrate had before hydrolysis.

The *inverting glycoside hydrolase mechanism* leads to a net inversion of the configuration at the anomeric carbon (C1] of the substrate after cleavage. This is performed via a single nucleophilic displacement mechanism, where the hydrolysis of a -glycosidic bond creates a product with the -configuration, and vice-versa [120].

12. Cellulose biodegradation

Although more than a dozen fungal species considered as cellulose degraders have been reported (including *T. viride, T. reesei, F. solani, A. niger, A. terreus, P. chrysosporium, B. adusta* and *P. sanguineus)* [3, 74]; and even with cellulases identified in nematodes (*Bursaphelenchus xylophilus,* a nematode infecting pine wood), yeast (*Aureobasidium pullulans)* and marine bacteria (*Saccharophagus degradans),* the search of new cellulases genes continues. This have led to the construction of metagenomic libraries and bioprospecting analysis from several environments: buffalo rumen, higher termite guts, bovine ruminal protozoan, decomposing poplar wood chips and hardwood forest leading to the identification of new genes and organisms with cellulolytic activities [107, 121-130].

To have a better impression of the latest developments regarding fungal carbohydrate-active enzymes, the following sections will discuss the enzymes needed for cellulose degradation.

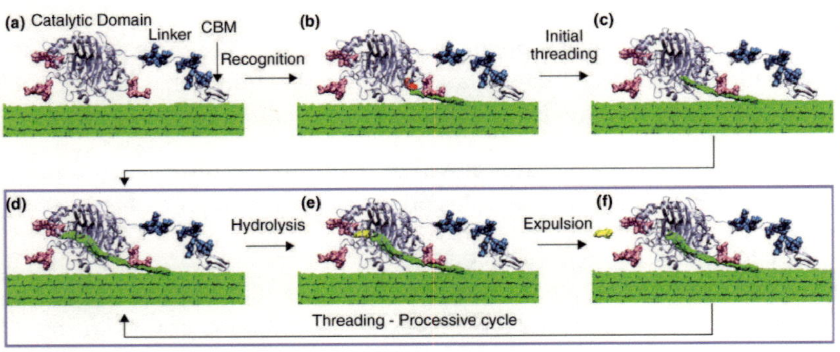

(*a*) Cel7A binding to cellulose, (*b*) recognition of a reducing end of a cellulose chain, (*c*) initial threading of the cellulose chain into the catalytic tunnel, (*d*) threading and formation of a catalytically active complex, (*e*) hydrolysis in a processive cycle and (*f*) product expulsion and threading of another cellobiose (shown in yellow in *e* and *f*). Image reproduced with publisher´s permission [131].

Figure 6. Activity on substrate of cellulase (exoglucanase, Cel7A) of *T. reesei.* The enzyme has a small carbohydrate-binding domain (CBD) of 36-amino acid, a long flexible linker with O-glycan (dark blue), and a large catalytic domain (CD) with N-linked glycan (pink) that can thread a single chain of cellulose into the catalytic tunnel of 50 Å.

13. Cellulases

Multiple types of modular cellulases formed by catalytic and carbohydrate binding domains have been discovered, including at least two exo-β-glucanases, or cellobiohydrolases (CBHs,CBH I and CBH II), four endoglucanases (EG; EG I, EG II, EG III, EG V), and one β-glucosidase (BG) [1].

Cellulases are O-glucoside hydrolases (GH, EC 3.2.1.), a widespread group of enzymes which hydrolyse the β-1,4 linkages or glycosidic bond between two or more carbohydrates or between a carbohydrate and a non-carbohydrate moiety. GH are classified into cellulases families on the basis of amino acid sequence similarity [31, 132].

A classification of glycoside hydrolases in families based on amino acid sequence similarities has been proposed a few years ago. Because there is a direct relationship between sequence and folding similarities, this classification reflects the structural features of these enzymes better than their sole substrate specificity, and helps to reveal the evolutionary relationships between these enzymes, which represent a convenient tool to deduce information of the mechanism [132-133].

Out of the currently existing 125 families, 15 correspond to cellulases (GHF 1,3, 5, 6, 7, 8, 9, 12, 44, 45, 48, 51, 74, 116, and 124), and 64 families group the cellulose binding domains (see http://www.cazy.org/). In [134] an excellent review of and classification system for many CBD families is provided.

The widely accepted mechanism for enzymatic cellulose hydrolysis involves synergistic actions by endoglucanases (EGL, EC 3.2.1.4], exoglucanases or cellobiohydrolases (CBH, EC 3.2.1.74; 1,4-β-D-glucan-glucanhydrolase and EC 3.2.1.91; 1,4-β-D-glucan cellobiohydrolase), and β-glucosidases (BGL, EC 3.2.1.21).

Endoglucanases hydrolyse accessible intramolecular β-1,4-glucosidic bonds of cellulose chains randomly to produce new chain ends; exoglucanases processively cleave cellulose chains at the reducing and non-reducing ends to release soluble cellobiose or glucose; and β-glucosidases hydrolyse cellobiose to glucose in order to eliminate cellobiose inhibition (13). These three hydrolysis processes occur simultaneously as shown in Figure 7.

The activity of cellulase enzyme systems is much higher than the sum of the activity of its individual subunits; a phenomenon known as synergism, so they have to be considered not just simply a conglomerate of enzymes with components from all three cellulase types, but as a mixture that efficiently hydrolyse cellulose fibres.

14. Endoglucanases

These enzymes cleave internal linkages in amorphous cellulose filaments, generating oligosaccharides with different sizes and creating new chain ends that can in turn be attacked by exoglucanases (135). The cellulolytic process is initiated by endoglucanases that randomly cleave internal linkages at amorphous regions of the cellulose fibre and creating new reducing and non reducing ends that are susceptible to the action of cellobiohydrolases [136].

Endoglucanases are monomeric enzymes with a molecular weight that ranges from 22 to 45 kDa, although some fungi such as *Sclerotium rolfsii* and *Gloeophyllum sepiarium* have endoglucanases twice this size [137]. In general, endoglucanases are not glycosylated; however, they sometimes may have relatively low amounts of carbohydrate (from 1 to 12%) [2]. Unlike other endoglucanases reported with optimum pH 4 to 5; the only known endoglucanase with a neutral pH optimum is that from the basidiomycete *Volvariella volvacea,* expressed in recombinant yeast. Basically, their optimum temperature ranges from 50 to 70 °C [138-139].

Exhaustively hydrolysing cellulose also requires the action of β-glucosidases (BGL) (EC 3.2.1.21), which hydrolyse cellobiose, releasing two molecules of glucose and thereby provide a carbon source that is easy to metabolize. Fungi causing white and brown rot, mycorrhizal fungi and plant pathogens produce these enzymes [2, 135].

According to [13], primary hydrolysis occurs on the surface of solid substrates and releases soluble sugars with a degree of polymerization (DP) up to 6 into the liquid phase upon hydrolysis by endoglucanases and exoglucanases. This depolymerisation step performed by endoglucanases and exoglucanases is the rate-limiting step for the whole cellulose hydrolysis process. The second hydrolysis involves primarily the hydrolysis of cellobiose to glucose by β-glucosidases, although some β-glucosidases also hydrolyse longer cellodextrins. The combined actions of endoglucanases and exoglucanases modify the cellulose surface characteristics over time, resulting in rapid changes in hydrolysis rates [32].

To assay endoglucanase activity, there are substrates that are used, such as carboxymethylcellulose (CMC), a soluble amorphous cellulose form that is an excellent substrate for endocellulases and its hydrolysis does not require a CBD [110].

15. Exoglucanases

Also known as cellobiohydrolases, these enzymes catalyse the successive hydrolysis of residues from the reducing and non-reducing ends of the cellulose, releasing cellobiose molecules as main product, which are hydrolysed by β-glucosidases. They account for 40 to 70% of the total component of the cellulase system, and are able to hydrolyse crystalline cellulose.

Exoglucanases have shown specificity on the ends of cellulose, such as*T. reesei* cellobiohydrolase (CBH) I and II that act on the reducing and non-reducing cellulose chain ends, respectively [112].

These enzymes are monomeric proteins with a molecular weight ranging from 50 to 65 kDa, although there are smaller variants (41.5 kDa) in some fungi, such as *Sclerotium rolfsii*. Low levels of glycosylation (around 12% to none at all) are found in these enzymes; and their optimum pH is 4 to 5, with an optimum temperature from 37 to 60 °C, depending on the specific enzyme-substrate combination [137, 140].

Exoglucanases form part of the cellulolytic machinery of the fungi causing white and soft rot and they are found only in some of the basidiomycetes causing the brown rot, such as *Fomitopsis palustris* [141].

Crystalline cellulose (Avicel, bacterial cellulose or filter paper), which is the main form of cellulose in most plant cell walls are good substrates for exoglucanase activity assay, because it has a low DP and relatively low accessibility; however, some endoglucanases can release considerable reducing sugars from Avicel [13].

16. β-glucosidases

β-D-glucosidases hydrolyse soluble cellobiose and other cellodextrins with a DP up to 6 to produce glucose in the aqueous phase in order to eliminate cellobiose inhibition [13].

These enzymes have molecular weights ranging from 35 to 640 kDa, and they can be monomeric or exist as homo-oligomers, as is the case β-glucosidase of the yeast *Rhodotorula minuta* [142]. Most β-glucosidases are glycosylated; in some cases, as that of the 300 kDa BGL from

Trametes versicolor, glycosylation may be superior to 90%. Their optimum pH ranges from 3.5 to 5.5, and their optimum temperature ranges from 45 to 75 °C (3). β-D-glucosidase activities can be measured using cellobiose, which is not hydrolysed by endoglucanases and exoglucanases [13].

17. Synergy between cellulases

Synergistic cooperation between cellulases is a prerequisite for efficient degradation of cellulose, but its molecular mechanisms are not fully understood. Synergistic action has been observed between two different cellobiohydrolases and between endoglucanases. However, more synergistic mechanisms have been proposed [143-144]:

Synergy endo-exo, occurs between endo and exoglucanases, where the action of endoglucanases provide free ends of the cellulose chain to the exoglucanases.

Synergy exo-exo, exoglucanases progressively act on reducing and non-reducing ends of the cellulose chain.

Synergy between exoglucanases and β-glucosidases, the latter process cellobiose produced as final product of the action of the exoglucanases.

Intramolecular synergy between catalytic domain and cellulose binding domain of cellulases.

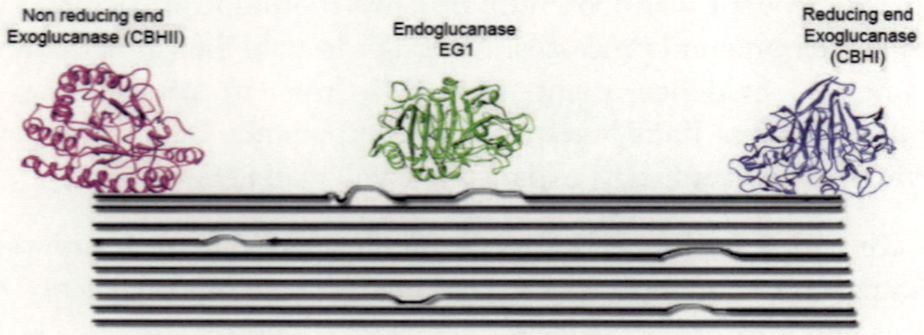

Figure 7. Cellulases activities. Exoglucanases act on reducing and non-reducing ends degrading crystalline cellulose, while Endoglucanase act on amorphous cellulose. Structures: CBHI (PBD, 1CB2), CBHII (PDB, 3CBH) and EGL (PDB, 1EG1).

As a whole system, plant cell wall polysaccharides should be degraded efficiently not only by synergy between cellulases but with participation of the other degrading enzymes as xylanases.

In (145) a synergistic mechanism between cellulases and xylanases in order to saccharify wheat straw for bioethanol production is reported. More recently, a new type of synergism between enzymes that employ oxidative reactions to break glycosidic bonds and hydrolytic enzymes was reported in chitin degradation [28].

Although a significant amount of information has been generated related to the action of cellulases and their mechanisms to degrading cellulose, the biodegradation of crystalline cellulose is still a slow process because the substrate is insoluble and poorly accessible to enzymes.

To overcome this situation scientists have optimized ratio of cellulolytic enzymes, and it was found that the best saccharification of crystalline cellulose is achieved with the enzyme blend: 60:20:20 (CBHI:CBHII:EGI) wherein a saturated level of BG was included to eliminate cellobiose inhibition [146]. In a different report, the impact of the cellulase mixture composition on cellulose conversion was modelled, and the findings suggested different optimum ratios for substrates with different characteristics, specifically degrees of polymerization and surface area [147].

Also, researchers have pointed out the use of proteins that relax plant cell wall structure as a complementary activity before action of cellulases in order to improve saccharification.

18. Plant cell remodelling proteins

In addition to lignocellulose-degrading enzymes, there are also enzymes involved in remodelling the cell wall, which could facilitate its later degradation.

18.1. Expansins

Expansins are pH-dependent wall-loosening proteins required for cell enlargement and expansion in many developmental processes. Although to date their precise mechanism of action remains unclear, evidence point toward a role in dissociating the cell wall polysaccharide complex that links together wall components, thus promoting slippage between wall polymers and, eventually, expansion in cell wall [148-149].

These proteins are coded by large multigene families present from bryophytes to angiosperms and also present in monocotyledonous plants (rice, maize), dicotyledonous plants (*Arabidopsis*), ferns and mosses.

Expansins have no hydrolytic activity (glucosidase) and therefore, it has been suggested to work by breaking hydrogen bonds between cellulose fibres or between cellulose and other polysaccharides (xyloglucans), using a non-enzymatic mechanism (Figure 8) [150-153].

Expansins have molecular weights ranging from 25 to 28 kDa and, like cellulases, have a two-domain modular structure and an approximately 20 amino acids-long amino-terminal signal peptide [149].

Domain I occupies the N-terminal part of the protein, and it has a DPBB (Double Psi Beta Barrel) structure. It is homologous to the catalytic domain of members of glycoside hydrolase family 45 (GH45), which includes mainly β-1,4-endoglucanases of fungal origin. The DPBB domain of members of this family adopts a six-stranded beta barrel structure forming a substrate-binding groove. Despite the presence of the GH45 catalytic domain in expansins, no hydrolytic activity has been detected for the latter [5].

http://dx.doi.org/10.5772/53719

Domain II, at the C-terminal end, is homologous to group II pollen allergens from grasses. Some authors have speculated that this might be a polysaccharide-binding domain, due to the presence of aromatic and polar amino acids on the protein surface, where two tryptophan and one tyrosine would form a planar platform of aromatic residues favouring this binding (149, 154). Domain II folds as a β-sandwich formed by two sheets of four antiparallel β strands each (Figure 8). In fact, a β-sandwich formed by 3 to 6 β strands per sheet is the most common fold in carbohydrate-binding modules of proteins binding substrates such as crystalline cellulose or chitin [155].

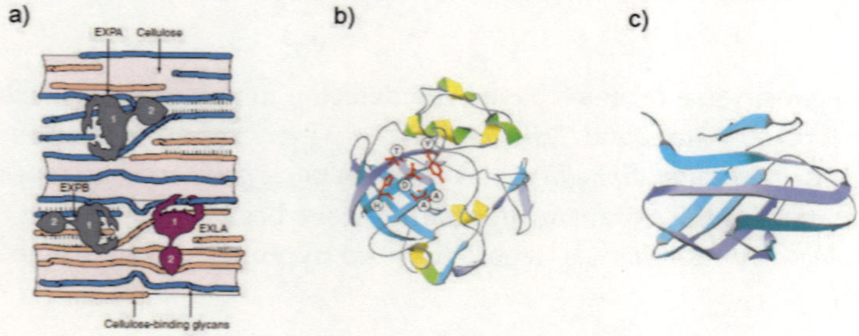

(a G2A protein from *Phleum pratense*; PDB 1WHO). In (a), the domain I forms a barrel; amino-acid residues that are conserved in expansins are indicated in the single-letter amino-acid code. Images reproduced with publisher BioMed permission [5].

Figure 8. a) Expansin proposed activity; b) Expansin domain I (the catalytic domain of a GH45 endoglucanase from *Humicolainsolens*; PBD, 2ENG); c) Expansin domain II

Expansins are classified in four families: α-expansins (EXPA), β-expansins (EXPB), α-expansin like-proteins (EXLA) and β-expansin like-proteins (EXLB) [5].

The *EXPA family* includes proteins participating in the relaxation and extension of plant cell walls through a pH-dependent mechanism; these proteins would participate in developmental processes such as organogenesis, the degradation of cell walls during the ripening of fruits and other processes where relaxation of the cell wall is crucial [156-159].

The *EXPB family* includes group I pollen allergens from grasses. These proteins are secreted by pollen and have been suggested to soften the tissues of the stigma and style to facilitate the penetration of the pollen tube [154].

EXPB proteins, unlike EXPA members, relax specifically the cell walls of grass cells, probably reflecting differences regarding the organization of cell walls between grasses and dicotyledonous plants. Although an HFD motif, that is known to form part of the active site of endoglucanases, has been found in domain I of EXPA and EXPB family members, they do not have hydrolytic activity [5, 160].

The *EXLA and EXLB families* do not have this sequence motif, which suggests that their mode of action differs to that of the other expansins. The EXLA and EXLB families are comprised of proteins identified by sequence analysis which, despite possessing the two- organization typical of expansins, have a number of divergent sequence features that separate them from the EXPA and EXPB families [161].

Another group included in the expansin superfamily is the *expansin-like X family* (EXLX), comprising proteins that exhibit weak sequence homology with the domains of EXPA and EXPB members, and identified in organisms other than plants, such as the mucilaginous fungus *Dictyostelium* and the bacteria *Bacillus subtilis,* and *Hahella chejuensis* [161-164].

The denomination of expansin or expansin-like is reserved for proteins exhibiting both domain I and domain II. Proteins with only one of these domains are not classified as expansins [161]. However other proteins with similar disrupting activity of the cell wall have been reported.

Expansins and expansin-like proteins have been detected in angiosperms such as *Arabidopsis thaliana, Oryza sativa, Zea mays* and *Triticum aestivum,* gymnosperms such as pine and poplar, ferns such as *Regnellidium diphyllum* and *Marsilea quadrifolia* and the moss *Physcomitrella patens.* Some members of the expansin superfamily have been found even in a potato-infecting nematode, *Globodera rostochiensis,* where they are hypothesized to promote the infection process [165-169].

18.2. Novel proteins with expansin-like activity

Proteins with expansin-like activity called swollenins and loosenins have been identified in ascomycete and basidiomycete fungi such as *T. reesei, A. fumigatus* and *B. adusta* [6-8, 170].

In [7], a swollenin gene from *T. reesei* denominated *swo1,* was cloned and expressed in *Saccharomyces cerevisiae,* coding for a protein that modifies the structure of cellulose in swollen regions of cotton fibres (hence the name) without releasing reducing sugars. Swo1 is a fungal expansin-like protein, containing a pollen allergen domain and a cellulose-binding domain.

Proteins with expansin activity could be used to improve the efficiency of cellulose bioconversion processes. For example, a swollenin purified from *A. fumigatus* has been used in combination with cellulases to facilitate the saccharification of microcrystalline cellulose (Avicel) [8]. In [163] also is described the synergism of an EXLX from *B. subtilis* in the enzymatic hydrolysis of cellulose and recently, and a new protein with expansin activity from the basidiomycete fungus *B. adusta,* denominated loosenin (LOOS1] was cloned and characterized [6].

Not only expansins, but also swollenins and loosenin represent good candidate as pretreatment to enhance sugar production from plant biomass. For example, loosenin activity was efficient to release reducing sugars (after cellulase treatment) from *Agave tequilana,* a crop extensively grown in some areas of Mexico, which shredded fibrous waste is usually burnt or left to decompose. Indeed, *A. tequilana* fiber became a susceptible substrate for a cocktail of commercial cellulases and xylanases in the presence of LOOS1. Loosenin shows optimum activity at the same pH as most cellulolytic enzymes, opening the possibility to use them as a mixture. This protein is able to relax the structure of cotton, enhancing up to 7.5-fold the rate of release of reducing sugars from agave fibre. Something similar was observed when a cucumber expansin was incubated with a compound of cellulose and xyloglucans of bacterial origin and occurred a rapid relaxation of the structure of this compound, suggesting that expansins modulate the binding between cellulose fibres and xyloglucans, relaxing or breaking the bonds keeping them together [171].

Given the optimum pH of LOOS1 (pH 5) and other expansin like proteins, they could be applied to processes of saccharification of natural substrates, facilitating the release of reducing sugars together with cellulases. For example, it might be used as an additive to obtain fermentable sugars from pretreated yellow poplar as reported in [172].

In [173], used swollenin as a pretreatment of cellulosic substrates and observed that even in non-saturating concentrations, a significant accelerated hydrolysis occurred. They also correlated particle size and crystallinity of the cellulosic substrates with initial hydrolysis rates, and it could be shown that the swollenin induced-reduction in particle size and crystallinity resulted in high cellulose hydrolysis rates.

It is not surprising that the idea of using plant expansins in saccharification processes has been patented [174-176].

The efficient enzymatic saccharification of cellulose has been a challenge over the past 50 years, mainly due to its crystallinity, which make it a recalcitrance substrate with a high potential to be used as a carbon source.

The bioconversion of cellulose to ethanol is the process where most interest has been focused. Fortunately, increasing of the loosened cellulose surface area by the use of non-hydrolytic proteins, a process known amorphogenesis, would allow access to hydrolytic enzymes making the saccharification process more efficient [177].

19. Conclusions

Cellulose biodegradation represents the major carbon flow from fixed carbon sinks to atmospheric CO_2, this process is very important in several agricultural and waste treatment processes. Also, cellulose contained in plant wastes could be used as a raw material to produce sustainable products and bioenergy to replace depleting fossil fuels. However, one of the most important and difficult technological challenges is to overcome the recalcitrance of natural cellulosic materials, which must be enzymatically hydrolysed to produce fermentable sugars. In order to achieve this goal, new enzymes with cellulolytic activities are being improved and organisms with novel properties have been found. Although the efforts are being directed to improve cellulolytic activity, proteins capable to relax plant cell structure (expansins, swollenins and loosenin) could be used as a biological pretreatment since they would be disrupting crystalline structure of cellulose making it more accessible to the enzymes and enhancing sugar releasing.

Author details

Rosa Estela Quiroz-Castañeda and Jorge Luis Folch-Mallol

*Address all correspondence to: rosa.quiroz@uaem.mx

Biotechnology Research Centre, Autonomous University of Morelos, Cuernavaca, Morelos, México

References

[1] Lynd LR, Weimer PJ, Van Zyl WH, Pretorius IS. Microbial cellulose utilization: fundamentals and biotechnology. Microbiology and molecular biology reviews. 2002;66(3):506-77.

[2] Baldrian P, Valášková V. Degradation of cellulose by basidiomycetous fungi. FEMS Microbiology Reviews. 2008;32(3):501-21.

[3] Dashtban M, Schraft H, Qin W. Fungal Bioconversion of Lignocellulosic Residues; Opportunities & Perspectives. International Journal of Biological Sciences. 2009;5(6): 578-94.

[4] Wada M, Nishiyama Y, Chanzy H, Forsyth T, Langan P. The structure of celluloses. Powder Diffr. 2008;23, No. 2, (2):92-5.

[5] Sampedro J, Cosgrove DJ. The expansin superfamily. Genome biology. 2005;6(12): 242.

[6] Quiroz-Castañeda R, Martinez-Anaya C, Cuervo-Soto L, Segovia L, Folch-Mallol J. Loosenin, a novel protein with cellulose-disrupting activity from Bjerkandera adusta. Microbial Cell Factories. 2011;10(1):8.

[7] Saloheimo M, Paloheimo M, Hakola S, Pere J, Swanson B, Nyyssonen E, Bhatia A, Ward M, Penttila M. Swollenin, a *Trichoderma reesei* protein with sequence similarity to the plant expansins, exhibits disruption activity on cellulosic materials. European Journal of Biochemistry. 2002 Sep;269(17):4202-11.

[8] Chen X-a, Ishida N, Todaka N, Nakamura R, Maruyama J, Takahashi H, Kitamoto K. Promotion of Efficient Saccharification with *Aspergillus fumigatus* AfSwo1 Towards Crystalline Cellulose. Applied and Environmental Microbiology. 2010;76(8):2556-61.

[9] Pauly M, Keegstra K. Cell-wall carbohydrates and their modification as a resource for biofuels. The Plant Journal. 2008;54(4):559-68.

[10] Field CB, Behrenfeld MJ, Randerson JT, Falkowski P. Primary Production of the Biosphere: Integrating Terrestrial and Oceanic Components. Science. 1998;281(5374): 237-40.

[11] Lieth H. Primary production of the major vegetation units of the world. In: Primary Productivity of the Biosphere In: Lieth H, Whittaker R, editors.: Springer-Verlag, New York and Berlin. ; 1975. p. 203-15. .

[12] Ragauskas AJ, Williams CK, Davison BH, Britovsek G, Cairney J, Eckert CA, Frederick WJ, Jr., Hallett JP, Leak DJ, Liotta CL, Mielenz JR, Murphy R, Templer R, Tschaplinski T. The Path Forward for Biofuels and Biomaterials. Science. 2006 January 27, 2006;311(5760):484-9.

[13] Zhang YHP, Lynd LR. Toward an aggregated understanding of enzymatic hydrolysis of cellulose: Noncomplexed cellulase systems. Biotechnology and Bioengineering. 2004;88(7):797-824.

[14] Holtzapple MT, in J., eds. Cellulose, Encyclopedia of Food Science, Food Technology, and Nutrition. Macrae R, Robinson, R. K., and Sadler, M. , editor: Academic Press, London, San Diego, CA, NY, Boston, MA, Sydney, Tokio, Toronto, ; 1993.

[15] Reddy N, Yang Y. Preparation and characterization of long natural cellulose fibers from wheat straw. J Agric Food Chem. 2007;55(21):8570-5. .

[16] Reddy N, Yang Y. Natural cellulose fibers from soybean straw. Bioresour Technol. 2009;100(14):3593-8. .

[17] Sun JX, Sun XF, Sun RC, Su YQ. Fractional extraction and structural characterization of sugarcane bagasse hemicelluloses. Carbohydrate Polymers. 2004;56(2):195-204.

[18] Fry S. Plant cell walls Encyclopedia of life sciences. 2001;DOI 10.1038/npg.els. 0001682. Chichester: Nature Publishing Group.

[19] Levy I, Shani Z, Shoseyov O. Modification of polysaccharides and plant cell wall by endo-1,4-[beta]-glucanase and cellulose-binding domains. Biomolecular Engineering. 2002;19(1):17-30.

[20] Fernandes AN, Thomas LH, Altaner CM, Callow P, Forsyth VT, Apperley DC, Kennedy CJ, Jarvis MC. Nanostructure of cellulose microfibrils in spruce wood. Proceedings of the National Academy of Sciences USA. 2011;108(47):1195-203.

[21] Endler A, Persson S. Cellulose Synthases and Synthesis in Arabidopsis. Molecular Plant. 2011;4(2):199-211.

[22] Festucci-Buselli RA, Otoni WC, Joshi CP. Structure, organization, and functions of cellulose synthase complexes in higher plants. Brazilian Journal of Plant Physiology. 2007;19:1-13.

[23] Koyama M, Helbert W, Imai T, Sugiyama J, Henrissat B. Parallel-up structure evidences the molecular directionality during biosynthesis of bacterial cellulose. Proceedings of the National Academy of Sciences USA. 1997;94(17):9091-5.

[24] Gautam SP, Bundela PS, Pandey AK, Jamaluddin, Awasthi MK, Sarsaiya S. Diversity of Cellulolytic Microbes and the Biodegradation of Municipal Solid Waste by a Potential Strain. International Journal of Microbiology. 2012;2012.

[25] Kimura S. Immunogold labeling of rosette terminal cellulose-synthesizing complexes in the vascular plant *Vigna angularis*. Plant Cell. 1999;11:2075-85.

[26] Park S, Baker JO, Himmel ME, Parilla PA, Johnson DK. Cellulose crystallinity index: measurement techniques and their impact on interpreting cellulase performance. Biotechnol Biofuels. 2010;3:10.

[27] Horn SJ, Vaaje-Kolstad G, Westereng B, Eijsink VG. Novel enzymes for the degradation of cellulose. Biotechnol Biofuels. 2012;5(1):45.

[28] Vaaje-Kolstad G, Westereng B, Horn SJ, Liu Z, Zhai H, Sørlie M, Eijsink VGH. An Oxidative Enzyme Boosting the Enzymatic Conversion of Recalcitrant Polysaccharides. Science. 2010;330(6001):219-22.

[29] Forsberg Z, Vaaje-Kolstad G, Westereng B, Bunaes AC, Stenstrom Y, MacKenzie A, Sorlie M, Horn SJ, Eijsink VG. Cleavage of cellulose by a CBM33 protein. Protein Sci. 2011;20(9):1479-83. .

[30] Fry SC. Cell Wall Polysaccharide Composition and Covalent Crosslinking. Annual Plant Reviews: Wiley-Blackwell; 2010. p. 1-42.

[31] Hildén L, Johansson G. Recent developments on cellulases and carbohydrate-binding modules with cellulose affinity. Biotechnology Letters. 2004;26(22):1683-93.

[32] Percival Zhang YH, Himmel ME, Mielenz JR. Outlook for cellulase improvement: Screening and selection strategies. Biotechnology Advances. 2006;24(5):452-81.

[33] Jager G, Wu Z, Garschhammer K, Engel P, Klement T, Rinaldi R, Spiess A, Buchs J. Practical screening of purified cellobiohydrolases and endoglucanases with alpha-cellulose and specification of hydrodynamics. Biotechnology for Biofuels. 2010;3(1): 18.

[34] Wilkie JS. Carl Nageli and the fine Structure of Living Matter. Nature. 1961;190(4782):1145-50.

[35] Meyer KH, Misch L. Positions des atomes dans le nouveau modèle spatial de la cellulose. Helvetica Chimica Acta. 1937;20(1):232-44.

[36] Atalla RH, Vanderhart DL. Native Cellulose: A Composite of Two Distinct Crystalline Forms. Science. 1984;223(4633):283-5.

[37] Kovalenko VI. Crystalline cellulose: structure and hydrogen bonds. Russian Chemical Reviews. 2010;79(3):231-41.

[38] Li Y, Lin M, Davenport JW. Ab initio studies of cellulose I: crystal structure, intermolecular forces, and interactions with water. The journal of physical chemistry. 2011;115:11533-9.

[39] Klemm D, Schmauder HP, Heinze T. Cellulose. In: Steinbüchel A, editor. Biopolymers Volume 6Polysaccharides II: Polysaccharides from Eukaryotes Münster, Germany: Wiley-VCH; 2004.

[40] Zugenmaier P. Conformation and packing of various crystalline cellulose fibers. Progress in Polymer Science. 2001;26(9):1341-417.

[41] Nishiyama Y, Sugiyama J, Chanzy H, Langan P. Crystal structure and hydrogen bonding system in cellulose I(alpha) from synchrotron X-ray and neutron fiber diffraction. J Am Chem Soc. 2003;125(47):14300-6.

[42] Nishiyama Y, Langan P, Chanzy H. Crystal structure and hydrogen-bonding system in cellulose Ibeta from synchrotron X-ray and neutron fiber diffraction. J Am Chem Soc. 2002;124(31):9074-82.

[43] Langan P, Nishiyama Y, Chanzy H. X-ray Structure of Mercerized Cellulose II at 1 Å Resolution. Biomacromolecules. 2001;2(2):410-6.

[44] Langan P, Sukumar N, Nishiyama Y, Chanzy H. Synchrotron X-ray structures of cellulose Iβ; and regenerated cellulose II at ambient temperature and 100 K. Cellulose. 2005;12(6):551-62.

[45] Wada M, Chanzy H, Nishiyama Y, Langan P. Cellulose IIII Crystal Structure and Hydrogen Bonding by Synchrotron X-ray and Neutron Fiber Diffraction. Macromolecules. [doi: 10.1021/ma0485585]. 2004;37(23):8548-55.

[46] Wada M, Heux L, Sugiya J. Polymorphism of cellulose I family: Reinvestigation of cellulose IV_I. Biomacromolecules 2004;5:1385-91.

[47] Mittal A, Katahira R, Himmel M, Johnson D. Effects of alkaline or liquid-ammonia treatment on crystalline cellulose: changes in crystalline structure and effects on enzymatic digestibility. Biotechnology for Biofuels. 2011;4(1):41.

[48] Hall M, Bansal P, Lee JH, Realff MJ, Bommarius AS. Cellulose crystallinity – a key predictor of the enzymatic hydrolysis rate. FEBS Journal. 2010;277(6):1571-82.

[49] Gardiner ES, Sarko A. Packing analysis of carbohydrates and polysaccharides. 16. The crystal structures of cellulose IV_I and IV_{II}. CanJ Chemistry. 1985;63:173-80. .

[50] Somerville C, Bauer S, Brininstool G, Facette M, Hamann T, Milne J, Osborne E, Paredez A, Persson S, Raab T, Vorwerk S, Youngs H. Toward a Systems Approach to Understanding Plant Cell Walls. Science. 2004 December 24, 2004;306(5705):2206-11.

[51] Ding S-Y, Himmel ME. The Maize Primary Cell Wall Microfibril: A New Model Derived from Direct Visualization. Journal of Agricultural and Food Chemistry. 2006;54(3):597-606.

[52] Doi R. Cellulases of mesophilic microorganisms: cellulosome & non-cellulosome producers. Annals of the New York Academy of Sciences. 2008;1125:267-79.

[53] Ransom-Jones E, Jones D, McCarthy A, McDonald J. The Fibrobacteres: an important phylum of cellulose-degrading bacteria. Microb Ecol 2012 Feb;63(2):267-81 2012.

[54] Schwarz W. The cellulosome and cellulose degradation by anaerobic bacteria. Applied Microbiology and Biotechnology. 2001;56:634–49.

[55] Dillon RJ, Dillon VM. The gut bacteria of insects: nonpathogenic interactions. Annu Rev Entomol. 2004;49:71-92.

[56] Arcand N, Kluepfel D, Paradis F, Morosoli R, Shareck F. Beta-mannanase of *Streptomyces lividans* 66: cloning and DNA sequence of the manA gene and characterization of the enzyme. Biochemical Journal. 1993;290(3):857-63.

[57] Khanna S, Gauri. Regulation, purification, and properties of xylanase from *Cellulomonas fimi*. Enzyme and Microbial Technology. 1993;15(11):990-5.

[58] Braithwaite KL, Black GW, Hazlewood GP, Ali BR, Gilbert HJ. A non-modular endo-beta-1,4-mannanase from *Pseudomonas fluorescens* subspecies cellulosa. Biochemical Journal. 1995;305(3):1005-10.

[59] Lin L, Thomson J. An analysis of the extracellular xylanases and cellulases of *Butyrivibrio fibrisolvens* H17c. FEMS Microbiology Letters. 1991;84(2):197-204.

[60] Murty MVS, Chandra TS. Purification and properties of an extra cellular xylanase enzyme of *Clostridium* strain SAIV. Antonie van Leeuwenhoek. 1992;61(1):35-41.

[61] Tomme P, Warren R, Gilke N. Cellulose hydrolysis by bacteria and fungi. Advances In Microbial Physiology. 1995;37:1–81.

[62] Eberhardt RY, Gilbert HJ, Hazlewood GP. Primary sequence and enzymic properties of two modular endoglucanases, Cel5A and Cel45A, from the anaerobic fungus *Piromyces equi*. Microbiology. 2000;146(8):1999-2008.

[63] Sánchez C. Lignocellulosic residues: Biodegradation and bioconversion by fungi. Biotechnology Advances. 2009;27(2):185-94.

[64] Wilson DB. Microbial diversity of cellulose hydrolysis. Curr Opin Microbiol. 2011;14(3):259-63. Epub 2011 Apr 29.

[65] Quiroz-Castañeda RE, Balcazar-Lopez E, Dantan-Gonzalez E, Martinez A, Folch-Mallol J, Martinez-Anaya C. Characterization of cellulolytic activities of *Bjerkandera adusta* and *Pycnoporus sanguineus* on solid wheat straw medium. Electronic Journal of Biotechnology [online]. 2009;12(4):Available from Internet: http://www.ejbiotechnology.cl/content/vol12/issue4/full/3/index.html.

[66] Koseki T, Yuichiro M, Shinya F, Kazuo M, Tsutomu F, Kiyoshi I, Yoshihit S, Haruyuki I. Biochemical characterization of a glycoside hydrolase family 61 endoglucanase from *Aspergillus kawachii*. Applied Microbiology and Biotechnology. 2008;77:1279–85.

[67] Lamed R, Naimark J, Morgenstern E, Bayer EA. Specialized cell surface structures in cellulolytic bacteria. J Bacteriol. 1987;169(8):3792-800.

[68] Kurzatkowski W, Torronen A, Filipek J, Mach RL, Herzog P, Sowka S, Kubicek CP. Glucose-induced secretion of *Trichoderma reesei* xylanases. Appl Environ Microbiol 1996;62(8):2859-65.

[69] Chao Y, Singh D, Yu L, Li Z, Chi Z, Chen S. Secretome characteristics of pelletized Trichoderma reesei and cellulase production. World J Microbiol Biotechnol. 2012;28(8):2635-41. Epub 012 May 12.

[70] Do Vale LH, Gomez-Mendoza DP, Kim MS, Pandey A, Ricart CA, Ximenes-Filho E, Sousa MV. Secretome analysis of the fungus Trichoderma harzianum grown on cellulose. Proteomics. 2012;29(10):201200063.

[71] Garcia-Kirchner O, Segura-Granados M, Rodriguez-Pascual P. Effect of media composition and growth conditions on production of beta-glucosidase by Aspergillus niger C-6. Appl Biochem Biotechnol. 2005;124:347-59.

[72] Tsujiyama S, Ueno H. Production of cellulolytic enzymes containing cinnamic acid esterase from Schizophyllum commune. J Gen Appl Microbiol. 2011;57(6):309-17.

[73] Ray A, Saykhedkar S, Ayoubi-Canaan P, Hartson SD, Prade R, Mort AJ. Phanerochaete chrysosporium produces a diverse array of extracellular enzymes when grown on sorghum. Appl Microbiol Biotechnol. 2012;93(5):2075-89.

[74] Quiroz-Castañeda R, Pérez-Mejía N, Martínez-Anaya C, Acosta-Urdapilleta L, Folch-Mallol J. Evaluation of different lignocellulosic substrates for the production of cellulases and xylanases by the basidiomycete fungi *Bjerkandera adusta* and *Pycnoporus sanguineus*. Biodegradation. 2010:1-8.

[75] Ji HW, Cha CJ. Identification and functional analysis of a gene encoding beta-glucosidase from the brown-rot basidiomycete Fomitopsis palustris. J Microbiol. 2010;48(6): 808-13. Epub 2011 Jan 9.

[76] Lee SS, Ha JK, Cheng KJ. The effects of sequential inoculation of mixed rumen protozoa on the degradation of orchard grass cell walls by anaerobic fungus Anaeromyces mucronatus 543. Can J Microbiol. 2001;47(8):754-60.

[77] Hodrova B, Kopecny J, Kas J. Cellulolytic enzymes of rumen anaerobic fungi Orpinomyces joyonii and Caecomyces communis. Res Microbiol. 1998;149(6):417-27.

[78] Griffith GW, Ozkose E, Theodorou MK, Davies DR. Diversity of anaerobic fungal populations in cattle revealed by selective enrichment culture using different carbon sources. Fungal Ecology. [doi: 10.1016/j.funeco.2009.01.005]. 2009;2(2):87-97.

[79] Li XL, Chen H, Ljungdahl LG. Monocentric and polycentric anaerobic fungi produce structurally related cellulases and xylanases. Appl Environ Microbiol. 1997;63(2): 628-35.

[80] Chen H-L, Chen Y-C, Lu M-Y, Chang J-J, Wang H-T, Wang T-Y, Ruan S-K, Wang T-Y, Hung K-Y, Cho H-Y, Ke H-M, Lin W-T, Shih M-C, Li W-H. A highly efficient beta-glucosidase from a buffalo rumen fungus Neocallimastix patriciarum W5. Biotechnology for Biofuels. 2012;5(1):24.

[81] Anderson I, Abt B, Lykidis A, Klenk H-P, Kyrpides N, Ivanova N. Genomics of Aerobic Cellulose Utilization Systems in Actinobacteria. PLoS ONE. 2012;7(6).

[82] Freier D MC, Wiegel J: . Characterization of *Clostridium thermocellum* JW20. . Appl Environ Microbiol 1988;54:204-11.

[83] Tuncer M, Ball AS. Degradation of lignocellulose by extracellular enzymes produced by Thermomonospora fusca BD25. Appl Microbiol Biotechnol. 2002;58(5):608-11. .

[84] Bredholt S, Sonne-Hansen J, Nielsen P, Mathrani IM, Ahring BK. Caldicellulosiruptor kristjanssonii sp. nov., a cellulolytic, extremely thermophilic, anaerobic bacterium. Int J Syst Bacteriol. 1999;3:991-6.

[85] Svetlichnyi V, Svetlichnaya T, Chernykh N, Zavarzin G. *Anaerocellum thermophilum* gen. nov., sp. nov., an extremely thermophilic cellulolytic eubacterium isolated from hot-springs in the valley of Geysers. . Microbiology. 1990;59:598–604.

[86] Yang SJ, Kataeva I, Hamilton-Brehm SD, Engle NL, Tschaplinski TJ, Doeppke C, Davis M, Westpheling J, Adams MW. Efficient degradation of lignocellulosic plant biomass, without pretreatment, by the thermophilic anaerobe "Anaerocellum thermophilum" DSM 6725. Appl Environ Microbiol. 2009;75(14):4762-9. .

[87] Doi R, Kosugi A. Cellulosomes: plant cell wall degrading enzyme complexes. Nature reviews microbiology. 2004;2:541-51.

[88] Bayer EA, Belaich JP, Shoham Y, Lamed R. The cellulosomes: multienzyme machines for degradation of plant cell wall polysaccharides. Annu Rev Microbiol. 2004;58:521-54.

[89] Blumer-Schuette SE, Kataeva I, Westpheling J, Adams MW, Kelly RM. Extremely thermophilic microorganisms for biomass conversion: status and prospects. Curr Opin Biotechnol. 2008;19(3):210-7.

[90] Fontes CMGA, Gilbert HJ. Cellulosomes: Highly Efficient Nanomachines Designed to Deconstruct Plant Cell Wall Complex Carbohydrates. Annual Review of Biochemistry. 2010;79(1):655-81.

[91] Sun Y, Cheng J. Hydrolysis of lignocellulosic materials for ethanol production: a review. Bioresource Technology. 2002;83(1):1-11.

[92] Henrissat B, Teeri TT, Warren RA. A scheme for designating enzymes that hydrolyse the polysaccharides in the cell walls of plants. FEBS Lett. 1998;425(2):352-4.

[93] Kuhad RC, Gupta R, Singh A. Microbial Cellulases and Their Industrial Applications. Enzyme Research. 2011;2011.

[94] Din N, Gilkes NR, Tekant B, Miller RC, Warren RAJ, Kilburn DG. Non-Hydrolytic Disruption of Cellulose Fibres by the Binding Domain of a Bacterial Cellulase. Nat Biotech. [10.1038/nbt1191-1096]. 1991;9(11):1096-9.

[95] Irwin D, Spezio M, Walker L, DB W. Activity studies of eight purified cellulases: specificity, synergism, and binding domain effects. Biotechnology and Bioengineering. 1993;42:1002–13.

[96] Feller G, Gerday C. Psychrophilic enzymes: hot topics in cold adaptation. Nat Rev Microbiol. 2003;1(3):200-8.

[97] Sonan G, Receveur-Brechot V, Duez C, Aghajari N, Czjzek M, Haser R, Gerday C. The linker region plays a key role in the adaptation to cold of the cellulase from an Antarctic bacterium. Biochemical Journal. 2007;407:293–302.

[98] Maki M, Leung KT, Qin W. The prospects of cellulase-producing bacteria for the bioconversion of lignocellulosic biomass. Int J Biol Sci. 2009;5(5):500-16.

[99] Li DC, Li AN, Papageorgiou AC. Cellulases from thermophilic fungi: recent insights and biotechnological potential. Enzyme Res. 2011;2011:308730.

[100] Miroshnichenko ML, Kublanov IV, Kostrikina NA, Tourova TP, Kolganova TV, Birkeland NK, Bonch-Osmolovskaya EA. Caldicellulosiruptor kronotskyensis sp. nov. and Caldicellulosiruptor hydrothermalis sp. nov., two extremely thermophilic, cellulolytic, anaerobic bacteria from Kamchatka thermal springs. Int J Syst Evol Microbiol. 2008;58(Pt 6):1492-6.

[101] Grogan DW. Evidence that beta-Galactosidase of Sulfolobus solfataricus Is Only One of Several Activities of a Thermostable beta-d-Glycosidase. Appl Environ Microbiol. 1991;57(6):1644-9.

[102] Zambare VP, Bhalla A, Muthukumarappan K, Sani RK, Christopher LP. Bioprocessing of agricultural residues to ethanol utilizing a cellulolytic extremophile. Extremophiles. 2011;15(5):611-8. .

[103] Kang HJ, Uegaki K, Fukada H, Ishikawa K. Improvement of the enzymatic activity of the hyperthermophilic cellulase from Pyrococcus horikoshii. Extremophiles. 2007;11(2):251-6.

[104] Bauer MW, Driskill LE, Callen W, Snead MA, Mathur EJ, Kelly RM. An endoglucanase, EglA, from the hyperthermophilic archaeon Pyrococcus furiosus hydrolyzes beta-1,4 bonds in mixed-linkage (1-->3),(1-->4)-beta-D-glucans and cellulose. J Bacteriol. 1999;181(1):284-90.

[105] Dutta T, Sahoo R, Sengupta R, Ray SS, Bhattacharjee A, Ghosh S. Novel cellulases from an extremophilic filamentous fungi Penicillium citrinum: production and characterization. J Ind Microbiol Biotechnol. 2008;35(4):275-82. .

[106] Kaper T, Brouns SJJ, Geerling ACM, De Vos WM, Van der Oost J. DNA family shuffling of hyperthermostable beta-glycosidases. Biochem J. 2002;368(2):461-70.

[107] Rees HC, Grant S, Jones B, Grant WD, Heaphy S. Detecting cellulase and esterase enzyme activities encoded by novel genes present in environmental DNA libraries. Extremophiles. 2003;7(5):415-21. .

[108] Ito S. Alkaline cellulases from alkaliphilic Bacillus: enzymatic properties, genetics, and application to detergents. Extremophiles. 1997;1(2):61-6.

[109] Shoseyov O, Shani Z, Levy I. Carbohydrate Binding Modules: Biochemical Properties and Novel Application. Microbiology and molecular biology reviews. 2006:283–95.

[110] Bhat M, Bhat S. Cellulose degrading enzymes and their potential industrial applications. Biotechnology Advances. 1997;15:583-620.

[111] Davies G, Henrissat B. Structures and mechanisms of glycosyl hydrolases. Structure. 1995;3(9):853-9.

[112] Teeri T, Koivula A, Linder M, Wohlfahrt G, Divne C, Jones T. *Trichoderma reesei* cellobiohydrolases: why so efficient on crystalline cellulose? Biochemical Society Transactions. 1998;26:173–8.

[113] Margolles-Clark E, Tenkanen M, Soderlund H, Pentilla.M. Acetyl xylan esterase from *Trichoderma reesei* contains an active site serine and a cellulose-binding domain. European Journal of Biochemistry. 1996;237:553–60.

[114] Wilson D, Irwin D. Genetics and Properties of Cellulases. Advances in Biochemical Engineering / Biotechnology. 1999;65:1-21.

[115] Srisodsuk M, Reinikainen T, Penttila M, Teeri T. Role of the interdomain linker peptide of *Trichoderma reesei* cellobiohydrolase I in its interaction with crystalline cellulose. Journal of Biological Chemistry. 1993;268:20756–61.

[116] Shen H, Schmuck M, Pilz I, Gilkes N, Kilburn D, Miller R, Warren R. Deletion of the linker connecting the catalytic and cellulose-binding domains of endoglucanase A (CenA) of *Cellulomonas fimi* alters its conformation and catalytic activity. Journal of Biological Chemistry. 1991;266:11335–40.

[117] Chundawat SP, Beckham GT, Himmel ME, Dale BE. Deconstruction of lignocellulosic biomass to fuels and chemicals. Annu Rev Chem Biomol Eng. 2011;2:121-45.

[118] Liu QP, Sulzenbacher G, Yuan H, Bennett EP, Pietz G, Saunders K, Spence J, Nudelman E, Levery SB, White T, Neveu JM, Lane WS, Bourne Y, Olsson ML, Henrissat B, Clausen H. Bacterial glycosidases for the production of universal red blood cells. Nat Biotechnol. 2007;25(4):454-64. .

[119] Rajan SS, Yang X, Collart F, Yip VL, Withers SG, Varrot A, Thompson J, Davies GJ, Anderson WF. Novel catalytic mechanism of glycoside hydrolysis based on the structure of an NAD+/Mn2+ -dependent phospho-alpha-glucosidase from Bacillus subtilis. Structure. 2004;12(9):1619-29.

[120] Dworkin M, Rosenberg E, Schleifer K. The Prokaryotes: Ecophysiology and biochemistry. New York, USA.: Springer; 2006.

[121] Nguyen NH, Maruset L, Uengwetwanit T, Mhuantong W, Harnpicharnchai P, Champreda V, Tanapongpipat S, Jirajaroenrat K, Rakshit SK, Eurwilaichitr L, Pongpattanakitshote S. Identification and characterization of a cellulase-encoding gene from the buffalo rumen metagenomic library. Biosci Biotechnol Biochem. 2012;76(6): 1075-84.

[122] van der Lelie D, Taghavi S, McCorkle SM, Li L-L, Malfatti SA, Monteleone D, Donohoe BS, Ding S-Y, Adney WS, Himmel ME, Tringe SG. The Metagenome of an Anaerobic Microbial Community Decomposing Poplar Wood Chips. PLoS ONE. [doi: 10.1371/journal.pone.0036740]. 2012;7(5):36740.

[123] Nimchua T, Thongaram T, Uengwetwanit T, Pongpattanakitshote S, Eurwilaichitr L. Metagenomic analysis of novel lignocellulose-degrading enzymes from higher termite guts inhabiting microbes. J Microbiol Biotechnol. 2012;22(4):462-9.

[124] Li LL, McCorkle SR, Monchy S, Taghavi S, van der Lelie D. Bioprospecting metagenomes: glycosyl hydrolases for converting biomass. Biotechnol Biofuels. 2009;2:10.

[125] Kellner H, Vandenbol M. Fungi Unearthed: Transcripts Encoding Lignocellulolytic and Chitinolytic Enzymes in Forest Soil. PLoS ONE. 2010;5(6):10971.

[126] Li LL, Taghavi S, McCorkle SM, Zhang YB, Blewitt MG, Brunecky R, Adney WS, Himmel ME, Brumm P, Drinkwater C, Mead DA, Tringe SG, Lelie D. Bioprospecting metagenomics of decaying wood: mining for new glycoside hydrolases. Biotechnol Biofuels. 2011;4(1):23.

[127] Findley SD, Mormile MR, Sommer-Hurley A, Zhang XC, Tipton P, Arnett K, Porter JH, Kerley M, Stacey G. Activity-based metagenomic screening and biochemical characterization of bovine ruminal protozoan glycoside hydrolases. Appl Environ Microbiol. 2011;77(22):8106-13.

[128] Chi Z, Chi Z, Zhang T, Liu G, Li J, Wang X. Production, characterization and gene cloning of the extracellular enzymes from the marine-derived yeasts and their potential applications. Biotechnology Advances. 2009;27:236–55

[129] Kikuchi T, Jones J, Aikawa T, Kosaka H, Ogura N. A family of glycosyl hydrolase family 45 cellulases from the pine wood nematode *Bursaphelenchus xylophilus*. FEBS Letters. 2004;572:201–5.

[130] Ekborg NA, Gonzalez JM, Howard MB, Taylor LE, Hutcheson SW, Weiner RM. Saccharophagus degradans gen. nov., sp. nov., a versatile marine degrader of complex polysaccharides. Int J Syst Evol Microbiol. 2005;55(Pt 4):1545-9.

[131] Beckham GT, Bomble YJ, Bayer EA, Himmel ME, Crowley MF. Applications of computational science for understanding enzymatic deconstruction of cellulose. Current Opinion in Biotechnology. [doi: 10.1016/j.copbio.2010.11.005]. 2011;22(2):231-8.

[132] Henrissat B. A classification of glycosyl hydrolases based on amino acid sequence similarities. Biochem J. 1991;280(Pt 2):309-16.

[133] Henrissat B, Bairoch A. New families in the classification of glycosyl hydrolases based on amino acid sequence similarities. Biochem J. 1993;293(Pt 3):781-8.

[134] Boraston AB, Bolam DN, Gilbert HJ, Davies GJ. Carbohydrate-binding modules: fine-tuning polysaccharide recognition. Biochem J. 2004 382(3):769-81.

[135] Aro N, Pakula T, Pentilla M. Transcriptional regulation of plant cell wall degradation by filamentous fungi. FEMS Microbiology Reviews. 2005;29(4):719-39.

[136] Lynd L, Cushman J, Nichols R, Wyman C. Fuel ethanol from cellulosic biomass. Science. 1991;15:1318–23.

[137] Sadana J, Lachke A, Patil R. Endo-(1-4)-beta-D-glucanases from *Sclerotium rolfsii* – purification, substrate specificity, and mode of action. Carbohydrate Research. 1984;133:297–312.

[138] Valášková V, Baldrian P. Degradation of cellulose and hemicelluloses by the brown rot fungus *Piptoporus betulinus* – production of extracellular enzymes and characterization of the major cellulases. Microbiology. 2006;152: 3613–22.

[139] Ding S, Ge W, Buswell J. Endoglucanase I from the edible straw mushroom, *Volvariella volvacea*. European Journal of Biochemistry. 2001;268(22):5687-95.

[140] Sadana J, Patil R. 1,4-beta-D-glucan cellobiohydrolase from *Sclerotium rolfsii*. Methods in Enzymology. 1988;160: 307–14.

[141] Song BC, Kim KY, Yoon JJ, Sim SH, Lee K, Kim YS, Kim YK, Cha CJ. Functional analysis of a gene encoding endoglucanase that belongs to glycosyl hydrolase family 12 from the brown-rot basidiomycete Fomitopsis palustris. J Microbiol Biotechnol. 2008;18(3):404-9.

[142] Onishi N, Tanaka T. Purification and properties of a galacto- and gluco-oligosaccharide-producing betaglycosidase from *Rhodotorula minuta* IFO879. . Journal of Fermentation and Bioengineering. 1996;82:439–43.

[143] Teeri T. Crystalline cellulose degradation: new insight into the function of cellobiohydrolases Trends in Biotechnology. 1997;15:160–7.

[144] Jalak J, Kurashin M, Teugjas H, Valjamae P. Endo-exo synergism in cellulose hydrolysis revisited. J Biol Chem. 2012;25:25.

[145] Tabka MG, Herpoël-Gimbert I, Monod F, Asther M, Sigoillot JC. Enzymatic saccharification of wheat straw for bioethanol production by a combined cellulase xylanase and feruloyl esterase treatment. Enzyme and Microbial Technology. 2006;39(4): 897-902.

[146] Baker JO, Ehrman CI, Adney WS, Thomas SR, Himmel ME. Hydrolysis of cellulose using ternary mixtures of purified celluloses. Appl Biochem Biotechnol. 1998;72:395-403.

[147] Levine SE, Fox JM, Clark DS, Blanch HW. A mechanistic model for rational design of optimal cellulase mixtures. Biotechnol Bioeng. 2011;108(11):2561-70. .

[148] Cosgrove DJ. Growth of the plant cell wall. Nature Reviews Molecular Cell Biology. 2005;6(11):850-61.

[149] Cosgrove DJ. Loosening of plant cell walls by expansins. Nature. 2000;407(6802): 321-6.

[150] McQueen-Mason S, Cosgrove DJ. Disruption of hydrogen bonding between plant cell wall polymers by proteins that induce wall extension. Proceedings of the National Academy of Sciences USA. 1994;91(14):6574-8.

[151] Lee Y, Choi D, Kende H. Expansins: ever-expanding numbers and functions. Current Opinion in Plant Biology. 2001;4(6):527-32.

[152] Wei W, Yanga C, Luoa J, Lua C, Wub Y, Yuana S. Synergism between cucumber alpha-expansin, fungal endoglucanase and pectin lyase. Journal of Plant Physiology. 2010;167:1204–10.

[153] Li Y, Jones L, McQueen-Mason S. Expansins and cell growth. Current Opinion in Plant Biology. 2003;6(6):603-10.

[154] Cosgrove DJ. Relaxation in a high-stress environment: the molecular bases of extensible cell walls and cell enlargement. Plant Cell. 1997;9(7):1031-41.

[155] Kerff F, Amoroso A, Herman R, Sauvage E, Petrella S, Filee P, Charlier P, Joris B, Tabuchi A, Nikolaidis N, Cosgrove DJ. Crystal structure and activity of *Bacillus subtilis* YoaJ (EXLX1), a bacterial expansin that promotes root colonization. Proceedings of the National Academy of Sciences USA. 2008;105(44):16876-81.

[156] Rose JKC, Lee HH, Bennett AB. Expression of a divergent expansin gene is fruit-specific and ripening-regulated. Proceedings of the National Academy of Sciences USA. 1997;94(11):5955-60.

[157] Civello PM, Powell ALT, Sabehat A, Bennett AB. An Expansin Gene Expressed in Ripening Strawberry Fruit. Plant Physiology. 1999;121(4):1273-9.

[158] Cho HT, Cosgrove DJ. Regulation of root hair initiation and expansin gene expression in *Arabidopsis*. Plant Cell. 2002;14(12):3237-53.

[159] Cosgrove D, Li L, Cho H, Hoffmann-Benning S, Moore R, Blecker D. The growing world of expansins. Plant Cell Physiology. 2002 43(12):1436-44.

[160] Cosgrove D, Bedinger P, Durachko D. Group I allergens of grass pollen as cell wall-loosening agents. Proceedings of the National Academy of Sciences USA. 1997;94(12):6559-64.

[161] Kende H, Bradford K, Brummell D, Cho HT, Cosgrove D, Fleming A, Gehring C, Lee Y, McQueen-Mason S, Rose J, Voesenek LA. Nomenclature for members of the expansin superfamily of genes and proteins. Plant Molecular Biology. 2004;55(3):311-4.

[162] Darley CP, Li Y, Schaap P, McQueen-Mason SJ. Expression of a family of expansin-like proteins during the development of *Dictyostelium discoideum*. FEBS Letters. 2003;546(2-3):416-8.

[163] Kim ES, Lee HJ, Bang WG, Choi IG, Kim KH. Functional characterization of a bacterial expansin from *Bacillus subtilis* for enhanced enzymatic hydrolysis of cellulose. Biotechnology and Bioengineering. 2009;102(5):1342-53.

[164] Lee HJ, Lee S, Ko HJ, Kim KH, Choi IG. An expansin-like protein from Hahella chejuensis binds cellulose and enhances cellulase activity. Molecules and cells. 2010;29(4):379-85.

[165] Carey RE, Cosgrove DJ. Portrait of the expansin superfamily in *Physcomitrella patens*: comparisons with angiosperm expansins. Annals of Botany. 2007;99(6):1131-41.

[166] Wu Y, Meeley R, Cosgrove D. Analysis and expression of the α-expansin and β-expansin gene families in maize. Plant Physiology. 2001;126:222-32.

[167] Li Y, Darley C, Ongaro V, Fleming A, Schipper O, Baldauf S, McQueen-Mason S. Plant expansins are a complex multigene family with an ancient evolutionary origin. Plant Physiology. 2002;128:854–64.

[168] Lin Z, Ni Z, Zhang Y, Yao Y, Wu H, Sun Q. Isolation and characterization of 18 genes encoding α- and β-expansins in wheat (*Triticum aestivum*). Molecular Genetics and Genomics. 2005;274(5):548-56.

[169] Kudla U, Qin L, Milac A, Kielak A, Maissen C, Overmars H, Popeijus H, Roze E, Petrescu A, Smant G, Bakker J, Helder J. Origin, distribution and 3D-modeling of Gr-EXPB1, an expansin from the potato cyst nematode *Globodera rostochiensis*. FEBS Letters. 2005;579(11):2451-7.

[170] Brotman Y, Briff E, Viterbo A, Chet I. Role of Swollenin, an Expansin-Like Protein from *Trichoderma*, in Plant Root Colonization. Plant Physiology. 2008;147(2):779-89.

[171] Whitney S, Gidley M, McQueen-Mason S. Probing expansin action using cellulose/hemicellulose composites. Plant Journal. 2000;22:327–34.

[172] Baker J, King M, Adney W, Decker S, Vinzant T, Lantz S, Nieves R, Thomas S, Li L-C, Cosgrove D, Himmel M. Investigation of the cell-wall loosening protein expansin as a possible additive in the enzymatic saccharification of lignocellulosic biomass. Applied Biochemistry and Biotechnology. 2000;84-86(1):217-23.

[173] Jager G, Girfoglio M, Dollo F, Rinaldi R, Bongard H, Commandeur U, Fischer R, Spiess A, Buchs J. How recombinant swollenin from Kluyveromyces lactis affects cellulosic substrates and accelerates their hydrolysis. Biotechnology for Biofuels. 2011;4(1):33.

[174] Cosgrove D, inventor The Penn State Research Foundation., assignee. Enhancement of accessibility of cellulose by expansins. . United States 2001.

[175] Cosgrove D, inventor The Penn State Research Foundation, assignee. Increased activity and efficiency of expansin-like proteins. United States2007.

[176] Cosgrove D, inventor The Penn State Research Foundation assignee. β-expansins as cell wall loosening agents, compositions thereof and methods of use 2004.

http://dx.doi.org/10.5772/53719

[177] Arantes V, Saddler J. Access to cellulose limits the efficiency of enzymatic hydrolysis: the role of amorphogenesis. Biotechnology for Biofuels. 2010;3(1):4.

Chapter 7

Optimization of Delignification and Enzyme Hydrolysis of Steam Exploded Oil Palm Trunk for Ethanol Production by Response Surface Methodology

Vittaya Punsuvon

Additional information is available at the end of the chapter

http://dx.doi.org/10.5772/54691

1. Introduction

The inevitable depletion of the world's petroleum supply and the increasing problem of greenhouse gas effects have resulted in an increasing worldwide interest in alternative nonpetroleum-base source of energy. As the transportation sector is practically entirely depending on oil and as it is responsible for half of the total CO_2 emission [1], the increasing in market share of renewable biofuels includes ethanol fuel. The uses of ethanol fuel will significantly reduce net carbon dioxide emission once it replaces fossil fuels because fermentation-derived ethanol is already a part of the global carbon cycle. However, to enhance the market position of the biofuel the production cost should be reduced. Nowadays, the raw material and enzyme production are the two main contributors to the overall costs, thus using high cellulose containing agricultural residues as feedstock agricultural could result in cost reduction. The techniques employed to produce bioethanol from agricultural residue materials or lignocellulosic materials are subjected to the same economical demands as the more traditional sugar and starch processes, as the price of bioethanol must be competitive with that of petrol. Conversion of lignocellulosic materials to monomeric sugars and finally ethanol must thus be performed at low cost, while still achieving high yields. This can be done by developing processes that require limited amounts of the material chemicals, yeast and enzymes. To convert lignocellulosic materials to monomeric sugars, they must pretreat by different methods, such as dilute acid, steam explosion, ammonia fiber explosion (AFEX) and dilute alkali. All of these methods can change lignocellulosic structure and enhance the enzymatic saccharification of cellulose to hexose sugar.

The bioconversion process from lignocellulose biomass to ethanol consists basically of three steps: pre-treatment, enzymatic hydrolysis and fermentation. Pretreatment is a necessary step to facilitate the enzymatic attack of lignocellulosic materials. Steam explosion is recognized as an efficient pre-treatment method in ethanol production [2]. The raw material is treated at high pressure steam followed by suddenly rapid reduction in pressure resulting in substantial breakdown of the lignocellulosic structure, hydrolysis of the hemicellulosic fraction, depolymerization of the lignin components and defibration [3]. Therefore, the accessibility of the cellulose components to degrade by enzymes is greatly increased. The process of ethanol production from lignocellulosic material is shown in Figure 1.

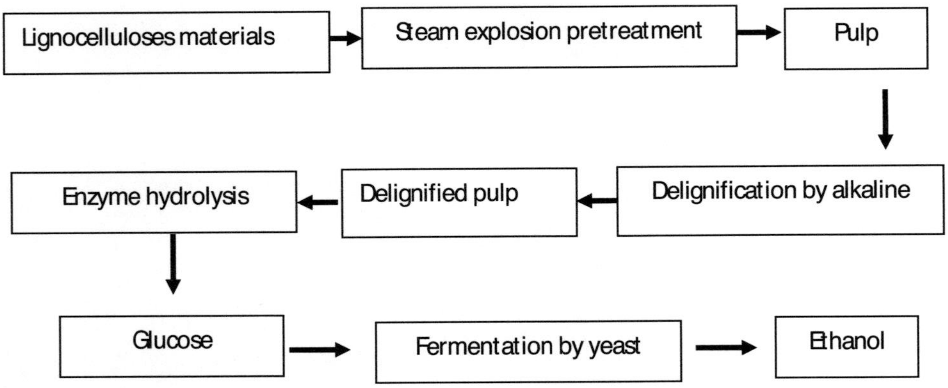

Figure 1. Flowchart of procedure for ethanol production

Numerous experimental studies of ethanol production from biomass have been carried out. A review article by Kaur *et al.* [4] examined the steam explosion of sugarcane bagasse as a pretreatment for ethanol production. In order to identify the optimum conditions of steam explosion, a range of operating temperatures at 188 – 243ºC and residence times at 0.5 - 44 min were applied. The results showed that pretreatment with steam explosion followed by enzyme hydrolysis had high efficiency in converting monosaccharide sugar to ethanol. Nunes *et al.* [5] reported the steam explosion pretreatment and enzymatic hydrolysis of Eucalyptus wood. The comparison under conditions of acid and non acid impregnation of wood before steam explosion was experimented. The results demonstrated the same solubilization effect of both experiments. Ballesteros *et al.* [6] reported that simultaneous saccharification and fermentation (SSF) process for ethanol production from various lignocellulosic woody (poplar and eucalyptus) and herbaceous (*Sorghum* sp. bagasse, wheat straw and *Brassica carinata* residue) materials had been assayed using the thermotolerant yeast strain. Biomass samples were previously treated in a steam explosion pilot plant to provide biomass with increased cellulose content relative to untreated materials and to enhance cellulase accessibility. SSF experiments were performed in laboratory conditions at 42ºC for 160 hours. The results showed that eucalyptus, wheat straw and sweet sorghum bagasse gave ethanol concentration at 17, 18 and 16.0 g/L respectively, in 72 hours of fermentation. Montane *et al.* [7] studied the steam explosion of wheat straw. A fractionation of wheat straw components in a two-step chemical pretreatment was proposed. Hemicellulose was hydrolyzed by dilute H_2SO_4, allowing a substantial recovery of xylose. Lignin was removed by means of a mild alkaline/oxidative solubilization procedure, involving no sulphite or chlorine and its derivatives. The use of diluted reagents and relatively low temperatures was both cheap and environmentally friendly. The pretreated material was nearly pure cellulose, whose enzyme hydrolysis proceeded fast with high yields, that leading to high glucose syrup of remarkable purity. Ballesteros *et al.* [8] investigated the enzyme hydrolysis of steam exploded herbaceous agricultural waste (*Brassica carinata*) at different particle sizes. The objective of this work was to evaluate the effect of particle size on steam explosion pretreatment of herbaceous lignocellulosic biomass. Hemicellulose and cellulose recovery and effectiveness of enzyme hydrolysis of the cellulosic residue was presented for the steam-exploded agricultural residue (*Brassica carinata*) with different particle sizes. The parameters tested were: particle size (2-5, 5-8 and 8-12 mm), temperature (190 and 210°C), and residence time (4 and 8 min). The composition analysis of filtrate and water insoluble fibre after pretreatment and enzyme digestibility data were presented. The results showed that larger steam exploded particle (8-12 mm) resulted in higher cellulose and enzyme digestibility. The use of small particles in steam explosion would not be desirable in optimizing the effectiveness of the process improving economy. Punsuvon *et al.* [9] studied the fractionation of chemical components of oil palm trunk by steam explosion. The results showed optimal conditions for pretreatment at temperature 214°C for 2 min of steam explosion. Ohgren *et al.* [10] reported the ethanol fuel production from steam-pretreated corn stover using SSF at higher dry matter content. This study was performed on steam-pretreated corn stover at 5, 7.5 and 10% water-insoluble solids (WIS) with 2 g/L hexose fermenting *Saccharomyces cerevisiae*. The results showed that SSF at 10% WIS gave 74% of ethanol yield based on the glucose content in the raw material. Ruiz *et al.* [11] studied the steam explosion pretreatment prior to enzymatic hydrolysis of sunflower stalks. The stalks were subjected to steam explosion pretreatment in the temperatures ranging between 180ºC and 230ºC. The steam-exploded pulp was further hydrolyzed by enzyme. The result showed that after 96 hours of enzymatic reaction, a maximum hydrolysis yield of 72% was obtained after pretreatment at 220ºC, corresponding to a glucose concentration of 43.7 g/L in hydrolysis media. With regard to the filtrate analysis, most of the hemicellulosic derived sugars released during the steam pre-treatment were in the oligomeric form. The highest recovery was obtained at 210ºC of pretreatment temperature. Moreover, the utilization of hemicellulosic-derived sugars as a fermentation substrate would improve the overall bioconversion of sunflower stalks into ethanol fuel.

1.1. Response Surface Methodology (RSM)

Response surface methodology is an empirical statistical technique employed for multiple regressions analysis by using quantitative data. It solves multivariable data which is obtained from properly designed experiments to solve multivariable equation simultaneously. The graphical representation of their function was called response surface, which is used to describe the individual and cumulative effect of the test variables and their subsequent effect on the response. The effect of the variables on the response is investigated using second-order polynomial regression equation. This equation, derived using RSM for the evaluation of the response variables, is as follows:

$$Y = b_o + \sum_{i=1}^{4} b_i x_i + \sum_{i=1}^{4} b_{ii} x_i^2 + \sum_{i<j=1}^{3} \sum^{4} b_{ij} x_{ij}$$

Where Y is the response, b_o, b, $b_{ii\ and}$ b_{ij} are regression coefficients for intercept, linear, quadratic and interaction terms, respectively. The x_i and x_{ij} are uncoded values for independent variables. An analysis of variance (ANOVA) is performed to determine the lack of fit and the effect of linear, quadratic and interaction terms on the response. Many researches have used RSM in optimization process as these examples. Roberto *et al.* [12] studied the dilute acid hydrolysis to recover xylose from rice straw in a semi-pilot reactor. Rice straw is consisted of pentose that could be used as a raw material for the production of many useful compounds. One of these was xylitol, with a potential application in the food and medical areas. The interest in biotechnological processes employing lignocellulosic residues was increased because this material was cheap, renewable and widespread sugar sources. The objective of the study was to determine the effects of H_2SO_4 concentration and reaction time on the production of sugars (xylose, glucose and arabinose) and on the reaction byproducts (furfural, HMF and acetic acid). Dilute sulfuric acid was used as a catalyst for the hydrolysis of rice straw at 121ºC in a 350-L batch hydrolysis reactor. Rationale for conducting this study was determined based on a central composite statistical design. Response surface methodology (RSM) was adopted to optimize the hydrolysis conditions aiming to attain high xylose selectivity. The optimum condition was 1% H_2SO_4 concentration for 27 min. This condition gave 77% of xylose yield and 5.0 g/g of selectivity. Kunamneni *et al.* [13] applied the response surface to optimize the enzymatic hydrolysis of maize starch for higher glucose production. Doses of pre-cooked α-amylase, post-cooked α-amylase, glucoamylase and saccharification temperature were examined to produce maximum conversion efficiency and all values were selected for optimization. Full factorial composite experimental design and response surface methodology were used in the experiment design and result analysis. The optimum values for the tested variables were: 2.243 U of pre-cooked α-amylase /mg solids, 3.383 U of post-cooked 3.383 U of α-amylase /mg solids, 2.243 U of glucoamylase /mg solids at a saccharification temperature of 55.1ºC. The maximum conversion efficiency of 96.25% was achieved. This method was efficient because only 28 experiments were necessary for the assessment and also the model adequacy was very satisfactory.

1.2. Oil palm trunk

The oil palm tree (Elaeis guineensis) is indigenous to the tropical forests in weat Africa. The oil palm tree has become one of the most valuable commercial cash a crop due to the palm oil is used as a raw material in many industries such as soap, cosmetic, detergent, vegeTable oil and biodiesel. Nowaday almost 80% of the world oil palm plantation is centered at Southeast Asia, with most of it occurring in Indonesia (5.44 x 10^6 hectares) and Malaysia (4.85 x 10^6 hectares). Additionally, there are 260,000 hectares planted in Thailand, with smaller areas in the Philippines and some recent planting in Cambodia and Myanmar [14]. Oil palm trunks are available only when the economic lifespan of the palm is reach at the time of replanting. The average age of replanting is approximately 25 years. The main economic criteria for felling are the height of the palm, reaching 13 m or above and the diameter of the felled trunk is around 45 cm to 65 cm. More than 15 million tons of oil palm trunks per year are replanted in the world [15]. The increase of oil

http://dx.doi.org/10.5772/54691

palm trunk every year can create massive pollutions thus development technology for value-added products are need for this raw material. There are many need uses of potential value-added products made from oil palm trunk such as particleboard, laminated board, plywood, fiberboard and furniture [16]. Oil palm trunk can also be used for making paper [17]. It can also be used as raw material in ethanol production, too [9].

The objectives of this research are performed according to central composite design (CCD) and response surface methodology (RSM) to optimize and compare the condition for delignification and hydrolysis of steam-exploded oil palm trunk prior to ethanol fermentation to understand the relationship between the critical factor involved in enzymatic degradation of pulp and conversion to ethanol.

2. Materials and methods

2.1. Raw material and microorganism

The steam-exploded pulp obtained from oil palm trunk was prepared by steam explosion treatment. An amount of 150 g of dry oil palm trunk chip sample was placed in 2.5 L batch digester (Nitto Koatsu Company, Japan). Heating was accomplished by direct steam injection into the digester and the temperature of steam at 214°C for 2 min. This condition was previous work by Punsuvon *et al.* [9]. It could briefly explained that oil palm trunk chip was steamed at temperatures varying between 214 and 220ºC for 2 and 5 minutes. The optimization of the pretreatment condition was 214ºC and 2 minutes that gave the highest glucose yield after enzyme hydrolysis. In this studied, the explosive discharge of the digester contents into a collecting tank was actuated by rapidly opening a value. The combined pulp slurry was collected and washed with hot water (80°C) at total volume of 2 L for 30 min. The pulp was filtered and dried at room temperature for using as raw material in alcohol production study.

Saccharomyces cervisiae TIRS 5339 obtained from TISTR, Thailand was used in this study. It was maintained on a medium containing 20.0 g/l glucose, 20.0 g/l peptone and 10.0 g/l yeast extract at 4ºC and subcultured every month at 30ºC. The growth medium of the yeast consisted of 10.0 g/l yeast extract, 6.4 g/l urea, 2.0 g/l KH_2PO_4, 1.0 g/l $MgSO_4$-7H2O and 2.0 g/l glucose at pH 5.5 [18].

2.2. Alkaline delignification of steam-exploded pulp

The water-insoluble cellulose pulp obtained from steam explosion was delignified with potassium hydroxide. The reactions were carried out in a beaker under various maintained temperature. Before RSM was applied on alkaline delignification, approximate conditions for glucose content in pulps, namely concentration of pulp, concentration of alkaline solution, reaction time and temperature were determined by varying one factor at time while keeping the other constant. The initial step of the preliminary experiment was to select an appropriate amount of concentration of pulp. Five different concentrations of pulp (3, 6, 9, 12, 15 %w/v) were examined. The other three factors, concentration of alkaline solution, reaction time and temperature, were kept constant at 20%w/w, 60 min and 80ºC, respectively. Based on the glucose content in pulp after delignification, the optimum concentration of pulp was chosen. The second step of the preliminary experiment was to determine the concentration of alkaline solution. The glucose content in pulp was analyzed using the optimum condition of pulp chosen in the previous step. The alkaline solution concentration varied from 2 to 30% w/w while holding reaction time and temperature at 60 min and 80ºC. The third step of the preliminary experiment was to determine the reaction time. Using the concentration of pulp, concentration of alkaline solution, reaction time from the previous steps, alkaline delignification was studied under various reaction times from 15 to 90 min. The final step was to select an appropriate temperature by using the concentration of pulp, concentration of alkaline solution, reaction time from the previous step. The temperature varied from 30 to 100ºC. Based on these results the five level of each process variable were determined for RSM. The independent variables of RSM experiments were shown in Table1. Delignified pulp was recovered by filtration, washed several times with distilled water, died and then analyzed for glucose content. These pulps were ready to be used as the substrate for enzymatic hydrolysis.

2.3. Enzyme hydrolysis

Delignified pulps were hydrolyzed by cellulase (Celluclast 1.2L, Novozymes A/S Denmark) in flasks. The hydrolysis was performed in 0.05M sodium citrate buffer (pH 4.8) at 150 rpm of shaking. The dependent variables of experiments were shown in Table 2. All enzymatic hydrolysis liquor was analyzed for glucose content by High Performance Liquid Chromatography (HPLC).

2.4. Inoculums and ethanol fermentation

S. cerevisaie was initiated in the maintenance medium at 30ºC. The yeast was grown for 48 h at 170 rpm on a rotary shaker at 30°C. A 2.5 % w/v inoculums was used for subsequent subcultures. Ethanol fermentation was evaluated at 30°C in 150–ml Erlenmeyer flasks containing 100 ml fermentation media. The yeast fermentation medium consisted of the hydrolysis liquor containing 50g/l glucose, 2.0 g/l $KHPO_4$, 1.0 g/l $MgSO_4.7H_2O$, 10.0 g/l yeast extracts and 6.4 g/l urea at pH 5.5. The flasks were sealed with a one-hole rubber stopper, which a glass tube was connected to an air lock filled with 40% sulfuric acid solution. Ethanol content from fermentation was analyzed by Gas Chromatography (GC).

Independent variable	Symbol	Coded variable levels				
		α	-1	0	1	α
Concentration of pulp, %w/v	X_1	3	6	8	12	15
Concentration of alkaline solution, %w/w	X_2	2	8	14	20	26
Reaction time, min	X_3	15	30	45	60	75
Temperature, ºC	X_4	35	50	65	80	95

Table 1. Independent variables and their levels for central composite design in optimization of alkaline delignification of steam-exploded pulp

Independent variable	Symbol	Coded variable levels				
		α	-1	0	1	α
Reaction time, h	X_1	10	30	50	70	90
Temperature, °C	X_2	28	35	42.5	50	57.5
Enzyme loading, Filter Paper Unit (FPU)/g substrate	X_3	5	30	55	80	105
Concentration of pulp, %w/v	X_4	1	2	3	4	5

Table 2. Independent variables and their levels for central composite design in optimization of enzyme hydrolysis of delignified pulp

2.5. Experimental design

A central composite design was employed the response, namely percentage of glucose for alkaline delignification, and percentage of glucose yield for enzyme hydrolysis. The independent variables of alkaline delignification were X_1, X_2, X_3 and X_4 representing concentration of pulp, %w/v, and concentration of alkaline solution, % w/w, reaction time, min and temperature, °C, respectively. The independent variables of enzyme hydrolysis were X_1, X_2, X_3 and X_4 representing reaction time, h, temperature, °C, enzyme loading, FPU/g substrate and concentration of pulp, %w/v, respectively. Each variable to be optimized was coded at five levels: - α, -1, 0, +1 and + α. This gives a range of these variables of alkaline delignification (Table 1) and enzyme hydrolysis (Table2). Six replication runs at the centre (0, 0, 0) of the design were performed to allow the estimation of the pure error.

2.6. Statistical analysis

The data obtained by carrying out the experiment according to central composite design were analyzed by SPSS package (version 12.0). The response surface was expressed at the following second-order polynomial equation:

$$Y=b_o+\sum_{i=1}^{4}b_ix_i+\sum_{i=1}^{4}b_{ii}x_i^2+\sum_{i<j=1}^{3}\sum^{4}b_{ij}x_{ij}$$

Where Y is the response (percent glucose, %), x_i and x_{ij} are uncoded independent variables, b_o is constant, b_i is linear term coefficients, b_{ii} is quadratic term coefficients. b_{ij} is cross-product term coefficients. SPSS package was used for regression analysis of variance (ANOVA) and response surface methodology was performed using STATISTICA Software. Response surface plots were developed using the fitted second order polynomial equation obtain from regression analysis holding one of the independent variables at a constant value corresponding to the stationary point and changing the other two variables.

2.7. Glucose determination

The glucose content of hydrolysis liquid was analyzed by High Performance Liquid Chromatography (HPLC, Shimadzu, Kyoto, Japan) with refractive index detector. An AMINEX HPX-87C carbohydrate analysis (Bio-Rad, Hercules, USA) was used as column. The mobile phase was deionied water with flow rate 0.6 ml/min. The injection volume was 20 μl and the column temperature was maintained at 80°C. The glucose content of the solid residue was determined based on monomer content that was measured after two steps of acid hydrolysis. The first step hydrolysis was performed with 72% (w/w) H_2SO_4 at 30ºC for 60 min. In the second step, the reaction mixture was diluted to 4% (w/w) H_2SO_4 with distilled water and subsequently autoclaved at 121°C for 1 h. This hydrolysis liquid was then analyzed for glucose content as described above. All analytical determinations were performed in duplication.

2.8. Ethanol determination

The ethanol content in fermented solution was analyzed by gas chromatography (GC). GC analysis was carried out with an Agilent Technologies (Santa clara, CA) 6890 gas chromatograph equipped with a flame ionization detector and a HP-5 (Bonded 5%phenyl, 95% dimethylpolysiloxane) capillary column (30 m x 0.32 mm ID, 0.25 μm film thickness). The temperature program was an initial temperature of 150ºC, increased to 190ºC at 10ºC/min then 15ºC/min to 250ºC, and held for 15 min. The injector and detector temperature were 250 ºC and 300ºC, respectively. Standard and sample were injected by using the split mode ratio of 50:1.

3. Results and discussion

3.1. Chemical components of oil palm trunk and steam exploded oil palm trunk pulp

The Figure 2 (a) showed oil palm trunk chip before steam explosion pretreatment and Figure 2 (b) showed steam exploded oil palm trunk pulp obtained after steam explosion.

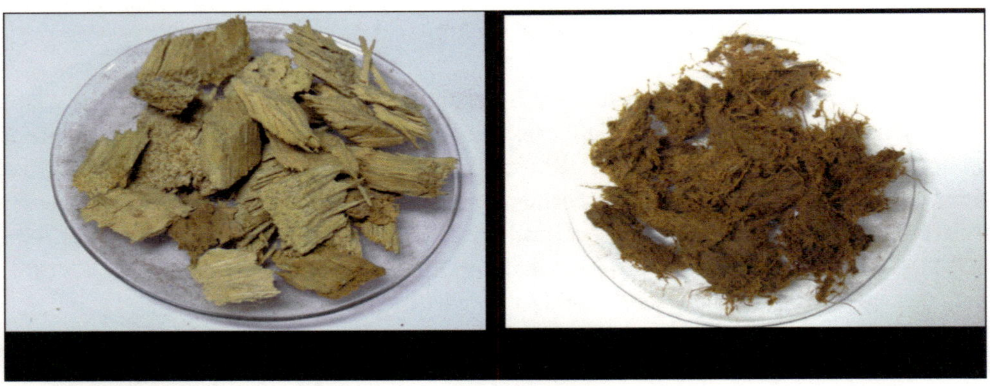

Figure 2. Oil palm trunk chip (a) and steam exploed oil palm trunk pulp (b)

http://dx.doi.org/10.5772/54691

The chemical components of steam exploded oil palm trunk pulp were 40.54% of cellulose, 9.36% of hemicelluloses, 38.46% of lignin and 8.56% of extractive in ethanol/benzene. This pulp obtained after steam explosion was used as raw material for optimization study in ethanol production.

3.2. Model fitting for optimization of alkaline delignification

The complete design matrix together with the values of both the experimental and predicted responses is given in Table3. Central composite design was used to develop correlation between the NaOH and KOH delignification variables to the percentages of glucose yield. The percentages of glucose yield were found to range between 41.4-49.8% for KOH delignification and 38.0-49.9% for NaOH delignification. Runs 17-23 at the center point were used to determine the experiment error. For both reponses of NaOH delignification and KOH delignification, the quadratic model was selected, as suggested by used software. The final empirical models in terms of coded factors are given by equation (1) and Equation (2) in Table 4. Where X_1, X_2, X_2 and X_4 were the coded values of test variables that represented pulp concentration, concentration of NaOH or concentration of KOH, reaction time and temperature, respectively. The variables X_1X_2, X_1X_3, X_1X_4, X_2X_3, X_2X_4 and X_3X_4 represented the interaction effects of pulp concentration and concentration of NaOH or concentration of KOH, pulp concentration and reaction time, pulp concentration and temperature, respectively. The quality of the model developed was evaluated based on the correlation coefficient R^2. The R^2 for the two obtained equations were found to be 0.875 and 0.890 in NaOH and KOH delignification, respectively. This indicates that 87.5% and 89.0% of the total variation in both delignifications were attributed to experimental variables studied. The R^2 of 0.875 and 0.890 were considered as the good fit of the models.

The adequacy of the two models was further justified through analysis of variance (ANOVA). The ANOVA for the quadratic models for the two reponses is listed in Table 5. The Fisher's test (F-test) carried out on experimental data make it possible to estimate the statistical significance of the proposed model. The F-test value of the models being 16.69 and 10.88, respectively for glucose in pulp obtained after NaOH and KOH lignifications, with a low probability value ($p<0.01$), we can conclude that they were statistically significant at 99.9% confidence level. It should be noted that p-value indicates the statistical significance of each parameter. It is based on hypothesis that a parameter is not significant, thus the more effect is significant. From Table 5, it was showed that the two models (both p-value<0.01) were adequate to predict the glucose in pulp obtained after NaOH and KOH lignifications within the range of studied variables.

Response surface contour plots of the RSM as a function of two factors at the time are helpful in understanding both the main and the interaction effects of these factors. The effects of concentration of pulp and concentration of NaOH on the percentage of glucose are shown in Figure 3 (a). Figure 3 showed that the concentration of pulp and NaOH could increased percent glucose in pulp after NaOH delignification. The concentration of pulp higher than 13.50% (w/w) had no significant effect on the amount of percent glucose in pulp. Response surface plot indicated the optimized condition at 12.50% w/v of pulp concentration and 21.50% (w/v) NaOH that gave 48.10% of percent glucose remained in pulp after NaOH delignification. Figure 3 (b) showed that increasing temperature and concentration of pulp could increased percent glucose in pulp after NaOH delignification. The temperature higher than 80ºC had no significant effect on the percent glucose in pulp after NaOH delignification. Response surface plot indicated the optimized condition at 12.5% w/w of pulp concentration and 80ºC that gave 47.95% of percent glucose remained in pulp after NaOH delignification. Figure 3 (c) showed that increasing time and concentration of pulp could increased percent glucose in pulp after NaOH delignification. The time longer than 67 min had no significant effect on the percent glucose in pulp after delignification. Response surface plot indicated the optimized condition at 12.5% w/v of pulp concentration and 65 min that gave 50.23% of percent glucose in pulp after NaOH delignification.

Response surface contour plots of the RSM on the effects of concentration of pulp and concentration of KOH on the percentage of glucose are shown in Figure 4 (a).Figure 4 (a) a showed that increasing concentration of pulp and KOH concentration could increased percent glucose in pulp. The pulp concentration higher than 12.5% (w/v) had no significant effect on percent glucose in pulp. Likewise, the KOH concentration higher than 23.4% (w/w) had no significant effect on percent glucose in pulp after KOH delignification. Response surface plot indicated the optimized condition at 10% (w/v) of pulp concentration and 23.5% (w/w) of KOH concentration that gave 49.04% of percent glucose in pulp after KOH delignification.

Figure 4 (b) showed that increasing concentration of pulp and temperature could increased percent glucose in pulp. The temperature higher than 80°C had no significant effect on percent glucose in pulp. Likewise, the concentration of pulp higher than 12% (w/v) had no significant effect on percent glucose in pulp after KOH delignification. Response surface plot indicated the optimized condition at 78°C of glucose in pulp after KOH delignification.

Figure 4 (c) showed that increasing time and concentration of pulp could increased percent glucose in pulp. The time longer than 60 min had no significant effect on percent glucose in pulp. Likewise, the concentration of pulp more than 12% (w/v) had no significant effect on percent glucose in pulp after KOH delignification. Response surface plot indicated the optimized condition at 12% (w/v) of pulp concentration and 70 min of reaction time that gave 48.35% of percent glucose in pulp after KOH delignification.

The summary result from combination of each response surface plot showed that the optimum condition for NaOH delignification was obtained from 11% (w/v) pulp concentration, 21% (w/w) NaOH concentration, 65 min reaction time and 78ºC temperature with the maximum glucose at 47.50% remaining in the pulp. The optimum condition for KOH delignification was 12% (w/v) pulp concentration, 23% (w/w) KOH concentration, 65 min of reaction time and 80ºC of temperature with the maximum glucose at 49.50% remaining in the pulp.

Run	X_1	X_2	X_3	X_4	Glucose (%), Delignification			
					Experimental		Predicted	
					KOH	NaOH	KOH	NaOH
1	1	1	-1	-1	48.3	45.2	48.7	45.1
2	1	1	-1	1	46.4	49.9	46.6	49.2
3	1	1	1	-1	42.0	43.6	42.3	43.4
4	1	1	1	1	49.2	49.0	49.6	49.8
5	1	-1	-1	-1	48.1	42.3	48.6	42.3
6	1	-1	-1	1	43.2	40.0	43.2	40.2
7	1	-1	1	-1	47.9	47.4	47.6	47.4
8	1	-1	1	1	49.3	49.5	49.3	49.5
9	-1	1	-1	-1	44.2	38.4	44.7	38.2
10	-1	1	-1	1	46.4	30.8	46.5	30.6
11	-1	1	1	-1	39.8	40.8	39.9	40.6
12	-1	1	1	1	47.2	42.1	47.6	42.7
13	-1	-1	-1	-1	44.1	48.1	44.6	48.5
14	-1	-1	-1	1	41.4	49.3	41.5	49.3
15	-1	-1	1	-1	46.3	38.0	46.7	38.1
16	-1	-1	1	1	49.0	48.2	49.0	48.7
17	0	0	0	0	48.3	43.3	48.2	43.6
18	0	0	0	0	47.4	44.9	47.3	44.3
19	0	0	0	0	48.6	44.0	48.0	44.5
20	0	0	0	0	46.4	44.9	46.9	44.6
21	0	0	0	0	46.3	44.3	46.0	44.3
22	0	0	0	0	48.5	43.8	48.2	43.2
23	0	0	0	0	48.1	43.2	48.3	43.6
24	α	0	0	0	49.5	39.8	49.6	39.7
25	-α	0	0	0	44.3	47.2	44.3	47.9
26	0	α	0	0	48.4	48.9	48.2	48.4
27	0	-α	0	0	45.7	43.9	45.0	43.7
28	0	0	α	0	49.8	48.6	49.5	48.8
29	0	0	-α	0	47.3	42.7	47.4	42.7
30	0	0	0	α	49.4	49.0	49.7	49.9
31	0	0	0	-α	43.7	44.7	43.3	44.9

Table 3. Design and response of the central composite design for glucose (%) obtained from NaOH and KOH delignification

Dependent variable	Predictive	R^2
Glucose (%) NaOH (Eqs.1)	$37.112 - 0.166 X_1 - 0.959X_2 + 0.511X_3 + 0.104X_4 - 0.074X_12$ $- 0.005X_22 + 2.706X_32 + 001X_42 + 0.133X_1X_2 - 0.011X_1X_3 + 0.001X_1X_4 - 0.005X_2X_3$ $- 0.005X_2X_4 - 0.006X_3X_4$	0.875
Glucose (%) KOH (Eqs.2)	$60.266 + 0.574X_1 - 1.398X_2 -0.320X_3 + 0.091X_4 - 0.030X_12$ $- 0.001X_22 + 0.006X_32 - 002X_42 + 0.133X_1X_2 - 0.008X_1X_3 + 0.006X_1X_4 - 0.004X_2X_3$ $+ 0.015X_2X_4 + 0.013X_3X_4$	0.890

Table 4. The linear regression of dependent for alkaline delignification of glucose

	Source	Degree of freedom	Sum of square	Mean square	F-value	P-value
NaOH	Model	14	10005.23	714.65	15.65	0.0078
delignification	Residual	16	728.64	45.54	-	-
	Lack of fit	10	530.43	53.04	1.60	-
	Pure error	6	198.21	33.04	-	-
	Total	30	10733.87			
	R^2	0.875				
KOH	Model	14	7935.29	566.81	10.88	0.0071
delignification	Residual	16	833.06	52.07	-	-
	Lack of fit	10	523.45	52.34	1.01	-
	Pure error	6	309.61	51.60	-	-
	Total	30	8768.35			
	R^2	0.890				

Table 5. Analysis of variance (ANOVA) for the fit of experimental data to response surface models

http://dx.doi.org/10.5772/54691

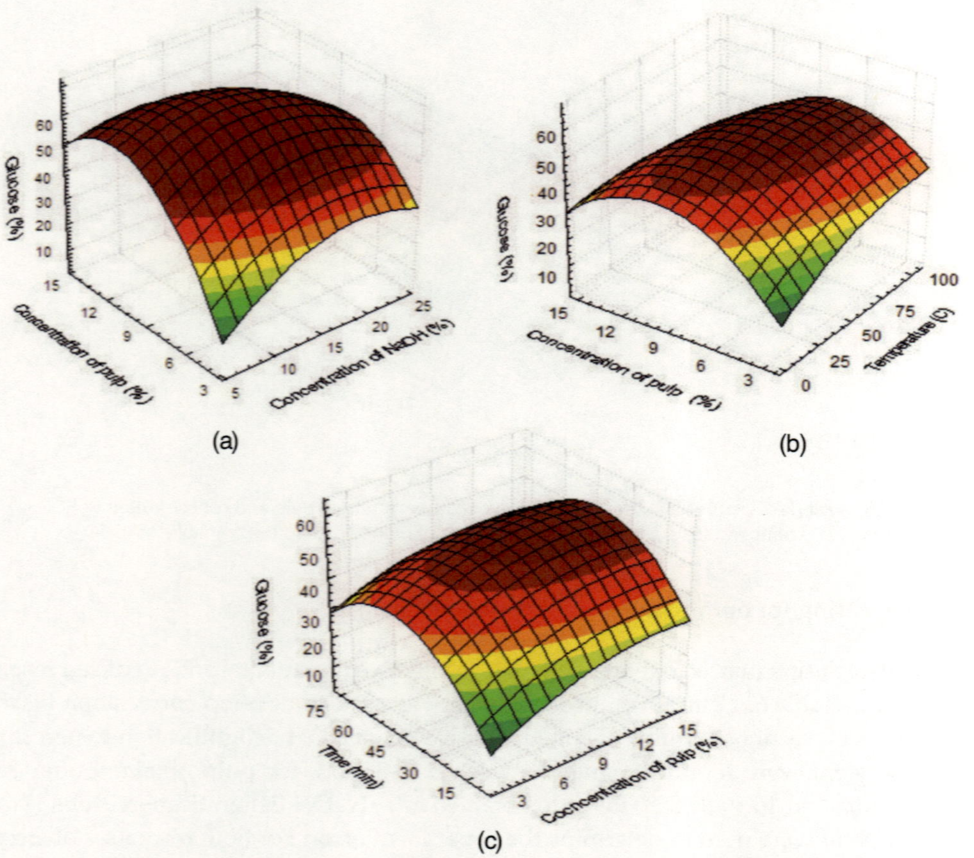

Figure 3. Response surface plots of glucose as a function of concentration of pulp and concentration of NaOH (a), concentration of pulp and temperature (b), time and concentration of pulp (c), other fixed variables

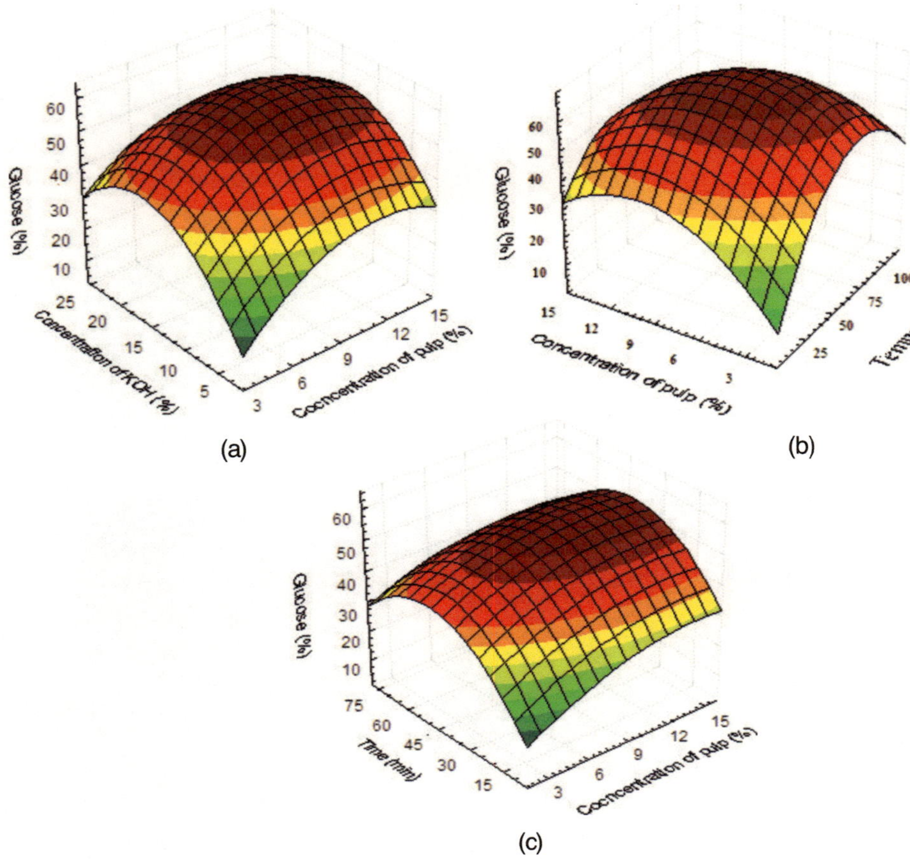

Figure 4. Response surface plots of glucose as a function of concentration of pulp and concentration of KOH (a), concentration of pulp and temperature (b), time and concentration of pulp (c), other fixed variables

3.3. Model fitting for optimization of enzyme hydrolysis

The complete design matrix together with values of both experimental and predicted responses is given in Table 6. Central composite design was used to develop correclation between enzyme hydrolysis of both pulps obtained after NaOH or KOH delignification to percentage of glucose yield were found to range between 6.30-80.10% for pulp obtained after KOH delignification and 10.30-89.80% for pulp obtained after NaOH delignification. Runs 17-23 at the center point were used to determine the experiment error. For both responses of enzyme hydrolysis, the quadratic model was selected, as suggested by used software. The final empirical models in terms of coded factors are given by Equation (3) and Equation (4) in Table 7. Where X_1, X_2, X_2 and X_4 were the coded values of test variables reaction time, temperature, enzyme loading and pulp concentration, respectively. The variables X_1X_2, X_1X_3, X_1X_4, X_2X_3, X_2X_4 and X_3X_4 represented the interaction effects of reaction time and temperature, reaction time and enzyme loading, reaction time and pulp concentration, temperature and enzyme loading, temperature and pulp concentration and enzyme loading and pulp concentration, respectively. The quality of the model developed was evaluated based on the correlation coefficient R^2. The R^2 for the two obtained equations were found to be 0.867 and 0.935 in enzyme hydrolysis of pulp obtained after NaOH and KOH delignification. This indicated that 86.70% and 93.50% of total variation in both enzyme hydrolysis were attributed to the experimental variables studied. The R^2 of 0.867 and 0.935 were considered as the good fit of the models.

The adequacy of the two models was further justified through analysis of variance as the same as delignification reaction in previous studied. The statistical result showed that the two models had p-values less than 0.01 that indicated these two models were adequate to predict the percentage of glucose yield in enzyme hydrolysis within the range of the studied variables.

In addition, response surface contour plots of the RSM on the effects of reaction time and concentration of pulp on the percentage of glucose in enzyme hydrolysis are shown in Figure 5 (a). Figure 5 (a) showed that increasing time and pulp concentration could increased percent glucose yield in hydrolyzed solution. The pulp concentration higher than 4.5% (w/v) had no significant effect on percent glucose yield in hydrolyzed in solution. Response surface plot indicated the optimized condition at 3% (w/v) of pulp concentration and 65 h of reaction time that gave 81.01% of percent glucose yield in hydrolyzed solution obtained after NaOH delignification. Figure 5 (b) showed that increasing time and temperature could increased percent glucose yield in hydrolyzed solution. The temperature higher than 50ºC had no significant effect on percent glucose yield in hydrolyzed solution. Response surface plot indicated the optimized condition at 44ºC of temperature and 65 h of reaction time that gave 81.33% of percent glucose yield hydrolyzed solution obtained after NaOH delignification. Figure 5 (c) showed that increasing time and enzyme loading could increased percent glucose yield in hydrolyzed solution. The enzyme loading higher than 65 (FPU/g substrate) had no significant effect on percent glucose yield in hydrolyzed solution. Response surface plot indicated the optimized condition at 54 (FPU/g substrate) of enzyme loading and 50 h of reaction time that gave 91.36% of percent glucose yield in hydrolyzed solution obtained after NaOH delignification. Figure 6 (a) showed that increasing time and pulp concentration could increased percent glucose yield in hydrolyzed solution. The reaction time longer than 75 h had no significant effect on percent glucose yield in hydrolyzed solution. Response surface plot indicated the optimized condition at 3% (w/v) of pulp concentration and 65 h of reaction time that gave 92.08% of percent glucose yield in hydrolyzed solution obtained after KOH delignification. Figure 6 (b) showed that increasing time and temperature could increased percent glucose yield in hydrolyzed solution. The temperature higher than 50ºC had no significant effect on percent glucose in hydrolyzed in solution. Response surface plot indicated the optimized condition at 50ºC of temperature and 65 h of reaction time that gave 88.91% (w/v) of percent glucose yield in hydrolyzed solution obtained after KOH delignification. Figure 6 (c) showed that increasing time and enzyme loading could increased percent glucose yield in hydrolyzed solution. The enzyme loading higher than 85 (FPU/g substrate) had no significant effect on percent glucose yield in hydrolyzed solution. Response surface plot indicated the optimized condition at 54.5 (FPU/g substrate) of enzyme loading and 66 h of reaction time that gave 88.44% of percent glucose yield in hydrolyzed in solution obtained after KOH delignification.

Run	X_1	X_2	X_3	X_4	Glucose (%), Enzyme hydrolysis			
					Experimental		Predicted	
					KOH	NaOH	KOH	NaOH
1	1	1	-1	-1	80.1	86.2	79.8	86.0
2	1	1	-1	1	69.5	75.4	69.0	75.3
3	1	1	1	-1	74.2	80.0	74.5	80.0
4	1	1	1	1	54.4	63.9	55.4	64.2
5	1	-1	-1	-1	37.6	46.7	37.4	46.2
6	1	-1	-1	1	16.8	23.0	16.1	23.2
7	1	-1	1	-1	30.3	35.4	30.9	35.1
8	1	-1	1	1	14.5	20.6	14.3	20.0
9	-1	1	-1	-1	74.4	80.3	74.4	80.8
10	-1	1	-1	1	63.8	71.6	63.3	70.2
11	-1	1	1	-1	70.1	76.4	70.4	76.3
12	-1	1	1	1	53.8	58.6	53.6	58.2
13	-1	-1	-1	-1	34.9	41.4	34.7	41.3
14	-1	-1	-1	1	12.2	18.8	12.5	18.8
15	-1	-1	1	-1	20.1	27.3	20.1	27.3
16	-1	-1	1	1	4.4	10.5	4.5	10.9
17	0	0	0	0	74.2	89.8	74.2	89.2
18	0	0	0	0	73.3	79.3	73.3	79.2
19	0	0	0	0	74.8	80.7	74.8	80.7
20	0	0	0	0	75.2	80.4	75.2	80.1
21	0	0	0	0	75.3	81.3	75.3	81.6
22	0	0	0	0	75.9	81.6	75.4	81.9
23	0	0	0	0	75.3	81.4	75.2	81.8
24	α	0	0	0	77.0	83.7	77.3	83.4
25	$-\alpha$	0	0	0	6.3	10.3	6.3	10.0
26	0	α	0	0	11.7	15.4	11.7	15.4
27	0	$-\alpha$	0	0	45.1	50.5	45.1	50.5
28	0	0	α	0	73.0	77.6	73.0	77.6
29	0	0	$-\alpha$	0	15.2	21.3	15.3	21.3
30	0	0	0	α	74.4	80.2	74.8	81.4
31	0	0	0	$-\alpha$	62.2	79.8	62.4	79.7

Table 6. Design and response of the central composite design for glucose (%) obtained from enzymatic hydrolysis

http://dx.doi.org/10.5772/54691

Dependent variable	Predictive	R^2
Glucose yield (%) NaOH (Eqs.3)	$-49.956 + 2.654X_1 + 20.429X_2 + 1.310X_3 - 6.477X_4 - 0.021X_1 2 - 0.228X_2 2 - 0.012X_3 2 - 1.134X_4 2 - 0.005X_1X_2 + 0.001X_1X_3 + 0.003X_1X_4 + 0.002X_2X_3 + 0.225X_2X_4 + 0.002X_3X_4$	0.867
Glucose yield (%) KOH (Eqs.4)	$-51.956 + 2.654X_1 + 20.332X_2 + 1.410X_3 - 6.477X_4 - 0.021X_1 2 - 0.224X_2 2 - 0.012X_3 2 - 1.835X_4 2 - 0.004X_1X_2 + 0.001X_1X3 + 0.003X1X4 + 0.002X2X 3 + 0.225X2X4 + 0.003X3X4$	0.935

Table 7. The linear regression of dependent for enzyme hydrolysis of glucose yield

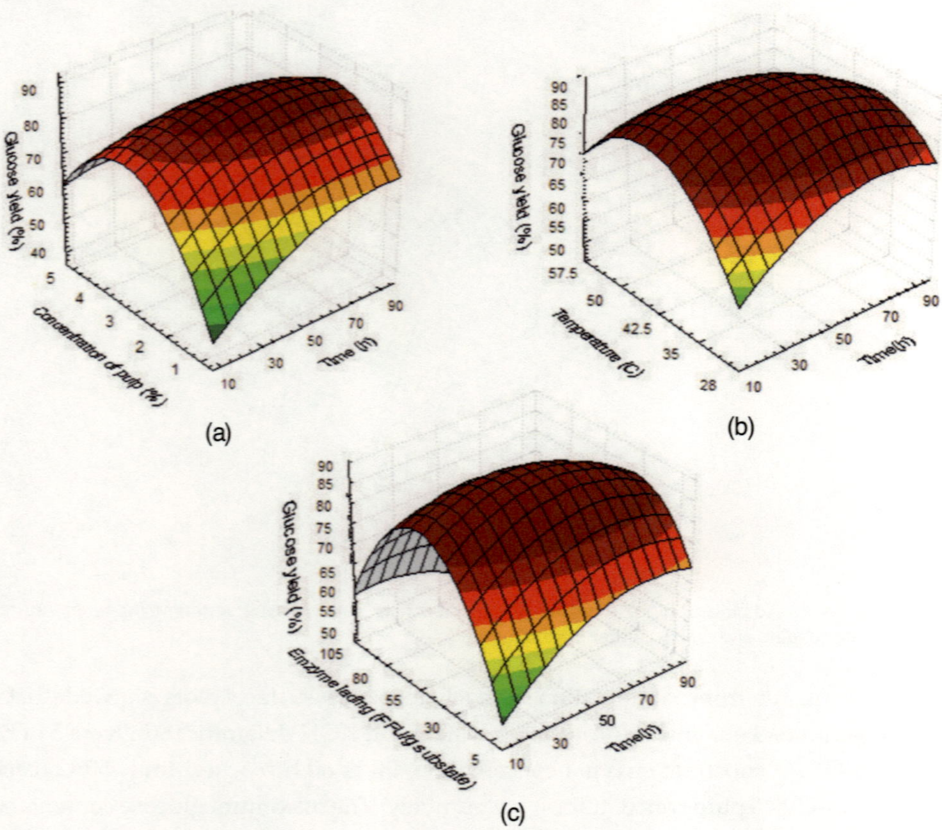

Figure 5. Response surface plots of glucose yield as a function concentration of pulp and time (a), temperature and time (b), enzymatic loading and time (c), other fixed variables

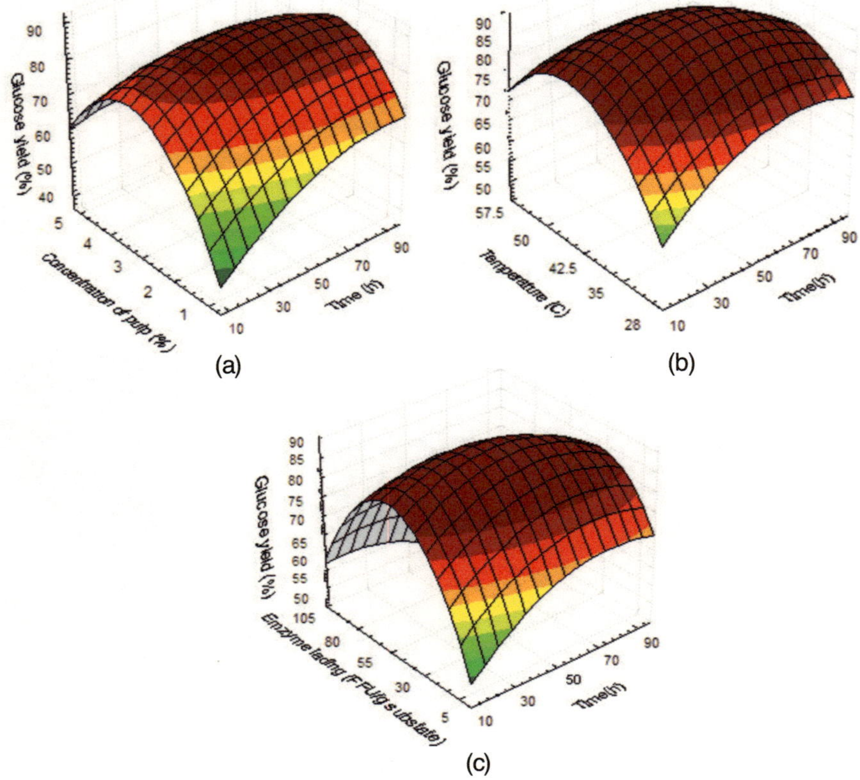

Figure 6. Response surface plots of glucose yield as a function concentration of pulp and time (a), temperature and time (b), enzymatic loading and time (c), other fixed variables

The summary result from combination of each response surface plots showed that the optimum condition of enzyme hydrolysis for NaOH and KOH delignification were 54 FPU/g substrate, 65 FPU/g substrate enzyme concentation, 50 h, 60 h reaction time, 50ºC reaction temperature and 2.5% pulp concentration, respectively. The maximum glucose contents were 47.50% and 48.00% from NaOH and KOH delignification, respectively. The maximum glucose yield obtained were 85%, 81% and 75% from NaOH and KOH delignification and without delignification respectively. Both optimization conditions in enzyme hydrolysis of delignification pulp by NaOH and KOH used to hydrolyze undelignification pulp, the result showed that percent glucose yield was lower than those from delignified pulp about 8-10%.

http://dx.doi.org/10.5772/54691

3.4. Ethanol fermentation

3.4.1. Production of ethanol from pure glucose

S. Cerevisiae TISTR 5339 was grown in YMB containing 50 g/l glucose, 20 g/l peptone and 10 g/l yeast extract. The experiment was performed at room temperature. The result of ethanol production was shown in Table 7.

Time (h)	Residual glucose (g/l)	Consumed glucose (g/g)	Ethanol (g/l)	Ethanol yield (%)
0	50.00	0.00	0.00	0.00
6	40.52	9.48	4.62	18.12
12	33.21	16.79	8.05	31.57
18	25.16	24.84	12.15	47.65
24	17.38	32.65	16.08	63.06
30	10.30	39.70	19.46	76.31
36	3.21	46.79	21.92	85.96
42	0.00	50.00	21.81	85.53
48	0.00	50.00	21.75	85.29
54	0.00	50.00	21.04	82.51

Table 8. Production of ethanol from 50 g/l of pure glucose by S. Cerevisiae TISTR 5339

$$\text{Ethanol yield} = \frac{\text{ethanol from experiment}}{\text{Theoretical ethanol}} \times 100$$

Table 8 showed the production of ethanol from 50 g/l of pure glucose by S. Cerevisiae TISTR 5339. The result showed that within 36 h of fermentation, the highest ethanol concentration was obtained at 21.92 g/l. The ethanol yield from calculation was 85.96%. This result indicated that S. Cerevisiae TISTR 5339 has 85.96% in capability to change pure glucose to ethanol.

3.4.2. Production of ethanol from hydrolyzed solution of NaOH and KOH delignified pulp and nondelignified pulp

The hydrolyzed solutions from three samples of pulp were concentrated to 50 g/l of glucose concentration. S. Cerevisiae TISTR 5339 was applied in the same amount as production of ethanol from pure glucose but pure glucose was replaced with the three hydrolyzed solutions. The fermentation solutions were analyzed by GC for ethanol concentration. The results were shown in Table 9.

Sample	Ethanol concentration (g/l)	Ethanol yield (%)
Nondelignification*	16.25	65.0
KOH delignification	16.42	65.7
NaOH delignification	16.35	65.4

*glucose obtained from steam exploded pulp that directly employed for ethanol production without delignification with NaOH or KOH.

Table 9. Comparison of resulting ethanol concentration (g/l) and yield (%) after 36 h fermentation

Comparison of resulting ethanol after 36 h fermentation of NaOH and KOH delignified pulp was shown. The obtained ethanol concentration and yield from non delignified pulp was 16.25 g/l and 65%, respectively. The fermentation of KOH delignified gave ethanol concentration and yield at 16.42 g/l and 65.7% whereas those obtained from NaOH delignified pulp were 16.35 g/l and 65.4%, respectively. Alkaline delignification process showed no significant influence on the fermentation process. When the ethanol yield obtained from pure glucose and hydrolyzed solution (Table 8 and Table 9) were compared, it showed that the three hydrolyzed solutions gave lower ethanol yield due to the three hydrolyzed solution contained toxic substances such as furfural, 5-hydroxy methyl furfural, phenolic compound and acetic acid derived from steam explosion process. All substances can inhibit the fermentation of S. Cerevisiae TISTR 5339.

4. Conclusion

From this study on optimization of alkaline delignification and enzyme hydrolysis on steam exploded oil palm trunk followed by yeast fermentation to produce ethanol, an encouraging results were obtained. We concluded that delignification of steam explode pulp with NaOH was more influenced than delignification with KOH on term of less alkaline concentration consuming and lower reaction temperature. The comparison enzyme hydrolysis condition for both delignified pulp showed that NaOH delignified pulp gave higher the percentage of glucose yield than KOH delignified pulp. After optimizing the both delignification (NaOH and KOH) and both enzyme hydrolysis parameter by RSM, the highest ethanol yield of 65% were obtained from both fermentations with 50 g/l of glucose raw material.

The empirical quadratic models successfully predicted the percentage of glucose in pulp after delignification and the percentage of glucose in hydrolyzed solution after enzyme hydrolysis and they used in the development of better estimation tools. In addition, response surface plot in three-dimension obtained from the empirical quadratic models can show the interaction effect of two variables on the studied response and the optimum values of the selected variables are obtained from response surface plot, too.

http://dx.doi.org/10.5772/54691

Acknowledgements

The financial support for this work was provided by Centre of Excellence-Oil Palm Kasetsart University, Bangkok, Thailand.

Author details

Vittaya Punsuvon[1,2*]

Address all correspondence to: fscivit@ku.ac.th

1 Department of Chemistry, Faculty of Science, Kasetsart University, Bangkok, Thailand

2 Center of Excellence-Oil Palm, Kasetsart University, Bangkok, Thailand

References

[1] Mieleuz JR. Ethanol production from biomass technology and commercialization status. Current Opinion Microbiology 2001;4 324–329.

[2] Rocha GJM, Goncalves AR, Oliveira BR, Olivares EG, Rossell CEV. Steam explosion pretreatment reproduction and alkaline delignification reactions performed on a pilot scale with sugarcane bagasse for bioethanol production. Industrial Crops and Product 2012;35 274-279.

[3] Wany K, Jiang JX, Xu F, Sun RC. Influence of steaming pressure on steam explosion pretreatment of Lespedeza stalks (Lespedeza crytobotrya): Part1. Characteristics of degraded cellulose. Polymer Degrad Stability 2009;94 1379-1388.

[4] Kaur WE, Gutierrez CV, Kinoshita CM. Steam explosion of sugarcane as a pretreatment for conversion to ethanol. Biomass and Bioenergy 1995;14 277-287.

[5] Nunes AP, Pourquie J. Steam explosion pretreatment and enzymatic hydrolysis of eucalyptus wood. Bioresource Technology 1996;57 107-110.

[6] Ballesteros M, Oliva JM, Negro MJ, Manzanares P, Ballesteros I. Ethanol from ligno-cellulosic material by simultaneous saccharification and fermentation process (SSF) with kluyveromyces marxianus CECT 10875. Process Biochemistry 2004;39 1843-1853.

[7] Montane D, Fattiol X, Salvado J, Jollez P, Chornet E. Fractation of wheat straw by steam explosion pretreatment and alkali delignification cellulose pulp and by product from hemicelluloses and lignin. Journal of wood Chemistry and Technology 1998;18(2) 171-191.

[8] Ballesteros I, Oliva JM, Negro MJ, Manzanares P, Ballesteros M. Enzyme hydrolysis of steam explode herbaceous agricultural waste (Brassica carnata) at different partical sizes. Process Biochemistry 2002:38 187-192.

[9] Punsuvon V, Vaithanomsat P, Pumiput S, Jantharanurak N, Anpannurak W. Fraction of chemical components of oil palm trunk by steam explosion for xylitol and ethanol production. In the proceedings of the 13th International Symposium of Woodfiber and Pulping Chemistry (ISWFPC), Auckland (New Zealand) 2005;301–308.

[10] Ohgren K, Bura R, Sadder J, Zacchi G. Effect of hemicelluloses and lignin removal on enzymatic hydrolysis of steam pretreated corn stover. Bioresource Technology 2007;98 2503-2510.

[11] Ruiz C, Cara CO, Manzauares P, Ballesteros M, Custro E. Evaluation of steam explosion pretreatment for enzymztic hydrolysis of sunflower stalks. Enzyme and Microbial Technology 2008;42 160-166.

[12] Roberto IC, Mussatto I, Rodrigues RCLB. Dilute-acid hydrolysis for optimization of xylose recovery from rice strew in a semi pilot reactor. Industrial Crops and Products 2003;17 171-176.

[13] Kunamneni A, Singh S. Response surface optimization of enzymatic hydrolysis of maize starch for higher glucose production, Biochemical Engineering Journal 2005;27 179-190.

[14] Malaysian Palm Oil Board Statistics 2010 (http://www.mpob.gov.my).

[15] Shuit SH, Tan KT, Lee KT, Kamaruddin AH. Oil palm biomass as sustainable energy source: A Malaysia case study energy 2009;34 1225-1235.

[16] Nordin K, Jamaludin MA, Ahmad M, Samsi HW, Salleh AH, Jallaludin Z. Minimizing the environmental burden of oil palm trunk residues through the development of laminated veneer lumber product. Manag. Environ. Qual. 2004;15(5) 484-490.

Section 3

Microbial Pretreatment of Lignocellulosics

Chapter 8

Pretreatment of Lignocellulosic Biomass Using Microorganisms: Approaches, Advantages, and Limitations

Thomas Canam, Jennifer Town, Kingsley Iroba,
Lope Tabil and Tim Dumonceaux

Additional information is available at the end of the chapter

http://dx.doi.org/10.5772/55088

1. Introduction

Much of Earth's recent geologic history is dominated by periods of extensive glaciation, with relatively low global mean temperatures and correspondingly low atmospheric CO_2 concentrations [1]. The current interglacial period stands out as an anomaly because the atmospheric CO_2 concentration has risen sharply above the range of approximately 180-280 parts per million by volume that has defined the past 420,000 years to reach levels that are nearly 40% higher than the biosphere has experienced over this time frame [2]. This rapid increase in CO_2 concentration is primarily due to the release of ancient fixed atmospheric CO_2 into the modern atmosphere through the combustion of fossil fuel resources over the past 200 years. Since it is clear from ice core records that atmospheric CO_2 concentration has a strong positive correlation to global temperature, it is expected that changes to global climate are forthcoming [3]. There are substantial uncertainties regarding the ability of terrestrial and oceanic carbon sinks to absorb this anthropogenic CO_2 on time scales that are relevant to human society [2], so the continued release of ancient CO_2 into the modern atmosphere at current rates carries with it an important risk of inducing climate changes of unknown amplitude along with a host of ancillary changes that are difficult to predict with certainty. This has led to the search for alternatives to fossil fuels to meet a rising global energy demand, and one such option is the use of extant organic matter to produce energy. This resource contains carbon that was fixed from the modern atmosphere, which means it does not result in a net increase in atmospheric CO_2 upon combustion.

Meeting the world's energy demands requires resources that are abundant and inexpensive to produce. Biomass from forestry and agricultural activities is certainly a candidate, as hundreds of millions of tonnes of agricultural waste from rice, wheat, corn, and other crops are produced worldwide, which could generate billions of litres of ethanol [4]. For ethanol, butanol, methane, and other biofuels to be produced economically, however, requires an integrated approach, with a number of value-added co-products produced in addition to the energy – a "biorefinery" that stands in analogy to petroleum refineries that produce both energy and a wide range of petroleum-based chemicals and products [5-7]. The biorefinery concept is hardly new, as the industrial-scale bacterial fermentation of starch to acetone and butanol (A-B) was developed a century ago. These A-B fermentations were done on an industrial scale in the West during World War I and persisted into the 1950's. They continued in Russia until late in the Soviet era, ultimately using corn cobs and other agricultural residues as input [8]. However, releasing the energy and co-product potential of plant-based material requires energy inputs and processing steps, as discussed below; this hinders the ability of biofuels to compete economically with petroleum resources, which have been exposed to millions of years of geological energy input to reach their current biochemical state.

Current paradigms for biofuels production include the production of ethanol by yeast or bacteria from glucose produced from soluble sugars and starch (1st generation ethanol) or from the cellulosic fraction of biomass (2nd generation ethanol). Due to a lack of competition with food production, the latter is typically seen as more sustainable on a long-term basis [9, 10]. An emerging option is the co-production of ethanol and hydrogen via consolidated bioprocessing [11]. In addition to the now little-used anaerobic A-B fermentations discussed above, another scheme for biofuels production from plant biomass involves the anaerobic production of methane by microbial consortia (anaerobic digestion) [12, 13]. The common link for all of these strategies is the exploitation and optimization of natural microbial activity to produce energy-rich molecules for combustion to produce energy. Direct thermochemical conversion of biomass via pyrolysis or gasification is also possible, although these strategies involve a large amount of energy input by heating the biomass to very high temperatures (normally >500°C) and are therefore independent of microbial activity [14].

Regardless of the means by which biofuels are produced by microbial activity from extant plant material, the same essential challenge must be faced: the substrate for biofuels production is the carbohydrate fraction, which must be made available to the microorganisms in order for the biochemical reactions to proceed efficiently. In the case of 1st generation ethanol, soluble sugars and starch are relatively easily converted to glucose that is fermented into ethanol by yeast. Strategies that utilize the non-food portion of crops, however, face a more formidable challenge. The resource from which energy is to be produced consists of three major biopolymers: cellulose (β(1,4)-linked glucose residues with a degree of polymerization up to ~15,000); hemicellulose (a heterogeneous, short-chained, branched carbohydrate with both 5- and 6-carbon sugars); and lignin (a complex aromatic polymer consisting of nonrepeating covalently linked units of coniferyl, sinapyl, and coumaryl alcohols). These polymers exist together in the plant as a composite, tightly interconnected molecule called lignocellulose [15]. Within lignocellulose, the lignin fraction in particular acts as a barrier to enzyme or microbial penetration, which greatly decreases the yields of fermentable sugars and negatively affects the overall process of energy production from these resources to the extent that it is uneconomical [5, 16]. To overcome this limitation, some form of pretreatment of the biomass is required for economical and efficient production of biofuels by any of the strategies described above [13, 16-21].

The purpose of this chapter is to review the various pretreatment options available for lignocellulosic biomass, with particular emphasis on agricultural residues and on strategies that exploit the natural metabolic activity of microbes to increase the processability of the biomass. These microbial-based strategies can be effective pretreatments on their own or, more probably, can be used in combination with thermomechanical pretreatments in order to provide a cost-effective means to make lignocellulosic substrates available for conversion to biofuels by microorganisms. The key advantages and disadvantages of this strategy will be presented along with a vision for how microbial pretreatment can be integrated into an economical biorefinery process for biofuels and co-product production.

2. Mechanical, thermomechanical and thermochemical pretreatments

An early, essential mechanical pretreatment step is comminution, or mechanical particle size reduction, to transform the biomass from its native state into a suitable substrate for further pretreatment and energy production [17]. This step is often not considered in the energy balance of biofuels processes, but it is important to keep in mind that particle size reduction involves energy input that can influence the effective energy yield of these processes [22]. While smaller particle sizes are often considered to be more desirable for yields of fermentable sugars, sizes smaller than about 0.4-0.5 mm provide no additional benefit [17, 23], and the process becomes economically unfeasible at even smaller particle sizes [22]. Methods for mechanical size reduction include wet milling, dry milling, ball milling or vibratory ball milling, and other forms of chipping and grinding of biomass [4, 17]. Regardless of the method employed, particle size reduction requires energy input; therefore, strategies that facilitate the production of biomass in the proper size range while minimizing energy input will provide positive benefits to the overall economics of biofuels processes.

A wide range of options is available for preparing ground biomass for further processing. One of the most common and simple technologies for rendering the carbohydrate fraction available for biofuels production is the application of a dilute solution of sulfuric acid (0.5%-2%) at temperatures of 140°C – 180°C with residence times of 10-30 minutes [24]. This process leaves a residue that is depleted in hemicellulose but retains most of the cellulose intact, making it an ideal substrate for enzymatic hydrolysis to yield fermentable sugars for ethanol production. There is a range of conditions for acid hydrolysis that will result in more or less carbohydrate remaining in the solid fraction, with the most severe conditions used to completely degrade the carbohydrate fraction for the determination of cell wall carbohydrate composition [25]. Harsher conditions (e.g. higher acid concentration and temperature), while resulting in a substrate that is highly digestible with enzymes to generate fermentable sugars, also result in a higher yield of compounds derived from pentoses (furfural), hexoses (5-hydroxymethylfurfural) and lignin (low molecular weight phenolic compounds) that are inhibitory to subsequent fermentation by ethanologenic yeasts [26]. The mathematical concept of combined severity, which combines the various factors that define acid hydrolysis conditions (e.g. temperature, residence time, pH), allows objective comparisons between different conditions that enables the determination of optimal conditions for a given substrate [26]; however, doubts have been raised about its accuracy [17].

Another highly effective pretreatment strategy is steam explosion, in which biomass is briefly heated to high temperatures (~200°C) under high pressure, then subjected to a rapid pressure drop that renders the biomass more penetrable by enzymes for subsequent hydrolysis [18]. In some cases, steam explosion is enhanced by the addition of an acid catalyst such as sulfuric acid [27]. For lignocellulosic agricultural residues, steam explosion under optimized conditions has been shown to be an effective pretreatment strategy for enzymatic saccharification [28]. Steam explosion has also been successfully used in combination with other physiochemical pretreatments such as acid/water impregnation of cereal straws [29]. Both of the latter studies resulted in the release of hemicellulose-derived pentose oligomers into the liquid fraction, and it was suggested that the use of ethanologenic strains capable of converting these pentoses into ethanol would further improve overall process efficiency [28]. Other assessments have suggested that the hemicellulose fraction would be more efficiently converted to other value-added products rather than ethanol using post-treatment enzyme addition or further acid hydrolysis [30].

Organosolv is a process by which the lignin fraction is chemically modified and essentially removed from biomass using high-temperature extraction with alcohols such as methanol or ethanol or other solvents, sometimes with dilute acid (e.g. hydrochloric or sulfuric acid) as a catalyst [17]. While organosolv processes require a solvent recovery step to be economical and efficient, they provide a robust means of generating three streams of potential products: an extracted, modified lignin component, a hemicellulose-enriched aqueous phase, and a residue that is highly enriched in cellulose and an excellent substrate for the production of biofuels by enzymatic saccharification followed by bacterial or yeast fermentation. Organosolv is one of the pretreatment options that results in a fraction containing chemically modified, low molecular weight lignin components. This stream has a good deal of product potential in addition to its possible use as a fuel for combustion to provide energy to the process [7, 31]. While organosolv is particularly suited to very lignin-rich feedstocks such as wood [32], there is increasing interest in using organosolv extractions for agricultural residues such as wheat straw and dedicated biofuels crops [33]. Goh et al. [34] optimized organosolv conditions for empty palm fruit bunch using combined severity calculations, with excellent results and the ability to accurately predict product stream yields.

Microwave pretreatment of biomass is another option that has been reported to improve subsequent enzymatic saccharification of rice straw [35]. Microwaves have the advantage of combining very rapid heating times with a lower energy input than conventional heating strategies. This irradiative pretreatment creates localized hotspots, which open up the lignocellulose composite molecule, thereby facilitating enzyme access for saccharification and biofuel production by fermentation [4]. A successful combination of microwave and chemical pretreatments in a microwave-acid-alkali-hydrogen peroxide sequence resulted in efficient enzymatic saccharification of rice straw [36]. A related pretreatment option that has been exploited to improve the enzymatic digestibility of switchgrass is the use of radio frequency heating in combination with alkali; this treatment has the key advantage of allowing a much higher solids content than conventional heating [37]. Irradiation of biomass can also enhance methane production by anaerobic digestion [12].

A number of other pretreatment options exist, including ammonia fiber explosion (AFEX), liquid hot water, alkalai/wet oxidative pretreament, and others; several recent reviews discuss these processes and their advantages and disadvantages in detail [4, 17-20, 23, 30, 38, 39]. Regardless of the strategy employed, a common feature of any pretreatment option is that energy input is required. Pretreatment is a major part of the overall operating expense and energy efficiency of any biofuels process, and, while essential, typically accounts for over 30% of the costs of biorefinery operation [40, 41]. Strategies to reduce these costs will have a major impact on the energy balance and economic sustainability of biorefineries.

3. Biological pretreatments

Microorganisms have evolved a capacity to modify and access lignocellulosic biomass to meet their metabolic needs. The exploitation of this capacity offers a natural, low-input means for preparing biomass for biofuels processes. Natural modification and degradation of the lignin component in particular can reduce the severity requirements of subsequent thermochemical pretreatment steps. For example, Itoh et al. [42] used a variety of lignin-degrading white-rot fungi to treat wood chips prior to extracting lignin by an organosolv method, and demonstrated that improved ethanol yields were obtained from the solid fraction along with a 15% savings in electricity use. Similarly, brown-rot fungal species *Coniophora puteana* and *Postia placenta* have been successfully used to improve glucose yields upon enzymatic saccharification of pine, acting as a complete replacement for thermomechanical pretreatments [43]. While it is clear that it is possible to exploit the metabolic capabilities of microorganisms to facilitate biofuels production, the very wide taxonomic array of microorganisms that modify or degrade lignocellulose presents a tremendous variety of choices for implementing such a strategy. Each approach carries its own advantages and challenges.

3.1. Microbial consortia

One approach for applying the power of microbial metabolism to the challenges of biofuel production involves ensiling, which is a commonly used means for enhancing the digestibility of forage and other biomass for ruminants [44, 45]. The process of ensiling exploits the capacity of naturally occurring bacteria, mostly Lactobacillaceae, to ferment the sugars within lignocellulosic residues and produce a substrate that is more easily digested by ruminal microorganisms. While these bacterial consortia lack the ability to substantially degrade the lignin component, the changes effected on the biomass can improve yields of fermentable sugars upon subsequent enzymatic hydrolysis. For example, ensiling a variety of agricultural residues, including wheat, barley, and triticale straws along with cotton stocks resulted in significant improvements in fermentable carbohydrate yields upon application of cellulose-degrading enzymes [46]. Due to limitations in the ability of ensilage to substantially modify the lignin component, this method is not normally a suitable stand-alone biological pretreatment. However, ensiling has been exploited as a means to preserve biomass for biofuels production and has been found to be a very effective, on-farm biomass pretreatment. A strain of *Lactobacillus fermentum* was highly effective in preserving sugar beet pulp cellulose and hemicellulose, and ensiling improved enzymatic saccharification by as much as 35% [47]. Ensiling has also been found to improve yields of methane in anaerobic digestion, with the added benefit of facilitating the longer-term storage of biomass (up to 1 year) while retaining the yield improvements [48, 49]. Improvements in methane yields of up to 50% have been observed with hemp and maize residue, while other crops showed little improvement [50]. However, other researchers have cautioned that the total solids loss may be overestimated for certain substrates, which may result in a misleading, apparent improvement in methane yields by ensiling [51]. Furthermore, while some studies noted above have shown that desirable carbohydrates can be preserved through ensiling, others have noted degradation of cellulose and hemicellulose of up to 10% in this relatively uncontrolled, complex process [46]. Nevertheless, ensiling does offer the substantial benefit of biomass preservation and, importantly, it utilizes existing technology and expertise and can be performed on-farm using unmodified farm equipment. Moreover, ensiling is a relatively low-input process that is anaerobic and therefore does not require mixing and aeration. For these reasons, ensiling could easily be incorporated into an overall biorefinery process at the earliest stages of energy production.

3.2. Lignin-degrading fungi

The earliest colonization of land by plants began around 450 million years ago. The evolutionary innovation that facilitated their spread and success in the non-marine environment was lignification, which provided protection from ultraviolet radiation, structural rigidity and eventually protection from coevolved pathogens and herbivores [52]. The complexity of the phenylpropanoid polymer also provided a carbon sink as land plants fixed atmospheric CO_2 into degradation-resistant lignin. The vast coal reserves whose combustion have contributed to the recent spike in atmospheric CO_2 concentrations trace their origins to the Carboniferous period (~350-300 million years ago), when lignin was not effectively decomposed [52]. Near the end of the Carboniferous period, saprophytic fungi of the class Agaricomycetes evolved the ability to degrade the lignin component of plant biomass, which contributed to a substantial decline in organic carbon burial to the extent that little coal formation occurs today [53, 54]. The large majority of fungal species that are capable of wood decay are known as "white-rot" fungi, which degrade all of the major wood polymers. Approximately 6% of wood decay species are "brown-rot" fungi, which evolved from white-rot fungi and selectively degrade the cellulose and hemicellulose fraction of wood, leaving a lignin-rich residue that is a major contributor to soil carbon in forest ecosystems [55].

http://dx.doi.org/10.5772/55088

The taxonomically broadly distributed white- and brown-rot fungi have developed a variety of means to access and degrade lignocellulose over their long evolutionary history, and their powerful metabolism has been exploited for industrial applications in recent decades. For example, lignin-degrading fungi were noted to have a brightening effect on kraft pulp derived from hardwoods, with savings in bleaching chemicals and potentially decreased environmental impact on paper mill operations [56]. This "biobleaching" was developed further using well-known fungi, such as *Trametes versicolor* [57, 58] and *Phanerochaete chrysosporium* [59, 60]. Similar approaches were used to decolorize and detoxify pulp mill effluent and black liquor [61-63]. In addition, white-rot fungi have been exploited for their ability to decrease energy requirements in pulp manufacturing. This process, known as biopulping, softens the woody substrate and substantially decreases mill electricity requirements for mechanical pulp manufacture [64, 65]. The required scale of industrial pulp manufacture and the applicability of white-rot fungi in providing manufacturing benefits led to the development of feasible means of applying white-rot fungi to biomass on an industrially-relevant scale [66]. This two-auger system featured a wood chip decontamination step and an inoculation step, followed by incubation at ambient temperatures in large chip piles with forced aeration. A series of outdoor trials of this method each featured the treatment of ~36 tonnes of softwood chips with the biopulping fungus *Ceriporiopsis subvermispora* for two weeks. The results were energy savings of around 30% in subsequent pulping, which is slightly higher than was observed in bench-scale trails [66].

3.2.1. Species and systems investigated

More recently, wood-degrading fungi have been investigated for their ability to assist in processing biomass for biofuels production. Again, with the tremendous variety of wood-rotting species and feedstocks available, there is a wide array of strategies reported for biological pretreatment. One very promising approach used rice straw as feedstock, treated with the white-rot fungus *Pleurotus ostreatus* (oyster mushroom) followed by AFEX [67]. This strategy resulted in significant reductions in the severity of the required pretreatment along with improved glucose yields upon enzyme treatment - and produced edible mushrooms as a by-product. Another study found that the incubation time required for *Pleurotus ostreatus* to improve enzymatic saccharification with rice hulls was decreased from 60 days to 18 days by pretreating the rice hulls with hydrogen peroxide prior to fungal inoculation [68]. Similarly, preconditioning of softwood using various white-rot fungi resulted in degradation and modification of the lignin, although significant cellulose loss was also observed [69]. Nevertheless, improved glucose yields were observed by enzymatic saccharification of softwood treated with *Stereum hirsutum* compared to untreated controls, which was attributed to an increase in the pore size of the substrate [69]. Other studies have exploited the selective lignin degradation ability of the white-rot fungus *Echinodontium taxodii* to enhance enzymatic saccharification of water hyacinth in combination with dilute acid pretreatment [70], or of woody substrates without subsequent thermochemical pretreatment [71]. Biological pretreatment has also been shown to improve biogas yields from agricultural residues via anaerobic digestion [72]. A tremendous variety of other approaches to biological pretreatment has been reported to be successful on many different lignocellulosic substrates [73, 74].

Exploitation of fungal metabolic activity for industrial purposes can take a variety of forms. For white- and brown-rot fungi, the mode of cultivation can have an effect on the results obtained, and the choice of cultivation conditions depends on the desired outcomes. In general, fungi can be cultivated under solid-state conditions (solid-state fermentation, or SSF), or using submerged fermentation (SmF). SSF involves culturing the fungus on the substrate under relatively low moisture conditions (~60-70%), while SmF uses liquid cultures of the fungus co-incubated with the normally insoluble substrate. Early pulp biobleaching experiments used SmF of white-rot fungi such as *Trametes versicolor*, which featured the advantage of shorter incubation times than SSF [75], but suffered the drawback that very large fermentation vessels would be required for industrial-scale treatments. Many white-rot fungi grow well and perform the desired metabolism under solid-state conditions. For example, species of the genera *Trametes*, *Phanerochaete*, and *Pycnoporus* preferentially removed color and chemical oxygen demand from olive mill wastewaters and pulp mill black liquors under SSF cultivation conditions [61, 76, 77]. SSF using white-rot fungi has also been used to modify the lignin in agricultural residues, such as wheat straw, for biofuels processes [78].

Despite relatively long incubation times, SSF offers an inexpensive and effective means of fungal cultivation that can also be used for the production of potentially valuable fungal enzymes [79-81]. Fungal enzymes produced by SSF have been used to enhance methane production by anaerobic digestion [82]. Alternatively, fungal lignocellulose modifying enzymes produced by SSF have been used to improve the ruminal digestibility of agricultural residues [83]. However, for SSF to work efficiently with white- or brown-rot fungi requires a decontamination step to allow the fungi to establish on the residues. In lab-scale studies, this is usually accomplished by autoclaving the residues prior to inoculation [84, 85]. While this is necessary at the research scale to establish with certainty the effects of the inoculated fungus on the substrate, autoclaving is in itself a form of pretreatment and is not feasible on an industrial scale. This is a limitation of SSF for application on the large scale that would be required for biological pretreatment of agricultural residues for biofuels production.

3.2.2. Enzymatic mechanisms of fungal lignocellulose degradation

The mechanisms that saprophytic wood degrading fungi have evolved to access their difficult growth substrate can be divided into two categories: oxidative mechanisms and hydrolytic mechanisms. These two groups of enzymes and chemicals act together in various combinations to effect the degradation of lignocellulose by different organisms.

3.2.2.1. Oxidative mechanisms

Due to the highly compact, complex nature of lignocellulose, enzymes cannot effectively penetrate this molecule to interact with their substrates. To overcome this limitation, wood-degrading fungi use chemical means to access the recalcitrant substrate. The production of reactive oxygen species (ROS) is a recurring theme in fungal lignocellulose degradation [86]. Specifically, since wood contains sufficient redox-active iron, fungal production of hydrogen peroxide will produce hydroxyl radicals via the Fenton reaction [86]. Hydroxyl radicals (•OH) are extremely powerful oxidizing agents that can catalyze highly non-specific reactions leading to the cleavage of covalent bonds in both lignin and cellulose [86]. Hydrogen peroxide is commonly produced through the action of fungal redox enzymes, such as glyoxal oxidase, pyranose-2 oxidase, and aryl-alcohol oxidase [15].

Another redox enzyme produced by a wide variety of wood-degrading fungi (as well as plants) is laccase, a multicopper oxidase. Laccase acts by removing a single electron from its substrate, which is typically a low-molecular weight compound (mediator) that can diffuse into the densely packed lignocellulose molecule and initiate free radical-mediated reactions leading to the depolymerisation of the substrate. The white-rot fungus *Pycnoporus cinnabarinus* uses laccase in combination with a secondary metabolite, 3-hydroxyanthranilic acid, to effect lignin depolymerisation [87, 88]. Laccase has been used in combination with a wide variety of chemical mediators to effect lignin degradation in wood pulp, with excellent results [89-91].

The presence of manganese in woody substrates is exploited by lignin-degrading fungi through the production of the enzyme manganese peroxidase (MnP). The importance of MnP in lignin degradation is illustrated by its presence in the genomes of white-rot fungi and absence in the non-lignin-degrading brown-rot fungi [92], as well as by the inability of MnP-deficient mutants of *Trametes versicolor* to delignify hardwood-derived kraft pulp [93]. MnP is a heme-containing enzyme with a catalytic cycle that is typical of heme peroxidases, but is uniquely selective for Mn^{2+} as its preferred electron donor [94]. The oxidation of Mn^{2+}, which is accompanied by the reduction of hydrogen peroxide to water, results in the formation of Mn^{3+}. The latter ion is a powerful, diffusible oxidant that is chelated by organic acids such as oxalate produced as a secondary metabolite of the fungus [94]. This highly reactive ion interacts with a wide variety of substrates, including phenols, non-phenolic aromatics, carboxylic acids, and unsaturated fatty acids, producing further ROS and resulting in lignocellulose bond cleavage through oxidative mechanisms [94]. Like laccase, MnP has found application as a delignifying enzyme for pine wood [95] as well as kraft pulp [96, 97]. Peroxidases related to MnP, including lignin peroxidase (LiP) and versatile peroxidase (VP) are also produced by a variety of wood-degrading fungi and play an important role in lignin degradation [98].

Cellobiose dehydrogenase (CDH) is a unique enzyme containing both a heme and a flavin cofactor [99]. CDH is produced by a wide range of fungal species, including both lignin-degrading organisms and fungi that are incapable of degrading lignin [100, 101]. CDH catalyzes the two-electron oxidation of a narrow range of β(1,4)-linked sugar molecules, principally cellobiose, and transfers these electrons to a very wide array of substrates, including metals such as ferric, cupric, or manganic ions, iron-containing proteins (e.g. cytochrome c), quinones, and other large and small molecules [102, 103]. The diversity of reduced substrates has led to much speculation regarding the role of CDH in lignocellulose degradation; roles have been postulated in the degradation of both cellulose [104] and lignin [105]. The reduction of cupric and ferric ions by CDH and the production of hydrogen peroxide by lignin-degrading fungi suggests that CDH may be involved in sustaining hydroxyl radical-based Fenton's chemistry, with many possible secondary reactions leading to lignocellulose bond cleavage [106]. The role of CDH in lignin-degrading basidiomycetes was addressed by generating mutants of *Trametes versicolor* that did not produce the enzyme, suggesting that CDH plays a role in cellulose degradation, with a more minor role in lignin degradation [107, 108]. Similarly, a recent study with the non- lignin-degrading ascomycete *Neurospora crassa* revealed that deletion of the gene encoding CDH resulted in vastly decreased cellulase activity, and that the oxidation of cellobiose was coupled to the reductive activation of copper-containing polysaccharide monooxygenases [109]. These studies strongly suggest a role for CDH in supporting cellulose catabolism by fungi, with the latter study in particular providing a highly plausible mechanism for the *in vivo* function of CDH.

3.2.2.2. Hydrolytic mechanisms

Complementing the degradative power of the redox chemistry catalyzed by the enzymes produced by lignocellulose-degrading fungi is a suite of enzymes that act by adding a water molecule to glycosidic bonds, resulting in bond cleavage and depolymerization. In contrast to the redox enzymes, these hydrolytic enzymes recognize and act on specific glycosidic linkages, releasing sugar molecules that can be utilized as an energy source to support fungal metabolism. Cellulose degradation is catalyzed by the synergistic action of three classes of hydrolytic cellulase enzymes: endo-(1,4)-β-glucanase (endocellulase), cellobiohydrolase (exocellulase), and β-glucosidase [110]. Endocellulases catalyze the cleavage of cellulose chains internally at amorphous regions, while exocellulases remove cellobiose units from the ends of cellulose chains. β-glucosidases are extracellular, cell wall-associated or intracellular enzymes that cleave cellobiose into glucose, which also supports exocellulase activity by relieving end-product inhibition [110]. The redundancy in cellulase genes in fungi is at least partially explained by the fact that different exocellulase enzymes preferentially attack the reducing or non-reducing end of a cellulose chain. This has the effect of exposing new sites for exocellulases of the opposite specificity and also generates new amorphous regions to be acted upon by endocellulases [110, 111]. Hemicellulose degradation is effected by the activity of a wide range of hydrolytic enzymes, including endo-xylanases; endo-α-L-arabinase; endo-mannanase, β-galactosidase, and an array of corresponding β-glucosidases [112]. In addition, covalent bonds within lignocellulose are hydrolyzed by cinnamoyl or feruloyl esterases, which cleave the ester bond between polymerized lignin subunits and the hemicellulose within the composite molecule [113, 114]. Complementary cellulase activity by these various "accessory enzymes" is shown on complex substrates by the improvement in enzymatic saccharification observed when enzymes such as xylanase, pectinase, and feruloyl esterase are added to cellulase cocktails [115, 116].

3.2.2.3. Fungal enzyme discovery, production, and application

Tremendous progress has been made in the last decade concerning the genetic mechanisms underlying plant biomass degradation and modification by microbes, specifically ascomycetous and basidiomycetous fungi. Key to these advancements was the complete genome sequencing of several biomass-degrading fungi, including *Phanerochaete chrysosporium, Phanerochaete carnosa, Postia placenta and Trametes versicolor*. The first basidiomycete genome to be sequenced and analyzed, *Phanerochaete chrysosporium*, revealed a tremendous diversity of genes encoding enzymes involved in wood degradation [117]. Among these genes were approximately 240 carbohydrate-active enzymes and several lignin and manganese-dependent peroxidases, which function to degrade the cellulosic/hemicellulosic and lignin components of the cell wall, respectively. This research provided the groundwork for more comprehensive analyses of the genome [118], transcriptome and secretome of *Phanerochaete chrysosporium* [119-122]. These studies highlighted hundreds of wood-degrading genes that were upregulated when *P. chrysosporium* was grown in cellulose-rich medium, including almost 200 genes encoding enzymes of unknown function [122].

Complementary to this research on white-rot fungi was a genome/transcriptome/proteome study on the brown-rot fungus *Postia placenta* [123]. Despite an abundance of similarities between *P. chrysosporium* and *P. placenta,* there were notably fewer glycoside hydrolases expressed by *P. placenta,* such as extracellular cellulases (e.g. endo-(1,4)-β-glucanases), highlighting the mechanistic differences between white- and brown-rot fungi. This work was followed by transcriptomic and proteomic studies investigating the biomass-degrading activity of *Phanerochaete carnosa* [124, 125]. Despite the overall similarity of the transcriptome composition among *P. carnosa* and *P. chrysosporium,* the most abundant transcripts in *P. carnosa* grown on wood substrates (hardwood and softwood) were peroxidases and oxidases involved in lignin degradation [124], whereas *P. chrysosporium* grown only on hardwood revealed only a few highly expressed lignin-degrading enzymes [121]. The differing expression of lignocellulosic enzymes in response to different woody substrates was also explored by examining gene expression patterns in both *P. placenta* and *P. chrysosporium* on hardwood and softwood species [126]. The results of this study strongly suggest that both species of fungi alter their gene expression patterns to degrade wood with different structural characteristics.

In addition to helping uncover the fundamental biochemical machinery involved in biomass degradation, these genomic, transcriptomic and proteomic studies of biomass-degrading fungi have also identified hundreds of target enzymes that could be utilized industrially for bioenergy production, with unique enzyme cocktails suited for specific substrates (e.g. hardwoods vs. softwoods). Several commercial enzymes are commonly used to degrade lignocellulosic residue into fermentable sugars (e.g. Celluclast and Novozyme 188). These 'omic' studies have identified hundreds of fungal glycoside hydrolases that may supplement or completely replace these industry standards. Pre-treatment strategies may also take advantage of the numerous lignin-modifying enzymes identified from biomass-degrading fungi, including lignin and manganese-dependent peroxidases, which have the potential to reduce the severity of thermomechanical and thermochemical pretreatment processes.

3.2.2.4. Exploiting fungal mutants for biological pretreatment

The explosion of 'omic' data for a wide variety of lignocellulose-degrading fungi [117, 123-125] along with the development of sophisticated tools for annotating fungal genomes [127] will continue to add to our understanding of the mechanisms of fungal decay of lignocellulose. Furthermore, increased knowledge of fungal decay mechanisms can aid in the development of strains with improved characteristics. For example, a major limitation to the application of fungal strains for biological pretreatment is the degradation of the desired carbohydrates (cellulose and hemicellulose) for fungal metabolism [73, 74]. Creating or selecting strains that lack the ability to degrade these carbohydrates while retaining the ability to degrade and modify lignin would provide a means to avoid this drawback of fungal pretreatment. Early studies with strains that were deficient in the production of cellulase met with only moderate success, with substantial degradation of cellulose observed [128, 129]. This is probably attributable to the high degree of redundancy in fungal cellulases, with large numbers of genes contributing to the hydrolytic degradation of cellulose and hemicellulose in various species [15]. More recently, we have applied a strain of *Trametes versicolor* that is unable to produce cellobiose dehydrogenase (CDH) to the pretreatment of canola residue, and found that the strain was proficient in lignin degradation but was unable to catabolize the cellulose [107]. Xylose within the substrate appeared to have been utilized to support the greatly decreased fungal growth compared to the wild-type strain. Furthermore, we found that the application of a fungal cell wall-degrading enzyme cocktail (glucanex; a concentrated supernatant of a SmF culture of *Trichoderma harzianum*) to the fungus-treated biomass resulted in the release of fungal cell wall-associated glucose [107]. Biological pretreatment with *T. versicolor* therefore had the overall effect of converting some of the xylose within the substrate to glucose, which is more easily fermented by ethanologenic yeasts.

Studies such as these also provide biological data regarding the role of the genes that are down-regulated in the mutant strains. This reverse genetics approach is a powerful method for investigating gene function, and in the current genomic era reverse genetics tools can often be applied in the known context of the entire genome of the fungus. Gene silencing by RNA interference (RNAi) is a common method for down-regulating genes in a variety of model systems [130, 131], and the recent demonstration of RNAi mechanisms in the model white-rot fungus *Phanerochaete chrysosporium* [132] suggests that RNAi could be used for targeted down-regulation of specific genes in species that are useful for biological pretreatment. The availability of convenient gene silencing transformation vectors for ascomycetes such as pSilent [133] and pTroya [134] as well as pHg/pSILBAγ for basidiomycetes [135] will greatly facilitate the investigation of gene function and may also result in the development of modified strains featuring enhanced properties for biological pretreatment of lignocellulosic substrates for biofuels production.

4. Conclusions and future outlook

Pretreatment of lignocellulosic materials with white- or brown-rot fungi can be incorporated into any strategy for the production of biofuels and bioproducts, with significant advantages including decreased energy requirements for subsequent steps, production of fewer fermentation-inhibiting substances, and the potential for the production of value-added co-product streams [73, 74]. With the wide variety of potential strains and substrates available, and the possibility to create or select new strains with more desirable properties, it seems likely that biological pretreatment can be used on nearly any biomass that is currently produced. One of the most important benefits of biological pretreatment is the resultant reduction in the severity of the subsequent thermomechanical or thermochemical pretreatment step that is required for efficient enzymatic saccharification. While this is a very important benefit, Keller et al. [136] identified six criteria for strains to be selected for biological pretreatment of agricultural waste:

little carbohydrate degradation, low costs for nutrients, a reasonable storage time, ability to compete with endogenous microbiota, decreased thermomechanical pretreatment severity, improved yields of glucose upon enzymatic saccharification, and a lack of production of compounds inhibitory to fermenting organisms. These criteria underscore the major limitations of biological pretreatment, the most important of which are the propensity of the organisms to degrade the carbohydrate component, their inability to establish growth on unsterilized biomass, and the relatively long incubation times that are required. These limitations are related to the ecological niche that these saprophytic fungi fill in nature. That is, they have evolved to access and utilize those plant carbohydrates that are difficult for other microorganisms to access. For this reason, these fungi typically appear at the end of an ecological succession of organisms that degrade decaying wood and are often ill equipped to compete with the faster-growing molds and bacteria that access the more easily degraded plant carbohydrates [137]. While it may be possible using reverse genetics tools and/or strain selection to limit carbohydrate degradation by pretreatment fungi [107], it is likely that such strains will be even less able to compete with endogenous microorganisms; therefore, establishment on recently harvested biomass will remain a challenge. Some sort of treatment of the biomass to suppress the growth of endogenous molds prior to inoculation with the pretreatment fungi will likely be necessary.

The unavoidable expense of the pre-inoculation treatment can be compensated by taking advantage of a potential benefit of biological pretreatment that has received very little attention: wood-degrading fungi may modify the lignin component sufficiently to provide positive benefits for particle compression of agricultural biomass during densification. Densification (briquetting or pelleting) of biomass aims to increase the bulk density of agricultural residues far beyond what is achievable by baling, and it is an essential step for providing biomass with sufficient caloric density for efficient transportation [138]. The production of biomass pellets provides a substrate that is suitable for conversion into biofuels through microbial processes or gasification [139, 140], or can be combusted directly to produce energy [141]. A wide variety of agricultural feedstocks is suitable for pelleting [142]; however, untreated biomass is very difficult to densify and, without pretreatment, produces weak, powdery pellets that are expensive to produce and cannot withstand the physical rigors of transportation. Lignin acts as a natural binder that provides strength and durability to biomass pellets, and pretreatment of the biomass is required in order to release lignin fragments during compaction and produce pellets with the desired characteristics [139, 143]. A number of options are available to prepare biomass for pelleting, with two very promising methods being microwave heating and radio frequency heating. Both of these methods provide a number of advantages over conventional heating, particularly regarding treatment times [144-146]. We have found that a very brief microwave treatment of a variety of agricultural feedstocks suppresses the growth of endogenous molds and bacteria sufficiently for inoculated white-rot fungi to establish growth on these substrates. Moreover, canola residue treated with *Trametes versicolor* produces pellets with excellent compaction characteristics and durability (Canam, Town, and Dumonceaux, unpublished). Such pellets would retain the thermochemical pretreatment benefits afforded by the fungal pretreatment in terms of enzymatic saccharification [107], but would offer vastly increased transportation efficiency in a full-scale biorefinery scenario.

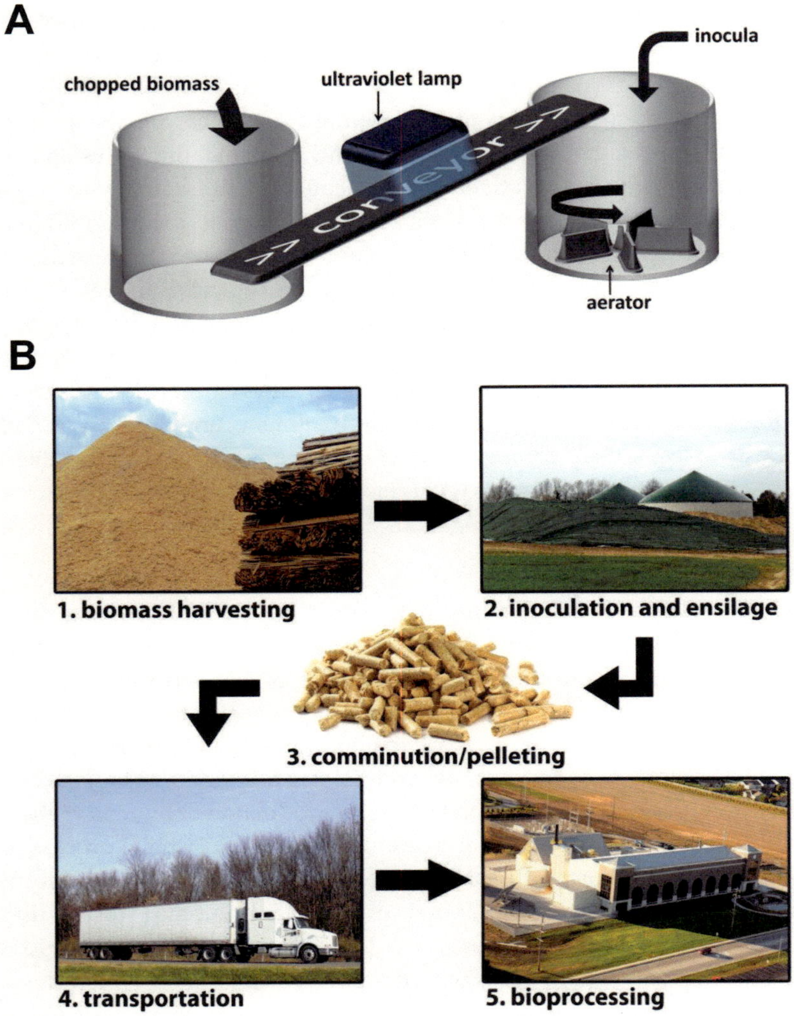

Figure 1. A possible scheme for incorporating biological pretreatment into biofuels manufacturing. A. Treatment to suppress the growth of endogenous microorganisms to allow establishment of the inoculated fungal culture. A variety of treatments could be utilized, including ultraviolet, microwave, or radio frequency treatment. This scheme is based on the successful biopulping inoculation strategy described by Scott et al. [66] B. Overall scheme for biofuels production including biological pretreatment. (B.5. photo courtesy of Jay Grabiec, Eastern Illinois University).

We can therefore envision a means by which fungal pretreatment might be incorporated into an overall process for producing energy from biomass by a variety of strategies (Figure 1). Biological pretreatment should be included at the earliest stages in order to take maximum advantage of its beneficial effects. Building on the successful bio-pulping model described by Scott et al. [66], chopped biomass would be briefly decontaminated by microwave or radio frequency heating within a conveyor; the objective of this is not sterilization or complete

http://dx.doi.org/10.5772/55088

thermal pretreatment of the biomass, but primarily growth suppression of endogenous microbiota. The lightly treated biomass would then be inoculated with a fungal suspension or formulation and transferred via auger to a pile analogous to a silage pile, but with aeration. The inoculated biomass would be incubated at ambient temperatures for several weeks to allow fungal growth. The fully infested biomass would then be milled to an appropriate size using standard equipment. After cooling and/or drying, pellets formed from the milled, pretreated biomass would be suitable for transport to a biorefinery for biofuels/bioproducts manufacture. In the absence of a viable product stream or a biorefinery, biomass pellets could be burned in a high-efficiency oven to exploit their calorific value [139]. Biological pretreatment would provide an array of benefits along this production chain, including decreased milling energy, decreased compression energy requirements, improved densification characteristics and the consequent reduction in transportation cost, decreased severity of thermochemical pretreatments, decreased production of fermentation inhibitors, improved yield of fermentable sugars upon enzymatic saccharification, and possibly co-products derived from the more easily extractable lignin phase. All of these benefits would be realized with a fairly minor energy input, and although it is difficult to avoid the long incubation times, SSF can be performed on time scales only slightly longer than the common on-farm practice of ensilage. Biological pretreatments can therefore, in theory, be performed on-farm or nearby, offering significant logistical and technical advantages when incorporated into an overall process for biofuels manufacture.

Author details

Thomas Canam[1], Jennifer Town[2,3], Kingsley Iroba[4], Lope Tabil[4] and Tim Dumonceaux[2,3*]

*Address all correspondence to: tim.dumonceaux@agr.gc.ca

1 Department of Biological Sciences, Eastern Illinois University, Charleston, IL, USA

2 Agriculture and Agri-Food Canada Saskatoon Research Centre, Saskatoon, Canada

3 Department of Veterinary Microbiology, University of Saskatchewan, Saskatoon, Canada

4 Department of Chemical and Biological Engineering, University of Saskatchewan, Saskatoon, Canada

References

[1] Wolff EW. Greenhouse gases in the Earth system: A palaeoclimate perspective. Philosophical Transactions of the Royal Society A: Mathematical, Physical and Engineering Sciences. 2011;369(1943):2133-47.

[2] Falkowski P, Scholes RJ, Boyle E, Canadell J, Canfield D, Elser J, et al. The Global Carbon Cycle: A Test of Our Knowledge of Earth as a System. Science. 2000;290(5490):291-6.

[3] Crowley TJ. Causes of climate change over the past 1000 years. Science. 2000;289(5477):270-7.

[4] Sarkar N, Ghosh SK, Bannerjee S, Aikat K. Bioethanol production from agricultural wastes: An overview. Renewable Energy. 2012;37(1):19-27.

[5] Menon V, Rao M. Trends in bioconversion of lignocellulose: Biofuels, platform chemicals & biorefinery concept. Progress in Energy and Combustion Science. 2012;38(4): 522-50.

[6] Ragauskas AJ, Williams CK, Davison BH, Britovsek G, Cairney J, Eckert CA, et al. The path forward for biofuels and biomaterials. Science. 2006;311(5760):484-9.

[7] Sannigrahi P, Pu Y, Ragauskas A. Cellulosic biorefineries-unleashing lignin opportunities. Curr Opin Environ Sustain. 2010;2(5-6):383-93.

[8] Zverlov VV, Berezina O, Velikodvorskaya GA, Schwarz WH. Bacterial acetone and butanol production by industrial fermentation in the Soviet Union: Use of hydrolyzed agricultural waste for biorefinery. Appl Microbiol Biotechnol. 2006;71(5): 587-97.

[9] Rude MA, Schirmer A. New microbial fuels: a biotech perspective. Curr Opin Microbiol. 2009;12(3):274-81.

[10] Khanna M, Hochman G, Rajagopal D, Sexton S, Zilberman D. Sustainability of food, energy and environment with biofuels. CAB Reviews: Perspectives in Agriculture, Veterinary Science, Nutrition and Natural Resources. 2009;4(28).

[11] Carere CR, Sparling R, Cicek N, Levin DB. Third generation biofuels via direct cellulose fermentation. Int J Molec Sci. 2008;9(7):1342-60.

[12] Chandra R, Takeuchi H, Hasegawa T. Methane production from lignocellulosic agricultural crop wastes: A review in context to second generation of biofuel production. Renewable and Sustainable Energy Reviews. 2012;16(3):1462-76.

[13] Frigon JC, Guiot SR. Biomethane production from starch and lignocellulosic crops: A comparative review. Biofuels, Bioproducts and Biorefining. 2010;4(4):447-58.

[14] Bahng MK, Mukarakate C, Robichaud DJ, Nimlos MR. Current technologies for analysis of biomass thermochemical processing: A review. Analytica Chimica Acta. 2009;651(2):117-38.

[15] Martinez AT, Ruiz-Duenas FJ, Martinez MJ, Del Rio JC, Gutierrez A. Enzymatic delignification of plant cell wall: from nature to mill. Curr Opin Biotechnol. 2009;20(3): 348-57.

[16] Margeot A, Hahn-Hagerdal B, Edlund M, Slade R, Monot F. New improvements for lignocellulosic ethanol. Curr Opin Biotechnol. 2009;20(3):372-80.

[17] Agbor VB, Cicek N, Sparling R, Berlin A, Levin DB. Biomass pretreatment: Fundamentals toward application. Biotechnol Adv. 2011;29(6):675-85.

[18] Chandra RP, Bura R, Mabee WE, Berlin A, Pan X, Saddler JN. Substrate pretreatment: the key to effective enzymatic hydrolysis of lignocellulosics? Adv Biochem Eng Biotechnol. 2007;108:67-93.

[19] Conde-Mejía C, Jiménez-Gutiérrez A, El-Halwagi M. A comparison of pretreatment methods for bioethanol production from lignocellulosic materials. Process Safety and Environmental Protection.

[20] Hendriks ATWM, Zeeman G. Pretreatments to enhance the digestibility of lignocellulosic biomass. Bioresour Technol. 2009;100(1):10-8.

[21] Kumar P, Barrett DM, Delwiche MJ, Stroeve P. Methods for pretreatment of lignocellulosic biomass for efficient hydrolysis and biofuel production. Industrial and Engineering Chemistry Research. 2009;48(8):3713-29.

[22] da Costa Sousa L, Chundawat SP, Balan V, Dale BE. 'Cradle-to-grave' assessment of existing lignocellulose pretreatment technologies. Curr Opin Biotechnol. 2009;20(3): 339-47.

[23] Chang VS, Burr B, Holtzapple MT. Lime pretreatment of switchgrass. Appl Biochem Biotechnol. 1997;63-65:3-19.

[24] Yang B, Lu Y, Wyman CE. Cellulosic ethanol from agricultural residues. In: Balscheck HP, Ezeji TC, Scheffran J, editors. Biofuels from Agricultural Wastes and Byproducts. Ames, Iowa: Wiley-Blackwell; 2010. p. 175-200.

[25] Sluiter A, Hames B, Ruiz R, Scarlata C, Sluiter J, Templeton D, et al. Determination of structural carbohydrates and lignin in biomass. NREL/TP-510-4261. 2008.

[26] Larsson S, Palmqvist E, Hahn-Hägerdal B, Tengborg C, Stenberg K, Zacchi G, et al. The generation of fermentation inhibitors during dilute acid hydrolysis of softwood. Enz Microb Technol. 1999;24(3-4):151-9.

[27] Ballesteros I, Negro MJ, Oliva JM, Cabanas A, Manzanares P, Ballesteros M. Ethanol production from steam-explosion pretreated wheat straw. Appl Biochem Biotechnol. 2006;129-132:496-508.

[28] Ruiz E, Cara C, Manzanares P, Ballesteros M, Castro E. Evaluation of steam explosion pre-treatment for enzymatic hydrolysis of sunflower stalks. Enz Microb Technol. 2008;42(2):160-6.

[29] Rosgaard L, Pedersen S, Meyer AS. Comparison of different pretreatment strategies for enzymatic hydrolysis of wheat and barley straw. Appl Biochem Biotechnol. 2007;143(3):284-96.

[30] Carvalheiro F, Duarte LC, GÃrio FM. Hemicellulose biorefineries: A review on biomass pretreatments. Journal of Scientific and Industrial Research. 2008;67(11):849-64.

[31] Doherty WOS, Mousavioun P, Fellows CM. Value-adding to cellulosic ethanol: Lignin polymers. Industrial Crops and Products. 2011;33(2):259-76.

[32] Bozell JJ, Black SK, Myers M, Cahill D, Miller WP, Park S. Solvent fractionation of renewable woody feedstocks: Organosolv generation of biorefinery process streams for the production of biobased chemicals. Biomass Bioenergy. 2011;35:4197-208.

[33] Buranov AU, Mazza G. Lignin in straw of herbaceous crops. Industrial Crops and Products. 2008;28(3):237-59.

[34] Goh CS, Tan HT, Lee KT, Brosse N. Evaluation and optimization of organosolv pretreatment using combined severity factors and response surface methodology. Biomass Bioenergy. 2011;35(9):4025-33.

[35] Ma H, Liu WW, Chen X, Wu YJ, Yu ZL. Enhanced enzymatic saccharification of rice straw by microwave pretreatment. Bioresour Technol. 2009;100(3):1279-84.

[36] Zhu S, Wu Y, Yu Z, Wang C, Yu F, Jin S, et al. Comparison of three microwave/chemical pretreatment processes for enzymatic hydrolysis of rice straw. Biosys Engin. 2006;93(3):279-83.

[37] Hu Z, Wang Y, Wen Z. Alkali (NaOH) pretreatment of switchgrass by radio frequency-based dielectric heating. Appl Biochem Biotechnol. 2008;148(1-3):71-81.

[38] Khanal SK, Rasmussen M, Shrestha P, Van Leeuwen H, Visvanathan C, Liu H. Bioenergy and biofuel production from wastes/residues of emerging biofuel industries. Wat Envrion Res. 2008;80(10):1625-47.

[39] Zhang X, Tu M, Paice MG. Routes to potential bioproducts from lignocellulosic biomass lignin and hemicelluloses. Bioenergy Res. 2011;4(4):246-57.

[40] Lynd LR. Overview and evaluation of fuel ethanol from cellulosic biomass: Technology, economics, the environment, and policy. Annu Rev Energy Environ. 1996;21(1): 403-65.

[41] Lynd LR, Elander RT, Wyman CE. Likely features and costs of mature biomass ethanol technology. Appl Biochem Biotechnol. 1996;57-58:741-61.

[42] Itoh H, Wada M, Honda Y, Kuwahara M, Watanabe T. Bioorganosolve pretreatments for simultaneous saccharification and fermentation of beech wood by ethanolysis and white rot fungi. J Biotechnol. 2003;103(3):273-80.

[43] Ray MJ, Leak DJ, Spanu PD, Murphy RJ. Brown rot fungal early stage decay mechanism as a biological pretreatment for softwood biomass in biofuel production. Biomass Bioenergy. 2010;34(8):1257-62.

[44] McEniry J, O'Kiely P, Clipson NJW, Forristal PD, Doyle EM. Bacterial community dynamics during the ensilage of wilted grass. J Appl Microbiol. 2008;105(2):359-71.

[45] Thompson DN, Barnes JM, Houghton TP. Effect of additions on ensiling and microbial community of senesced wheat straw. Appl Biochem Biotechnol. 2005;121(1-3): 21-46.

[46] Chen Y, Sharma-Shivappa RR, Chen C. Ensiling agricultural residues for bioethanol production. Appl Biochem Biotechnol. 2007;143(1):80-92.

[47] Zheng Y, Yu C, Cheng YS, Zhang R, Jenkins B, VanderGheynst JS. Effects of ensilage on storage and enzymatic degradability of sugar beet pulp. Bioresour Technol. 2011;102(2):1489-95.

[48] Herrmann C, Heiermann M, Idler C. Effects of ensiling, silage additives and storage period on methane formation of biogas crops. Bioresour Technol. 2011;102(8): 5153-61.

[49] Nizami AS, Korres NE, Murphy JD. Review of the integrated process for the production of grass biomethane. Environmental Science and Technology. 2009;43(22): 8496-508.

[50] Pakarinen A, Maijala P, Jaakkola S, Stoddard F, Kymäläinen M, Viikari L. Evaluation of preservation methods for improving biogas production and enzymatic conversion yields of annual crops. Biotechnol Biofuels. 2011;4(20).

[51] Kreuger E, Nges I, Björnsson L. Ensiling of crops for biogas production: Effects on methane yield and total solids determination. Biotechnol Biofuels. 2011;4(44).

[52] Weng JK, Chapple C. The origin and evolution of lignin biosynthesis. New Phytologist. 2010;187(2):273-85.

[53] Floudas D, Binder M, Riley R, Barry K, Blanchette RA, Henrissat B, et al. The paleozoic origin of enzymatic lignin decomposition reconstructed from 31 fungal genomes. Science. 2012;336(6089):1715-9.

[54] Hittinger CT. Evolution: Endless rots most beautiful. Science. 2012;336(6089):1649-50.

[55] Eastwood DC, Floudas D, Binder M, Majcherczyk A, Schneider P, Aerts A, et al. The plant cell wall-decomposing machinery underlies the functional diversity of forest fungi. Science. 2011;333(6043):762-5.

[56] Kirk TK, Yang HH. Partial delignification of unbleached kraft pulp with ligninolytic fungi. Biotechnol Lett. 1979;1(9):347-52.

[57] Roy BP, Archibald F. Effects of kraft pulp and lignin on Trametes versicolor carbon metabolism. Appl Environ Microbiol. 1993;59(6):1855-63.

[58] Selvam K, Saritha KP, Swaminathan K, Manikandan M, Rasappan K, Chinnaswamy P. Pretreatment of wood chips and pulp with Fomes lividus and Trametes versicolor

to reduce chemical consumption in paper industries. Asian J Microbiol Biotechnol Environ Sci. 2006;8(4):771-6.

[59] Jiménez L, López F, Martínez C. Biological pretreatments for bleaching wheat-straw pulp. Process Biochem. 1994;29(7):595-9.

[60] Jiménez L, Martínez C, Pérez I, López F. Biobleaching procedures for pulp from agricultural residues using Phanerochaete chrysosporium and enzymes. Process Biochem. 1997;32(4):297-304.

[61] Da Re V, Papinutti L. Black liquor decolorization by selected white-rot fungi. Appl Biochem Biotechnol. 2011;165(2):406-15.

[62] Kumar A, Shrivastava V, Pathak M, Singh RS. Biobleaching of pulp and paper mill effluent using mixed white rot fungi. Biosci Biotechnol Res Asia. 2010;7(2):925-31.

[63] Mittar D, Khanna PK, Marwaha SS, Kennedy JF. Biobleaching of pulp and paper mill effluents by Phanerochaete chrysosporium. J Chem Technol Biotechnol. 1992;53(1): 81-92.

[64] Scott GM, Akhtar M, Kirk TK, editors. An Update on Biopulping Commericialization. Proceedings of the 2000 TAPPI Pulping/Process and Product Quality Process; 2000; Boston, MA.

[65] Sena-Martins G, Almeida-Vara E, Moreira PR, Polónia I, Duarte JC, editors. Biopulping of pine wood chips for production of kraft paper board. Proceedings of the 2000 TAPPI Pulping/Process and Product Quality Process; 2000; Boston, MA.

[66] Scott GM, Akhtar M, Lentz MJ, Kirk TK, Swaney R. New technology for papermaking: Commercializing biopulping. Tappi J. 1998;81(11):220-5.

[67] Balan V, Da Costa Sousa L, Chundawat SPS, Vismeh R, Jones AD, Dale BE. Mushroom spent straw: A potential substrate for an ethanol-based biorefinery. J Ind Microbiol Biotechnol. 2008;35(5):293-301.

[68] Yu J, Zhang J, He J, Liu Z, Yu Z. Combinations of mild physical or chemical pretreatment with biological pretreatment for enzymatic hydrolysis of rice hull. Bioresour Technol. 2009;100(2):903-8.

[69] Lee JW, Gwak KS, Park JY, Park MJ, Choi DH, Kwon M, et al. Biological pretreatment of softwood Pinus densiflora by three white rot fungi. J Microbiol. 2007;45(6): 485-91.

[70] Ma F, Yang N, Xu C, Yu H, Wu J, Zhang X. Combination of biological pretreatment with mild acid pretreatment for enzymatic hydrolysis and ethanol production from water hyacinth. Bioresour Technol. 2010;101(24):9600-4.

[71] Yu H, Guo G, Zhang X, Yan K, Xu C. The effect of biological pretreatment with the selective white-rot fungus Echinodontium taxodii on enzymatic hydrolysis of softwoods and hardwoods. Bioresour Technol. 2009;100(21):5170-5.

[72] Muthangya M, Mshandete AM, Kivaisi AK. Two-stage fungal pre-treatment for improved biogas production from sisal leaf decortication residues. Int J Molec Sci. 2009;10(11):4805-15.

[73] Isroi, Millati R, Syamsiah S, Niklasson C, Cahyanto MN, Lundquist K, et al. Biological pretreatment of lignocelluloses with white-rot fungi and its applications: A review. BioResources. 2011;6(4):5224-59.

[74] Chen S, Zhang X, Singh D, Yu H, Yang X. Biological pretreatment of lignocellulosics: potential, progress and challenges. Biofuels. 2010;1(1):177-99.

[75] Paice MG, Archibald FS, Bourbonnais R, Jurasek L, Reid ID, Charles T, et al. Enzymology of kraft pulp bleaching by Trametes versicolor. In: Jeffries TW, Viikari L, editors. Enzymes for Pulp and Paper Processing. Washington: American Chemical Society; 1996. p. 151-64.

[76] Alaoui SM, Merzouki M, Penninckx MJ, Benlemlih M. Relationship between cultivation mode of white rot fungi and their efficiency for olive oil mill wastewaters treatment. Elec J Biotechnol. 2008;11(4).

[77] Aloui F, Abid N, Roussos S, Sayadi S. Decolorization of semisolid olive residues of "alperujo" during the solid state fermentation by Phanerochaete chrysosporium, Trametes versicolor, Pycnoporus cinnabarinus and Aspergillus niger. Biochem Engin J. 2007;35(2):120-5.

[78] Dinis MJ, Bezerra RMF, Nunes F, Dias AA, Guedes CV, Ferreira LMM, et al. Modification of wheat straw lignin by solid state fermentation with white-rot fungi. Bioresour Technol. 2009;100(20):4829-35.

[79] Rodríguez Couto S, Moldes D, Liébanas A, Sanromán A. Investigation of several bioreactor configurations for laccase production by Trametes versicolor operating in solid-state conditions. Biochem Engin J. 2003;15(1):21-6.

[80] Rodríguez Couto S, Sanromán MA. Application of solid-state fermentation to ligninolytic enzyme production. Biochem Engin J. 2005;22(3):211-9.

[81] Winquist E, Moilanen U, Mettälä, A., Leisola M, Hatakka A. Production of lignin modifying enzymes on industrial waste material by solid-state cultivation of fungi. Biochem Engin J. 2008;42(2):128-32.

[82] Bochmann G, Herfellner T, Susanto F, Kreuter F, Pesta G. Application of enzymes in anaerobic digestion. 2007. p. 29-35.

[83] Graminha EBN, GonÃ§alves AZL, Pirota RDPB, Balsalobre MAA, Da Silva R, Gomes E. Enzyme production by solid-state fermentation: Application to animal nutrition. Anim Feed Sci Technol. 2008;144(1-2):1-22.

[84] Kuhar S, Nair LM, Kuhad RC. Pretreatment of lignocellulosic material with fungi capable of higher lignin degradation and lower carbohydrate degradation improves

substrate acid hydrolysis and the eventual conversion to ethanol. Can J Microbiol. 2008;54(4):305-13.

[85] Rasmussen ML, Shrestha P, Khanal SK, Pometto Iii AL, van Leeuwen J. Sequential saccharification of corn fiber and ethanol production by the brown rot fungus Gloeophyllum trabeum. Bioresour Technol. 2010;101(10):3526-33.

[86] Hammel KE, Kapich AN, Jensen Jr KA, Ryan ZC. Reactive oxygen species as agents of wood decay by fungi. Enz Microb Technol. 2002;30(4):445-53.

[87] Eggert C, Temp U, Dean JF, Eriksson KE. A fungal metabolite mediates degradation of non-phenolic lignin structures and synthetic lignin by laccase. FEBS Lett. 1996;391(1-2):144-8.

[88] Eggert C, Temp U, Eriksson KE. Laccase is essential for lignin degradation by the white-rot fungus Pycnoporus cinnabarinus. FEBS Lett. 1997 Apr 21;407(1):89-92.

[89] Bajpai P, Anand A, Bajpai PK. Bleaching with lignin-oxidizing enzymes. Biotechnology Annual Review2006. p. 349-78.

[90] Bourbonnais R, Paice MG. Enzymatic delignification of kraft pulp using laccase and a mediator. Tappi J. 1996;79(6):199-204.

[91] Call HP, Mucke I. History, overview and applications of mediated lignolytic systems, especially laccase-mediator-systems (Lignozym®-process). J Biotechnol. 1997;53(2-3): 163-202.

[92] Ruiz-Duenas F, Martınez A. Microbial degradation of lignin: How a bulky recalcitrant polymer is efficiently recycled in nature and how we can take advantage of this. Microb Biotechnol. 2009;2:164-77.

[93] Addleman K, Dumonceaux T, Paice MG, Bourbonnais R, Archibald FS. Production and characterization of Trametes versicolor mutants unable To bleach hardwood kraft pulp. Appl Environ Microbiol. 1995 10;61(10):3687-94.

[94] Hofrichter M. Review: Lignin conversion by manganese peroxidase (MnP). Enz Microb Technol. 2002;30(4):454-66.

[95] Hofrichter M, Lundell T, Hatakka A. Conversion of milled pine wood by manganese peroxidase from Phlebia radiata. Appl Environ Microbiol. 2001;67(10):4588-93.

[96] Feijoo G, Moreira MT, Alvarez P, Lú-Chau TA, Lema JM. Evaluation of the enzyme manganese peroxidase in an industrial sequence for the lignin oxidation and bleaching or eucalyptus kraft pulp. J Appl Polymer Sci. 2008;109(2):1319-27.

[97] Paice MG, Reid ID, Bourbonnais R, Archibald FS, Jurasek L. Manganese peroxidase, produced by Trametes versicolor during pulp bleaching, demethylates and delignifies kraft pulp. Appl Environ Microbiol. 1993;59(1):260-5.

[98] Hammel KE, Cullen D. Role of fungal peroxidases in biological ligninolysis. Curr Opin Plant Biol. 2008;11(3):349-55.

[99] Henriksson G, Johansson G, Pettersson G. A critical review of cellobiose dehydrogenases. J Biotechnol. 2000;78(2):93-113.

[100] Harreither W, Sygmund C, Augustin M, Narciso M, Rabinovich ML, Gorton L, et al. Catalytic properties and classification of cellobiose dehydrogenases from Ascomycetes. Appl Environ Microbiol. 2011;77(5):1804-15.

[101] Zamocky M, Ludwig R, Peterbauer C, Hallberg BM, Divne C, Nicholls P, et al. Cellobiose dehydrogenase - A flavocytochrome from wood-degrading, phytopathogenic and saprotropic fungi. Curr Protein Peptide Sci. 2006;7(3):255-80.

[102] Henriksson G, Ander P, Pettersson B, Pettersson G. Cellobiose dehydrogenase (cellbiose oxidase) from Phanerochaete chrysosporium as a wood degrading enzyme. Studies on cellulose, xylan and synthetic lignin. Appl Microbiol Biotechnol. 1995;42(5):790-6.

[103] Roy BP, Dumonceaux T, Koukoulas AA, Archibald FS. Purification and characterization of cellobiose dehydrogenases from the white rot fungus Trametes versicolor. Appl Environ Microbiol. 1996 12;62(12):4417-27.

[104] Mansfield SD, De Jong E, Saddler JN. Cellobiose dehydrogenase, an active agent in cellulose depolymerization. Appl Environ Microbiol. 1997;63(10):3804-9.

[105] Roy BP, Paice MG, Archibald FS, Misra SK, Misiak LE. Creation of metal-complexing agents, reduction of manganese dioxide, and promotion of manganese peroxidase-mediated Mn(III) production by cellobiose:quinone oxidoreductase from Trametes versicolor. J Biol Chem. 1994;269(31):19745-50.

[106] Mason MG, Nicholls P, Wilson MT. Rotting by radicals--the role of cellobiose oxidoreductase? Biochem Soc Trans. 2003;31(Pt 6):1335-6.

[107] Canam T, Town JR, Tsang A, McAllister TA, Dumonceaux TJ. Biological pretreatment with a cellobiose dehydrogenase-deficient strain of Trametes versicolor enhances the biofuel potential of canola straw. Bioresour Technol. 2011;102:10020–7.

[108] Dumonceaux T, Bartholomew K, Valeanu L, Charles T, Archibald F. Cellobiose dehydrogenase is essential for wood invasion and nonessential for kraft pulp delignification by Trametes versicolor. Enz Microb Technol. 2001;29(8-9):478-89.

[109] Phillips CM, Beeson WT, Cate JH, Marletta MA. Cellobiose dehydrogenase and a copper-dependent polysaccharide monooxygenase potentiate cellulose degradation by Neurospora crassa. ACS Chem Biol. 2011 Dec 16;6(12):1399-406.

[110] Baldrian P, Valaskova V. Degradation of cellulose by basidiomycetous fungi. FEMS Microbiol Rev. 2008;32(3):501-21.

[111] Gilkes NR, Kwan E, Kilburn DG, Miller RC, Warren RAJ. Attack of carboxymethylcellulose at opposite ends by two cellobiohydrolases from Cellulomonas fimi. J Biotechnol. 1997;57(1-3):83-90.

[112] Shallom D, Shoham Y. Microbial hemicellulases. Curr Opin Microbiol. 2003;6(3): 219-28.

[113] Crepin VF, Faulds CB, Connerton IF. Functional classification of the microbial feruloyl esterases. Appl Microbiol Biotechnol. 2004;63(6):647-52.

[114] Mathew S, Abraham TE. Ferulic acid: An antioxidant found naturally in plant cell walls and feruloyl esterases involved in its release and their applications. Crit Rev Biotechnol. 2004;24(2-3):59-83.

[115] Berlin A, Gilkes N, Kilburn D, Bura R, Markov A, Skomarovsky A, et al. Evaluation of novel fungal cellulase preparations for ability to hydrolyze softwood substrates - Evidence for the role of accessory enzymes. Enz Microb Technol. 2005;37(2):175-84.

[116] Berlin A, Maximenko V, Gilkes N, Saddler J. Optimization of enzyme complexes for lignocellulose hydrolysis. Biotechnol Bioengin. 2007;97(2):287-96.

[117] Martinez D, Larrondo LF, Putnam N, Gelpke MD, Huang K, Chapman J, et al. Genome sequence of the lignocellulose degrading fungus Phanerochaete chrysosporium strain RP78. Nat Biotechnol. 2004 Jun;22(6):695-700.

[118] Vanden Wymelenberg A, Minges P, Sabat G, Martinez D, Aerts A, Salamov A, et al. Computational analysis of the Phanerochaete chrysosporium v2.0 genome database and mass spectrometry identification of peptides in ligninolytic cultures reveal complex mixtures of secreted proteins. Fungal Genet Biol. 2006;43(5):343-56.

[119] Abbas A, Koc H, Liu F, Tien M. Fungal degradation of wood: initial proteomic analysis of extracellular proteins of Phanerochaete chrysosporium grown on oak substrate. Curr Genet. 2005;47(1):49-56.

[120] Ravalason H, Jan G, Mollé D, Pasco M, Coutinho P, Lapierre C, et al. Secretome analysis of Phanerochaete chrysosporium strain CIRM-BRFM41 grown on softwood. Appl Microbiol Biotechnol. 2008;80(4):719-33.

[121] Sato S, Feltus F, Iyer P, Tien M. The first genome-level transcriptome of the wood-degrading fungus Phanerochaete chrysosporium grown on red oak. Curr Genet. 2009;55(3):273-86.

[122] Vanden Wymelenberg A, Gaskell J, Mozuch M, Kersten P, Sabat G, Martinez D, et al. Transcriptome and secretome analyses of Phanerochaete chrysosporium reveal complex patterns of gene expression. Appl Environ Microbiol. 2009 June 15, 2009;75(12): 4058-68.

[123] Martinez D, Challacombe J, Morgenstern I, Hibbett D, Schmoll M, Kubicek CP, et al. Genome, transcriptome, and secretome analysis of wood decay fungus Postia placen-

ta supports unique mechanisms of lignocellulose conversion. Proc Natl Acad Sci USA. 2009 February 10, 2009;106(6):1954-9.

[124] MacDonald J, Doering M, Canam T, Gong Y, Guttman DS, Campbell MM, et al. Transcriptomic responses of the softwood-degrading white-rot fungus Phanerochaete carnosa during growth on coniferous and deciduous wood. Appl Environ Microbiol. 2011;77(10):3211-8.

[125] MacDonald J, Master ER. Time-dependent profiles of transcripts encoding lignocellulose-modifying enzymes of the white rot fungus Phanerochaete carnosa grown on multiple wood substrates. Appl Environ Microbiol. 2012 March 1, 2012;78(5): 1596-600.

[126] Vanden Wymelenberg A, Gaskell J, Mozuch M, Splinter BonDurant S, Sabat G, Ralph J, et al. Significant alteration of gene expression in wood decay fungi Postia placenta and Phanerochaete chrysosporium by plant species. Appl Environ Microbiol. 2011;77(13):4499-507.

[127] Levasseur A, Piumi F, Coutinho PM, Rancurel C, Asther M, Delattre M, et al. FOLy: an integrated database for the classification and functional annotation of fungal oxidoreductases potentially involved in the degradation of lignin and related aromatic compounds. Fungal Genet Biol. 2008 May;45(5):638-45.

[128] Karunanandaa K, Fales SL, Varga GA, Royse DJ. Chemical composition and biodegradability of crop residues colonized by white-rot fungi. J Sci Food Agric. 1992;60(1):105-12.

[129] Akin DE, Sethuraman A, Morrison WH, III, Martin SA, Eriksson KEL. Microbial delignification with white rot fungi improves forage digestibility. Appl Environ Microbiol. 1993;59(12):4274-82.

[130] Mahmood ur R, Ali I, Husnain T, Riazuddin S. RNA interference: The story of gene silencing in plants and humans. Biotechnol Adv. 2008;26(3):202-9.

[131] Nakayashiki H, Nguyen QB. RNA interference: roles in fungal biology. Curr Opin Microbiol. 2008;11(6):494-502.

[132] Matityahu A, Hadar Y, Dosoretz CG, Belinky PA. Gene silencing by RNA interference in the white rot fungus Phanerochaete chrysosporium. Appl Environ Microbiol. 2008;74(17):5359-65.

[133] Nakayashiki H, Hanada S, Nguyen BQ, Kadotani N, Tosa Y, Mayama S. RNA silencing as a tool for exploring gene function in ascomycete fungi. Fungal Genet Biol. 2005;42(4):275-83.

[134] Shafran H, Miyara I, Eshed R, Prusky D, Sherman A. Development of new tools for studying gene function in fungi based on the Gateway system. Fungal Genet Biol. 2008;45(8):1147-54.

[135] Kemppainen MJ, Pardo AG. pHg/pSILBAγ vector system for efficient gene silencing in homobasidiomycetes: Optimization of ihpRNA - Triggering in the mycorrhizal fungus Laccaria bicolor. Microb Biotechnol. 2010;3(2):178-200.

[136] Keller FA, Hamilton JE, Nguyen QA. Microbial pretreatment of biomass: potential for reducing severity of thermochemical biomass pretreatment. Appl Biochem Biotechnol. 2003;105 -108:27-41.

[137] Lakovlev A, Stenlid J. Spatiotemporal patterns of laccase activity in interacting mycelia of wood-decaying basidiomycete fungi. Microb Ecol. 2000;39(3):236-45.

[138] Panwar V, Prasad B, Wasewar KL. Biomass residue briquetting and characterization. J Eng Engineering. 2011;137(2):108-14.

[139] Granada E, López González LM, Míguez JL, Moran J. Fuel lignocellulosic briquettes, die design and products study. Renewable Energy. 2002;27(4):561-73.

[140] Kaliyan N, Morey RV. Natural binders and solid bridge type binding mechanisms in briquettes and pellets made from corn stover and switchgrass. Bioresour Technol. 2010;101(3):1082-90.

[141] Alaru M, Kukk L, Olt J, Menind A, Lauk R, Vollmer E, et al. Lignin content and briquette quality of different fibre hemp plant types and energy sunflower. Field Crops Res. 2011;124(3):332-9.

[142] Karunanithy C, Wang Y, Muthukumarappan K, Pugalendhi S. Physiochemical Characterization of Briquettes Made from Different Feedstocks. Biotechnol Res Int. 2012;2012:165202.

[143] Adapa PK, Tabil LG, Schoenau GJ. Compression characteristics of non-treated and steam-exploded barley, canola, oat, and wheat straw grinds. Appl Engin Agric. 2010;26(4):617-32.

[144] Ramaswamy H, Tang J. Microwave and radio frequency heating. Food Sci Technol International. 2008;14(5):423-7.

[145] Iroba K, Tabil L, editors. Densification of radio frequency pretreated lignocellulosic biomass barley straw. ASABE Annual International Meeting; 2012; St. Joseph, MI, USA: American Society of Agricultural and Biological Engineers.

[146] Kashaninejad M, Tabil LG. Effect of microwave-chemical pre-treatment on compression characteristics of biomass grinds. Biosys Engin. 2011;108(1):36-45.

Chapter 9

Microbial Degradation of Lignocellulosic Biomass

Wagner Rodrigo de Souza

Additional information is available at the end of the chapter

http://dx.doi.org/10.5772/54325

1. Introduction

The search for renewable sources of energy requires a worldwide effort in order to decrease the harmful effects of global climate change, as well as to satisfy the future energy demands. In this context, biofuels are emerging as a new source of energy derived from biomass. The production of biofuels could decrease effectively the impact of pollutants in the atmosphere, in addition to assisting in the management for tons of biomass waste. Biomass (plant matter) can be referred to "traditional biomass", which is used in inefficient ways such as the highly pollutant primitive cooking stoves (wood), and "modern biomass" that refers to biomass produced in a sustainable way and used for electricity generation, heat production and transportation of liquid fuels [1]. In addition to these definitions, The International Energy Agency (IEA) defines biomass as any plant matter that could be used directly as fuel or converted into fuels, electricity or heat. Therefore, in order to provide useful management of biomass, it is clear that one needs to learn how to extract energy from plants.

Plant cells are mainly composed by lignocellulosic material, which includes cellulose, hemicellulose and lignin (lignocellulosic complex). The hydrolysis of lignocellulose to glucose is a major bottleneck in cellulosic biofuel production processes [2]. In nature, microorganisms, especially fungi, are able to degrade the plant cell wall through a set of acting synergistically enzymes. This phenomenon leads to glucose being released in a free form, which can enter the metabolism of the microorganism, providing its energy. A great challenge is to modify the architecture of the plant cell walls and/or the ability of the microorganisms to degrade it, by modifying their genomes. For instance, researchers can generate genetically engineered microorganisms able to degrade efficiently the polymers in the plant cells, producing sugars monomers that can be fermented directly by yeasts, generating ethanol. This chapter will describe the composition of plant cell walls and how microorganisms cope with the lignocellulosic material. The main focus will be on fungi cell wall degrading enzymes (CWDEs) and their genetic regulation. The aim of this chapter is to guide scientists in order to genetically improve microorganisms that can be able to efficiently degrade the plant biomass. Also, new strains could be great producers of CWDEs, providing enzymatic cocktails that can be introduced commercially. Finally, the future perspectives will demonstrate how far we are from cellulosic ethanol and other biomass-derived chemical compounds, regarding to research of microorganisms.

2. Plant cell wall polysaccharides and lignin

Plant cell wall [6] polysaccharides are the most abundant organic compounds found in nature. These compounds consist mainly of polysaccharides such as cellulose, hemicelluloses and pectin, as well as the phenolic polymer lignin. Together, the polysaccharides and lignin provide high complexity and rigidity to the plant cell wall. Cellulose, the major constituent of plant cell wall consists of ß-1,4 linked D-glucose units that form linear polymeric chains of about 8000-12 000 glucose units. In its crystalline form, cellulose consists of chains that are packed together by hydrogen bonds to form highly insoluble structures, called microfibrils. In addition to the crystalline structure, cellulose contains amorphous regions within the microfibrils (noncrystalline structure) [3].

Hemicelluloses are heterogeneous polysaccharides consisted by different units of sugars, being the second most abundant polysaccharides in plant cell wall. Hemicelluloses are usually classified according to the main residues of sugars present in the backbone of the structural polymer. Xylan, the most abundant hemicellulose polymer in cereals and hardwood, is composed by ß-1,4-linked D-xylose units in the main backbone, and can be substituted by different side groups such as D-galactose, L-arabinose, glucuronic acid, acetyl, feruloyl, and *p*-coumaroyl residues [3]. Another two major hemicelluloses in plant cell wall are galacto(gluco)mannans, which consist of a backbone of ß-1,4-linked D-mannose (mannans) and D-glucose (glucomannans) residues with D-galactose side chains, and xyloglucans that consist of a ß-1,4-linked D-glucose backbone substituted by D-xylose [3]. Moreover, in xyloglucan polymer, L-arabinose and D-galactose residues can be attached to the xylose residues, and L-fucose can be attached to galactose residues. The diversity of side groups that can be attached to the main backbone of xyloglucans confers to this polymers high structural complexity and variability [4]. Figure 1 represents schematic views of the three major hemicellloses.

Pectins are another family of plant cell wall heteropolysaccharides, containing a backbone of α-1,4-linked D-galacturonic acid. The polymers usually contain two different types of regions. The so-called "smooth" region of pectins contains residues of D-galacturonic acids that can be methylated or acetylated. In the other region, referred as "hairy" region, the backbone of D-galacturonic acids residues is interrupted by α-1,2-linked L-rhamnose residues. In the hairy region, long side chains of L-arabinose and D-galactose residues can be attached to the rhamnose residues [5]. Figure 2 depicts a schematic representation of the hairy region of pectins.

http://dx.doi.org/10.5772/54325

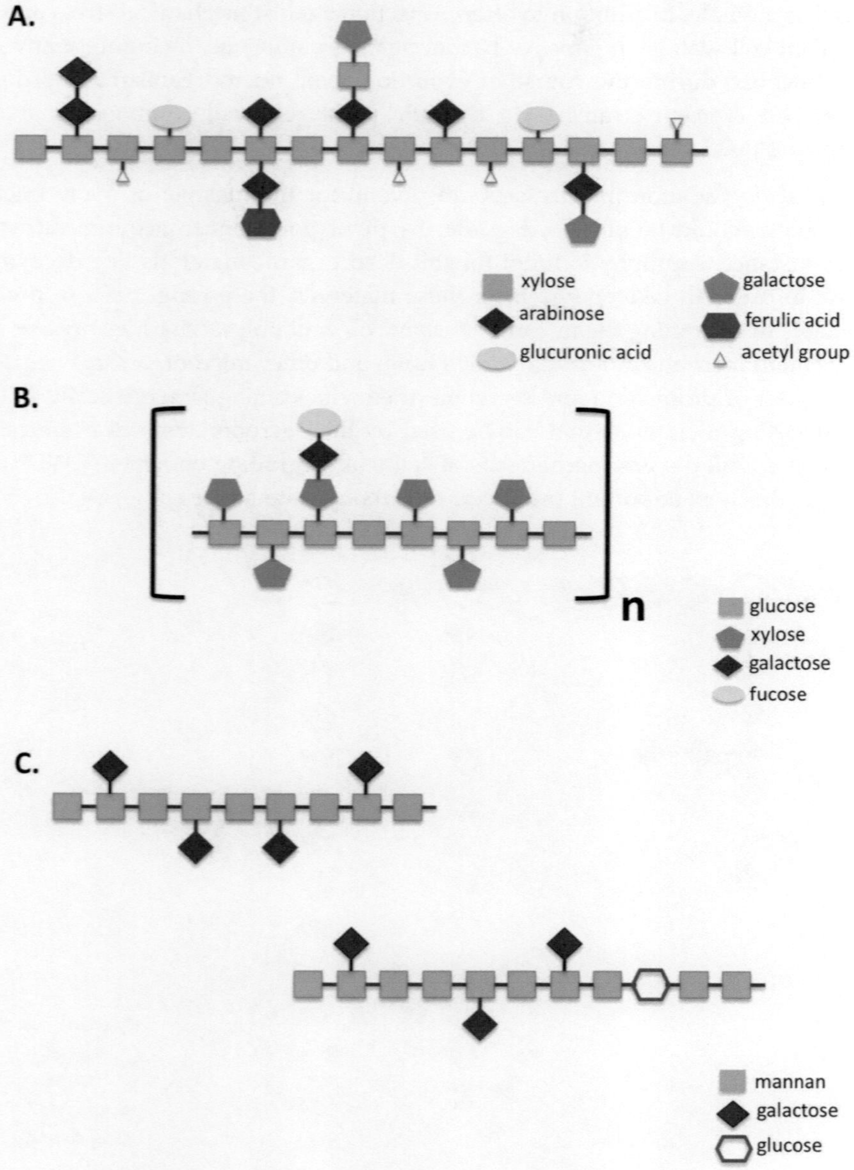

Figure 1. Schematic representation of the three major hemicellulose structures. A. Xylan. B. Xyloglucan. C. Galactomannan (upper left) and Galactoglucomannan (lower right).

Lignin is a phenolic polymer that confers strength to plant cell wall. Lignin is a highly insoluble complex branched polymer of substituted phenylpropane units, which are joined together by ether and carbon-carbon linkages, forming an extensive cross-linked network within the cell wall. The cross-linking between the different polymers described above confers the complexity and rigidity of the plant cell wall, which is responsible for the protection of plant cell as a whole. In addition to offer protection against mechanical stress and osmotic lysis, the plant cell wall is an effective barrier against pathogens, including many microorganisms. However, during the course of evolution some microorganisms, in order to survive, developed efficient strategies to degrade plant cell wall components, mainly the polysaccharides [6].

Plant cell wall degradation mechanisms are pivotal for the lifestyle of many microorganisms, once they should be able to degrade the plant polymers to acquire nutrients from plants. For instance, saprophytic fungi inhabit dead organic materials like decaying wood and leaves. In order to take energy from these materials, these fungi need to produce enzymes capable of degrading the majority of plant cell wall polysaccharides present in the biomass. The main mechanism through which fungi and other microorganisms degrade plant biomass consists of production and secretion of enzymes acting synergistically in the plant cell wall, releasing monomers that can be used by the microorganism as chemical energy. The next section will discuss mechanisms of cell wall degrading enzymes (CWDE) production by fungi, the most important producers of carbohydrate-active enzymes.

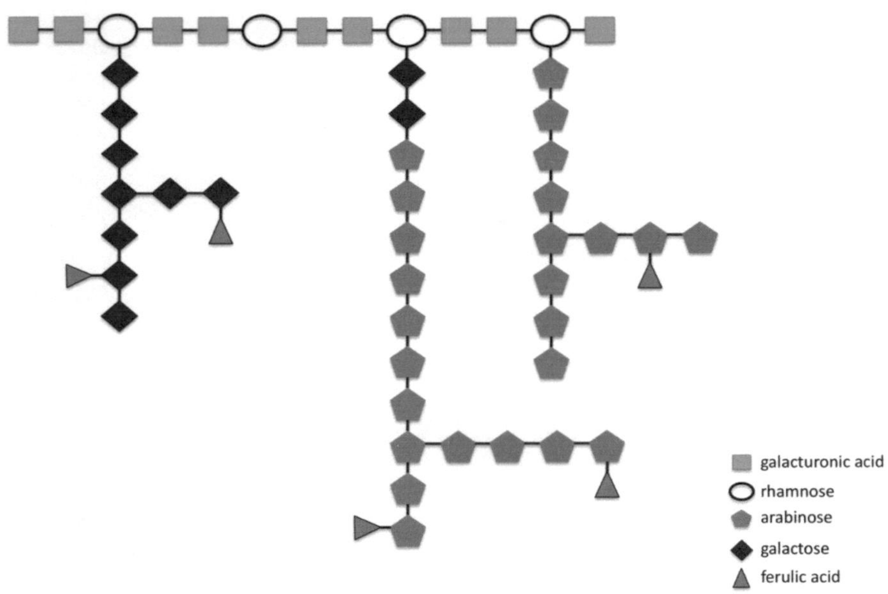

Figure 2. Schematic view of the hairy region of pectin.

3. Microbial degradation of plant cell wall polysaccharides

In order to survive, microorganisms developed, during the course of evolution, physiological mechanisms to cope with a variety of environmental factors. The acquirement of nutrients represents a challenge for all living organisms, especially for microorganisms. Saprophytism, one of the most common lifestyle of microorganisms, involves living in dead or decaying organic matter, mainly composed by plant biomass. In this context, microorganisms developed cellular mechanisms in order to take energy from plant biomass, and one of this mechanisms involves the production and secretion of carbohydrate-active enzymes. These enzymes degrade the plant cell wall, releasing sugars monomers that can be used as substrates for the metabolism of the microorganism. The microbial use of plant biomass is pivotal for life on Earth, because it is responsible for large portions of carbon flux in the biosphere. In addition, plant cell wall-degrading enzymes (CWDEs) have a broad range of industrial applications, such as within the food and feed industry and for sustainable production of many chemicals and fuels.

The capacity to degrade lignocellulose is mainly distributed among fungi and bacteria. Cellulolytic bacteria can be found in different genus such as *Clostridium, Ruminococcus, Caldicellulosiruptor, Butyrivibrio, Acetivibrio, Cellulomonas, Erwinia, Thermobifida, Fibrobacter, Cytophaga,* and *Sporocytophaga.* Bacterial degradation of cellulolytic material is more restrict to biomass containing low amounts of lignin, once bacteria are poor producers of lignanases. Plant biomass produced in aquatic environment, containing little amounts of lignin, is typically degraded by bacteria, which are better adapted for an aquatic environment than fungi [7]. Cellulolytic bacteria can also be found in digestive tracts of herbivore animals [8]. Fungal cellulose utilization is distributed within the entire kingdom, from the protist-like *Chytridomycetes* to the advanced *Basidiomycetes*.

Concerning to lignin degradation, many white-rot basidiomycetes and some actinomycetes are able to produce lignin-degrading enzymes, especially peroxidases. For instance, *Phanerochaete chrysosporium* and *Phlebia radiata* are well known producers of extracellular peroxidases [9], as well as *Coriolus tersicolor,* which was shown to produce the intracellular haem peroxidase upon the induction by phenolic compounds [10]. A white-rot basidiomycete, *Rigidoporous lignosus,* is known to secrete two oxidative enzymes, laccase and Mn peroxidase, responsible for solubilizing the lignin in a synergistic way [11].

The fungi *Hypocrea jecorina* (*Trichoderma reesei*) is the most important organism used in cellulase production [12, 13] and it has been the focus of cellulases research for over 50 years. Degradation of cellulose is performed by cellulases, a high specific class of enzymes able to degrade the cellulose glycosidic bonds. The filamentous fungi *Aspergillus niger* is known to produce a wide range of hemicellulose-degrading enzymes and it has been used for many industrial applications. As discussed above, hemicellulose is a complex class of polysaccharides composed by different units of sugars. In order to degrade hemicellulose, the organism should be able to produce a large set of enzymes (hemicellulases), acting in a synergistic way to hydrolyze such complex substrate. Therefore, the ascomycetes *T. reesei* and *A. niger* are considered the most important microorganisms for cellulase/hemicellulase production, and constitute the source of these enzymes for industrial applications, including the production of biofuels from plant biomass.

The plant cell wall-degrading machinery of aerobic and anaerobic microorganisms differs significantly, regarding to its macromolecular organization. The cellulase/hemicellulase apparatus of anaerobic bacteria is frequently assembled into a large multienzyme complex, named cellulosomes [14, 15]. This complex contains enzymes with a variety of activities such as polysaccharide lyases, carbohydrate esterases and glycoside hydrolases [16-18]. Basically, the catalytic components of the cellulosomes include a structure named dockerins, which are noncatalytic modules that bind to cohesin modules, located in a large noncatalytic protein acting as scaffold [15]. The protein-protein interaction between dockerins and cohesins allows the integration of the hydrolytic enzymes into the complex [19, 20]. It has been demostrated that scaffoldins are also responsible for the anchoring of the whole complex onto crystalline cellulose, through a noncatalytic carbohydrate-binding module (CBM) [21]. The main studies concerning cellulosomes are being focused on anaerobic bacteria, especially from *Clostridium* species, but a range of other anaerobic bacteria and fungi were shown to produce cellulosomal systems. These include anaerobic bacteria such as *Acetivibrio cellulolyticus, Bacteroides cellulosolvens, Ruminococcus albus, Ruminococcus flavefaciens,* and the anaerobic fungi of the genera *Neocalimastix, Pyromices* and *Orpinomyces* [14, 15]. Cellulosome-based complexes design and construction is a promising approach for the improvement of hydrolytic activity systems. Cellulosomes able to integrate fungal and bacterial enzymes from nonaggregating systems could be generated to increase hydrolytic activities and consequently the biomass saccharification [22]. In addition, genetic manipulations could be used in order to introduce genes responsible for the synthesis of cellulosome into microorganisms able to ferment simple sugars but that do not have a functional plant cell wall-degrading machinery [23]. Alternatively, microorganisms naturally synthesizing cellulosomes could be engineered to increase their capacity to produce ethanol from lignocellulose [15]. Recently, using the architeture of cellulosomes as template, self-assembling protein complexes were successfully designed and constructed. These protein complexes were termed xylanosomes, and were designed specifically for hemicellulose hydrolysis, but demonstrated synergy with cellulases, suggesting a possible use of these nanostructures in cellulose hydrolysis as well [24].

4. Ethanol production from the fermentable feedstock from lignocellulosic biomass

Fermentative production of ethanol is largely performed nowadays through the use of starch or sucrose provided by agricultural crops such as wheat, corn or sugarcane. In Brazil, for instance, the ethanol production through yeast fermentation of substrates from sugarcane is a well-known and consolidated process. However, the improvement of fermentative processes towards utilization of lower-value substrates such as lignocellulosic residues is emerging as a valuable approach for reducing the production cost and consequently increasing the use of ethanol as biofuel. In sugarcane mills, for instance, a large quantity of sugarcane bagasse, which is a great source of lignocellulosic residue, is produced as a by-product of the industrial process. The sugarcane bagasse can be used as a lower-value substrate to produce the so-called second generation ethanol, in other words the ethanol generated from lignocellulosic material. The conversion of lignocellulose to

http://dx.doi.org/10.5772/54325

ethanol requires challenging biological processes that includes: (i) delignification in order to release free cellulose and hemicellulose from the lignocellulosic material; (ii) depolymerization of the carbohydrates polymers from the cellulose and hemicellulose to generate free sugars; and (iii) fermentation of mixed hexose and pentose sugars to finally produce ethanol [25]. Glucose presents approximately 60% of the total sugars available in cellulosic biomass. The yeast *Saccharomices cerevisiae* is the most important microorganism able to ferment glucose (hexose), generating ethanol [26]. However, the presence of pentose sugars such as xylose and arabinose represents a challenge for the fermentation of these sugars in lignocellulosic biomass, once *S. cerevisiae* is not able to efficiently ferment C5 sugars. The naturally occurring microorganisms able to ferment C5 sugars include *Pichia stipitis, Candida shehatae,* and *Pachysolen tannophilus* [27]. From these microorganisms, the yeast *P. stipitis* has the highest ability to perform xylose fermentation, producing ethanol under low aeration rates. It appears that ethanol yields and productivity from xylose fermentation by *P. stipitis* are significantly lower than glucose fermentation by *S. cerevisiae* [28]. Therefore, genetic improvement of yeasts is a valuable tool to obtain strains able to ferment pentoses, hexoses and, in addition, produce ethanol with a high yield and a high ethanol tolerance as well. Genetically engineered organisms with C5 fermenting capabilities already include *S. cerevisiae, Escherichia coli, Zymomonas mobilis* and *Candida utilis* [28-31]. Studies on fungi degradation of lignocellulosic material could yield promising candidate genes that could be subsequently used in engineering strategies for improved cellulosic biofuel production in these yeast strains.

In summary, many microorganisms are able to produce and secrete hemicellulolytic enzymes, but fungi are pointed as the most important microorganisms concerning the biomass degradation. The significance of secreted enzymes in the life of these organisms and the biotechnological importance of filamentous fungi and their enzymes prompted an interest towards understanding the mechanisms of expression and regulation of the extracellular enzymes, as well as the characterization of the transcription factors involved. The next sections of this chapter will discuss the fungal enzyme sets for lignocellulosic degradation and the gene expression regulation of these enzymes.

5. Fungal enzyme sets for lignocellulosic degradation

Fungi play a central role in the degradation of plant biomass, producing an extensive array of carbohydrate-active enzymes responsible for polysaccharide degradation. The enzyme sets for plant cell wall degradation differ between many fungal species, and our understanding about fungal diversity with respect to degradation of plant matter is essential for the improvement of new strains and the development of enzymatic cocktails for industrial applications.

Carbohydrate-active enzymes are usually classified in different families, based on amino acid sequence of the related catalytic module. An extensive and detailed database presenting these hydrolytic enzymes can be found at www.cazy.org (CAZymes, Carbohydrate-

Active En**Z**ymes) [32]. The fungal carbohydrate-active enzymes commonly (but not always) present a carbohydrate-binding module (CBM), which promotes the association of the enzyme with the substrate. A review on fungal enzymes related to plant biomass degradation describes that such enzymes are assigned to at least 35 glycoside hydrolase (GH) families, three carbohydrate esterase (CE) families, and six polysaccharide lyase (PL) families [33]. Although the classification of CWDEs into families facilitates our view about a specific enzyme, the activities of these enzymes are quite complicated to classify, because some families can contain several enzymatic activities. This is especially important because CAZymes usually act in a synergistic way, complementing the substrate specificity of each other, in order to degrade complex polysaccharide matrices. For instance, GH5 comprises many catalytic activities, such as endoglucanases, exoglucanases and endomannanases [34]. This section will describe the fungal enzymatic set required for the main polysaccharides present in the plant biomass: cellulose, hemicellulose and pectin. A brief description of enzymes required for lignin degradation will be depicted. Because most of the research in cellulase/hemicellulase field is performed using the fungi *T. reesei* and *A. niger*, the focus of our discussion about CWDEs will be conducted based on these microorganisms, although some aspects related to other fungal species could be mentioned to demonstrate the diversity of carbohydrate-active enzymes.

6. Cellulose degradation

Cellulose, a polysaccharide consisted of linear β-1,4-linked D-glucopyranose chains, requires three classes of enzymes for its degradation: β-1,4-endoglucanases (EGL), exoglucanases/cellobiohydrolases (CBH), and β-glucosidase (BGL). The endoglucanases cleave cellulose chains internally mainly from the amorphous region, releasing units to be degraded by CBHs and/or BGLs. The cellobiohydrolases cleave celobiose units (the cellulose-derived disaccahride) from the end of the polysaccharide chains [6]. Finally, β-glucosidases hydrolise cellobiose to glucose, the monomeric readily metabolisable carbon source for fungi [35]. These three classes of enzymes need to act synergistically and sequentially in order to degrade completely the cellulose matrix. After endo- and exo-cleaving (performed by EGLs and CBHs, respectively), the BGLs degrade the remaining oligosaccharides to glucose. A schematic view of cellulose degradation is depicted in the Figure 3.

The most efficient cellulose-degrading fungi is *Trichoderma reesei*. The highly efficient degradation of cellulose by *T. reesei* is mainly due to the highly effectiveness of cellulases acting synergistically in this specie, although *T. reesei* does not have the biggest number of cellulases in the fungi kingdom [36]. The *T. reesei* has five characterized EGLs, two highly expressed CBHs and two characterized BGLs, the latter being expressed at low levels [37, 38] reviewed in [33]. In addition to being expressed at very low levels in *T. reesei*, the BGLs are strongly subjected to product inhibition [39]. These features reduce the utilization of *T. reesei* for in vitro saccharification of cellulose substrates and, in industrial applications, cellulase mixtures from *T. reesei* are often supplemented with BGLs from Aspergilli, which are highly expressed and tolerant to glucose inhibition [33].

http://dx.doi.org/10.5772/54325

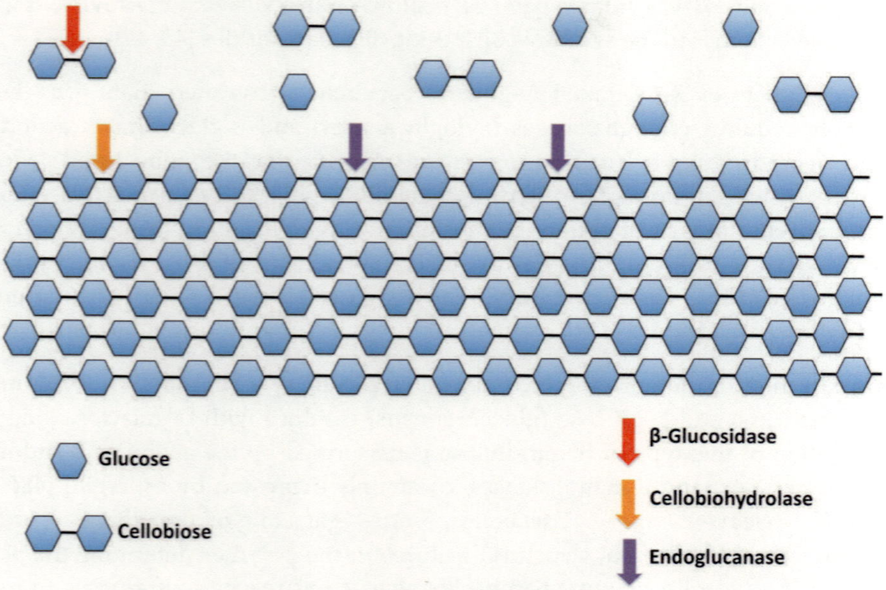

Figure 3. Schematic view of cellulose degradation. Endoglucanases hydrolise cellulose bonds internally, while cellobiohydrolases cleave celobiose units from the ends of the polysaccharide chains. The released cellobiose units (disaccharide) are finally hydrolyzed by β-glucosidases, releasing glucose, the main carbon source readily metabolisable by fungi.

7. Hemicellulose degradation

Hemicellulose is a complex polysaccharide matrix composed of different residues branched in three kinds of backbones, named xylan, xyloglucan and mannan. The complexity of hemicellulose requires a concerted action of endo-enzymes cleaving internally the main chain, exo-enzymes releasing monomeric sugars, and accessory enzymes cleaving the side chains of the polymers or associated oligosaccharides, leading to the release of various mono- and disaccharides depending on hemicellulose type.

Xylan, a polymer composed by ß-1,4-linked D-xylose units, is degraded through the action of ß-1,4-endoxylanase, which cleaves the xylan backbone into smaller oligosaccharides, and ß-1,4-xylosidase, which cleaves the oligosaccharides into xylose. Fungal ß-1,4-endoxylanase are classified as GH10 or GH11 [40], differing from each other in substrate specificty [41]. Endoxylanases belonging to family GH10 usually have broader substrate specificity than endoxylanases from family GH11 [33]. GH10 endoxylanases are known to degrade xylan backbones with a high degree of substitutions and smaller xylo-oligosaccharides in addition to degrade linear chains of 1,4-linked D-xylose residues. Thus, GH10 endoxylanases are necessary to degrade completely substituted xylans [42]. ß-Xylosidases are highly specific for small unsubstituted xylose olygosaccharides and they are important for the complete degradation of xylan. Some ß-xylosidases have been shown transxylosylation activity, suggesting a role for these enzymes in the synthesis of specific oligosaccharides [43, 44].

Xyloglucan consists of ß-1,4-linked D-glucose backbone substituted mainly by D-xylose and therefore requires endoglucanases (xyloglucanases) and ß-glucosidases action in order to be degraded. Some endoglucanases are specific for the substituted xyloglucan backbone, and they are not able to hydrolise cellulose [45]. Xyloglucan-active endoglucanases have specific modes of action. For instance, a xyloglucanase from *T. reesei* cleaves at branched glucose residues, whereas the GH12 xyloglucanase from *A. niger* prefers xylogluco-oligosaccharides containing more than six glucose residues with at least one non-branched glucose residue [46]-[47].

Mannans, also referred to galacto(gluco)mannans, consist of a backbone of ß-1,4-linked D-mannose (mannans) and D-glucose (glucomannans) residues with D-galactose side chains. The degradation of this type of hemicellulose is performed by the action of ß-endomannanases (ß-mannanases) and ß-mannosidases, commonly expressed by aspergilli [48]. The ß-mannanases cleave the backbone of galacto(gluco)mannans, releasing mannooligosaccharides. Several structural features in the polymer determine the ability of ß-mannanases to hydrolise the mannan backbone, such as the ratio of glucose to mannose and the number and distribution of substituents on the backbone [49]. It has been shown that ß-mannanase is most active on galactomannans with a low substitution of the backbone [50], and the presence of galactose residues on the mannan backbone significantly prevents ß-mannanase activity [51]. The main products of ß-mannanase activity on mannan are mannobiose and mannotriose. ß-Mannosidases act on the nonreducing ends of mannooligosaccharides, releasing mannose. As shown by substrate specificity studies, ß-Mannosidase is able to completely release terminal mannose residues when one or more adjacent unsubstituted mannose residues are present [52].

The complete degradation of hemicellulose is only achieved after release of all substitutions present on the main backbone. The high degree of substitution in the hemicellulose polymers requires the action of various accessory enzymes able to release all these substitutions from the polysaccharide. At least nine different enzyme activities distributed along 12 GH and 4 CE families are required to completely degrade the hemicellulose substituents [33].

Arabinose is one of the most common sugar residues in hemicellulose and is present in arabinose-substituted xyloglucan and (arabino-)xylan. The release of arabinose from the polymer is performed by α-arabinofuranosidases and arabinoxylan arabinofuranohydrolases. α-Arabinofuranosidases are mainly found in GH 51 and 54 families, and the differences in the substrate specificity between these enzymes could be exemplified by two arabinofuranosidases of *A. niger*, AbfA and AbfB. AbfA (GH 51) releases L-arabinose from arabinan and sugar beet pulp, while AbfB (GH 54) also releases L-arabinose from xylan [48]. The arabinoxylan arabinofuranohydrolases act in the L-arabinose residues of arabinoxylan, specifically against the α-1,2- or α-1,3-linkages [53]. Moreover, arabinoxylan arabinofuranohydrolases appear to be sensitive to the substitutions of adjacent D-xylose residues. AxhA, an arabinox-

http://dx.doi.org/10.5772/54325

ylan arabinofuranohydrolase from *A. niger*, is not able to release arabinobiose from xylan or substitued L-arabinose from D-xylose residues adjacent to D-glucuronic acid residues [54].

Another type of substituent present in hemicellulose is D-xylose. Hydrolases responsible for the release of D-xylose residues from the xyloglucan backbone are referred to α-xylosidases. These enzymes can differ with respect to the type of glycoside they can hydrolize. For instance, α-xylosidase II (AxhII) from *Aspergillus flavus* hydrolyzes xyloglucan oligosaccharides and AxhIII is most active on *p*-nitrophenyl α-L-xylose residues and does not hydrolyze xyloglucan [55, 56].

There are many other possible substituents in hemicellulose, such as L-fucose, α-linked D-galactose, D-glucuronic acid, acetyl group and *p*-coumaric and ferulic acids (Figure 1). A list containing the respective fungal accessory enzyme responsible for the release of each of these residues is shown in Table 1. An overview of fungal enzymatic complex for hemicellulose degradation is shown in Figure 4.

Hemicellulose polymer	Residue released	Enzyme
Xyloglucan/xylan	L-arabinose	α-arabinofuranosidases
		arabinoxylan arabinofuranohydrolases
Xyloglucan	D-xylose	α-xylosidases
Xyloglucan	L-fucose	α-fucosidases
Xylan/galactomannans	D-galactose	α-galactosidases
Xylan	D-glucuronic acid	α-glucuronidases
Xylan	acetyl group	acetyl xylan esterases
Xylan	*p*-coumaric acid	*p*-coumaroyl esterases
Xylan	ferulic acid	feruloyl esterases

Table 1. Fungal accessory enzymes for the cleavage of hemicellulose-derived residues.

8. Pectin degradation

Pectins are composed of a main backbone of α-1,4-linked D-galacturonic acid, and consist of two regions: the "smooth" region and the "hairy" region. The "smooth" region contains residues of D-galacturonic acids that can be methylated or acetylated, while in the "hairy" region, the backbone of D-galacturonic acids residues is interrupted by α-1,2-linked L-rhamnose residues. Moreover, in the hairy region, long side chains of L-arabinose and D-galactose residues can be attached to the rhamnose residues (Figure 2). As observed for cellulose and hemicellulose, degradation of pectins also requires a set of hydrolytic enzymes to degrade completely the polymer. Glycoside hydrolases (GHs) and polysaccharide lyases (PLs) are the two classes of hydrolytic enzymes required for pectin backbone degradation.

The GHs involved in pectin backbone degradation include endo- and exo-polygalacturonases, which cleave the backbone of smooth regions, while the intricate hairy regions are further degraded by endo- and exo-rhamnogalacturonases, xylogalacturonases, α-rhamnosidases, unsaturated glucuronyl hydrolases, and unsaturated rhamnogalacturonan hydrolases [33]. Endo- and exo-polygalacturonases are able to cleave α-1,4-glycosidic bonds of α-galacturonic acids. Rhamnogalacturonases cleave α-1,2-glycosidic bonds between D-galacturonic acid and L-rhamnose residues in the hairy region of the pectin backbone [57]. An endo-xylogalacturonase from *Aspergillus tubingensis* has been shown to cleave the xylose-substituted galacturonic acid backbone [58]. The other GHs required for the degradation of main chain of pectin, α-rhamnosidases, unsaturated glucuronyl hydrolases, and unsaturated rhamnogalacturonan hydrolases, are not well-characterized biochemically [33].

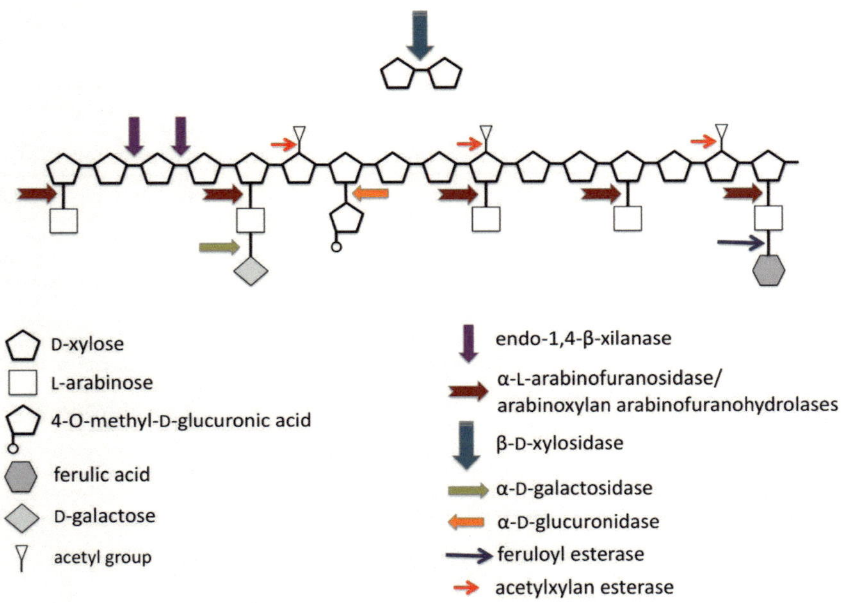

Figure 4. Schematic view on a hemicellulolytic system, degradation of arabinoxylan is depicted. The arrows represent each enzyme active for a determined substrate.

The fungal PLs pectin and pectate lyases hydrolyze α-1,4-linked D-galacturonic acid residues in the smooth regions of pectin backbone [59]. Pectin lyases have preference for substrates with a high degree of methylesterification, whereas pectate lyases prefer substrates with a low degree of esterification. Moreover, pectate lyases require Ca^{2+} ions for catalysis while pectin lyases lack such ion requirement to catalysis [60]. The PL rhamnogalacturonan lyase cleaves within the hairy region of pectin and appears to be structurally different from pectin and pectate lyases. As presented by nailing reviews [33, 48], the pectin structures xylogalacturonan and rhamnogalacturonan require a repertoire of accessory enzymes to remove the side chains, providing access for the main-chain pectinolytic enzymes. The accessory enzymes endoarabinanases, exoarabinanases, β-endogalactanases, and several esterases are specific for pectin degradation, while α-arabinofuranosidases, β-galactosidases, and β-xylosidases are also required for hemicellulose degradation.

9. Lignin degradation

Lignin, a highly insoluble complex branched polymer of substituted phenylpropane units joined by carbon-carbon and ether linkages, provides an extensive cross-linked network within the cell wall, and it is known to increase the strength and recalcitrance of the plant cell wall. Microbial lignin degradation is often complicated, once the microbe needs to cope with three major challenges related to lignin structure: (i) enzymatic system to degrade the lignin polymer needs to be essentially extracellular, because lignin is a large polymer, (ii) the mechanism of enzymatic degradation should be oxidative and not hydrolytic, since the lignin structure comprises carbon-carbon and ether bonds, and (iii) lignin stereochemistry is irregular, requiring enzymes with less specificity than hydrolytic enzymes required for cellulose/hemicellulose degradation [61]. The most well characterized enzymes able to degrade the lignin polymer are lignin peroxidase (LiP), laccase (Lac), manganese peroxidase (MnP), versatile peroxidase, and H_2O_2-generating enzymes such as glyoxal oxidase (GLOX) and aryl alcohol oxidase (AAO).

Lignin and manganese peroxidases (LiP and MnP, respectively) catalyse a variety of oxidative reactions dependent on H_2O_2. LiP oxidizes non-phenolic units of lignin (mainly Cα-Cβ bonds) by removing one electron and creating cation radicals that decompose chemically [62]. MnPs differ significantly from LiPs, once they cannot oxidize directly non-phenolic lignin-related structures [63]. In order to oxidize non-phenolic lignin-related components, the oxidizing power of MnPs is transferred to Mn^{3+}, a product of the MnP reaction: 2 Mn(II) + $2H^+ + H_2O_2 \rightarrow$ 2 Mn(III) + $2H_2O$ [64]. In this way, Mn^{3+} diffuses into the lignified cell wall, attacking it from the inside [63].

Laccases oxidize phenolic compounds and reduce molecular oxygen to water. Lac catalyses the formation of phenoxyl radicals and their unspecific reactions, leading finally to Cα-hydroxyl oxidation to ketone, alkyl-aryl cleavage, demethoxylation and Cα-Cβ cleavage in phenolic substructures [61]. Versatile peroxidases (VPs) are able to oxidize phenolic and non-phenolic aromatic compounds, as well as Mn^{2+} [64].

In order to degrade lignin, microbes require sources of extracellular H_2O_2, to support the oxidative turnover of LiPs and MnPs responsible for ligninolysis. The hydrogen peroxide is provided by extracellular oxidases that reduce molecular oxygen to H_2O_2, with the synergistic oxidation of a cosubstrate. The most well characterized extracellular H_2O_2-generating enzymes are glyoxal oxidase (GLOX) and aryl alcohol oxidase (AAO).

Most studies on enzymatic lignin degradation rely on white-rot fungi, which can mineralise lignin to CO_2 and H_2O in pure cultures [65, 66]. Among these fungi are *Phanero-*

chaete chrysosporium, *Ceriporiopsis subvermispora*, *Phlebia subserialis*, and *Pleurotus ostreatus*, which are able to metabolize the lignin in a variety of lignocellulosic biomass [62, 67, 68]. In addition, other species of fungi, such as *Postia placenta* (a brown-rot fungus), and some bacteria (such as *Azospirillum lipoferum* and *Marinomonas mediterranea*), are able to metabolize lignin. The saprotrophic homobasidiomycete *Pycnoporus cinnabarinus* is recognized by its high lignocellulolytic potential [69] overproducing high redox potential laccase, and a variety of studies have been performed in order to increase the ability of this specie to produce laccases for biotechnological applications, including heterologous expression in other species such as *A. niger* [70, 71, 72]. In addition, white-rot fungi such as *Cyathus cinnabarinus* and *Cyathus bulleri* demonstrated potential to degrade lignin [73, 74].

In summary, microbial degradation of lignocellulosic material requires a concerted action of a variety of enzymes arranged in an enzymatic complex, depending on the biomass to be degraded. The gene expression, production and secretion of plant cell wall-degrading enzymes demand energy from the microbial cells and therefore the overall process is highly regulated. There is an intense cross-talk in induction of expression of the genes encoding different classes of enzymes. The control of the regulation of CWDEs production could be the key for the development of new microbial strains that efficiently produce and secrete CWDEs. The regulation of genes encoding polysaccharide-degrading enzymes will be the subject of the next section of this chapter.

10. Regulation of cell-wall degrading enzymes production in fungi

The production of CWDEs by fungi is an energy-consuming process. The fine-tuned regulation of genes encoding CWDEs ensures that these enzymes will be produced only under conditions in which the fungus requires plant polymers as carbon source. Readily metabolizable carbohydrates repress the synthesis of enzymes related to catabolism of alternative carbon sources such as plant cell wall polysaccharides. In this way, preferential utilization of the most favored carbon source prevails, and one of the regulatory mechanisms involved in this adaptation is carbon catabolite repression (CCR). The CCR is activated by many carbon sources, depending on the lifestyle of the microorganism, but usually glucose is the most repressive molecule [75]. Nowadays, the search for microorganisms able to efficiently degrade lignocellulosic biomass is pivotal for the establishment of sustainable production of biomass-derived ethanol and other biocompounds. In this context, CCR appears as a major challenge to overcome, once this mechanism is responsible for enzymatic exclusion of less preferred carbon source such as lignocellulose-derived sugars. Hence, the comprehension of molecular mechanisms behind CCR, as well as the transcriptional control of cell wall derived enzymes are prerequisite in order to develop new microbial strains for lignocellulose degradation. In this section, the induction of expression of cellulases and hemicellulases, the transcriptional control of genes encoding CWDEs and the overall mechanism behind CCR will be discussed.

http://dx.doi.org/10.5772/54325

11. Induction of cellulases

Although the biochemistry of the process behind lignocellulosic degradation has been studied in detail, the mechanism by which filamentous fungi sense the substrate and initiate the overall process of hydrolases production is still unsolved. Some researchers have been proposed that a low constitutive level of cellulase expression is responsible for the formation of an inducer from cellulose, amplifying the signal. Another group of scientists suggest that the fungus initiates a starvation process, which could in turn activates cellulase/hemicellulase expression. Also, it is possible that an inducing sugar derived from carbohydrates released somehow from the fungal cell wall could be the derepressing molecules for hydrolase induction. Despite of the fact that the true mechanism behind natural cellulase/hemicellulase induction is still lacking, some individual molecules are known to induce these hydrolases.

The fungus *Trichoderma ressei* is an impressive producer of cellulases and most of studies concerning the regulation of cellulase genes have been performed in this specie. The most powerful inducer of cellulases in *T. reesei* is sophorose, a disaccharide composed of β-1,2-linked glucose units. Sophorose appears to be formed from cellobiose through transglycosylation activity of β-glucosidase [76 - 78]. In addition to *T. reesei,* sophorose is known to induce cellulase expression in *A. terreus* and *P. purpurogenum* [79 - 80].

Cellobiose (two β-1,4-linked glucose units) appears to induce cellulase expression in many species of fungi. Cellobiose is formed as the end product of cellobiohydrolases activity, and it has been show to induce cellulase expression in *T. reesei, Volvariella volvacea, P. janthinellum* and *A. nidulans* [81 - 84]. However, studies concerning the inducing effect of cellobiose on cellulase expression are controversial [6]. For instance, cellobiose can be transglycosylated by β-glucosidases, producing sophorose, which could be the true inducer of cellulases. Besides, β-glucosidases are able to cleave the cellobiose into glucose, which may cause repression by CCR. Therefore, the outcome in cellobiose cultures appears to be dependent on the balance between hydrolysis and transglycosylation, as well as the subsequent uptake of the generated sugars and the intracellular signals they initiate.

Lactose (1,4-*O*- β-D-galactopyranosyl-D-glucose) is a disaccharide that has been shown economically viable to induce cellulase expression in *T. reesei.* Interestingly, lactose is not a component of plant cell wall polymers and the mechanism through which this sugar induces cellulase expression appears to be complex. In filamentous fungi, lactose is cleaved by extracellular β-galactosidase into glucose and galactose. Lactose induction of cellulase genes requires the β-anomer of D-galactose, which can be converted to fructose by an alternative pathway in addition to the Leloir pathway [85]. In this alternative pathway, D-xylose reductase (encoded by *xyl1*) is the enzyme catalyzing the first step [86].

Moreover, induction of cellulase genes could be achieved in *T. reesei* cultures after addition of various other oligosaccharides such as laminaribiose, gentiobiose, xylobiose, L-sorbose and δ-cellobiono-1,5-lactone. L-arabitol and different xylans also have been show to induce expression of cellobiohydrolase 1 (*cbh1*) in *T. reesei* (reviewed in reference [6].

12. Induction of hemicellulases and pectinases

Hemicellulase expression has been studied mostly in Aspergilli and *T. reesei*. However, a comparison of the genome sequences of *T. reesei* [87] and *Aspergillus niger* [88] demonstrated that *A. niger* is more versatile in the range of hemicellulases, and therefore this specie has become a very useful model fungus for basic studies on CWDEs in recent years. It is known that the presence of the hemicelluloses xylan, xyloglucan, arabinan and mannan usually induces a high production of hemicellulases, though the mechanism of sensing is still lacking, as assigned for cellulase induction. Usually, small hemicellulose-derived molecules are able to induce the expression of a wide range of hemicellulases.

The monosaccharide D-xylose is a well-known inducer of xylanolytic enzymes in *Aspergillus* species. In *A. niger*, D-xylose appears to induce other hydrolase genes rather than xylanase genes, such as the accessory enzymes α-glucuronidase (*aguA*), acetylxylan esterase (*axeA*) and feruloyl esterase (*faeA*) [89, 90]. Some results have been demonstrated that xylose can act as a repressor carbon source of hemicellulase induction at high concentrations [91], whereas other studies demonstrated that utilization of a high D-xylose concentration was beneficial for the induction of hemicellulase-encoding genes [92]. In addition to xylose, xylobiose and D-glucose-β-1,2-D-xylose have been demonstrated to induce expression of xylanolytic genes in *A. terreus* [80].

The genes encoding enzymes responsible for the degradation of arabinoxylan in *A. niger* were induced by arabinose and L-arabitol. These genes encode enzymes such as arabinofuranosidases (*afbA* and *afbB*) and arabinoxylan arabinofuranosidases (*axhA*) [6]. High intracellular accumulation of L-arabitol in *A. nidulans* mutant strains was able to induce higher amounts of arabinofuranosidases and endoarabinases than in the wild type strain [93]. The arabinolytic system appears to be independent from xylanolytic system in *A. niger*, as demonstrated by the isolation of *ara* mutants of *A. niger* defective for the induction of the genes encoding arabinolytic enzymes, but not for the induction of enzymes belonging to xylanolytic system [94].

Regarding to pectinolytic enzymes, D-galacturonic acid, polygalacturonate and sugar beet pectin have been shown to induce virtually all the genes encoding for pectin degradation enzymes in *A. niger*, suggesting that these genes are under the control of a general pectinolytic regulatory system responding to D-galacturonic acid or a metabolic product derived from it in *A. niger* [48, 90].

In *T. ressei*, the induction of hemicellulases was observed during growth in the presence of cellulose, xylan, sophorose, xylobiose and L-arabitol [6]. A xylanase gene (*xyn2*) was induced by sophorose and xylobiose [95, 96]. Xylobiose was able to induce genes involved in xylan degradation in *T. reesei*, such as the xylanase genes *xyn1* and *xyn2* and the β-xilosidase gene *bxl1*. In addition, sophorose induced some genes encoding enzymes that cleave the side chains of xylan such as *agl1* and *agl2* (α-galactosidase genes), and *glr1*, encoding a α-glucuronidase gene [97].

Complex mixtures of polysaccharides have been shown to induce a wide range of cellulases/ hemicellulases genes in *A. niger*. For instance, sugarcane bagasse, a by-product of sugar/

ethanol factories, was able to induce a variety of cell-wall degrading enzyme genes in *A. niger* [98]. This is especially important for the development of second-generation ethanol (cellulosic ethanol) technology in countries producing ethanol from sugarcane, which could be able to use a very cheap raw material to produce the biofuel.

13. Induction of ligninases

The ligninolytic system of many fungi appears to be induced under nutrient deprivation, mainly nitrogen, carbon and sulphur. Therefore, the expression of most of the ligninolytic genes is regarded as a stress response to nutrient deprivation. Also, the presence of Mn(II) is required for induction of manganese peroxidase (MnPs) genes in the white-rot fungi *P. chrysosporium* [99 - 101]. Besides nutrient depletion and the presence of Mn(II), MnP genes were found to be expressed under heat shock in nitrogen limited cultures [102], and in the presence of H_2O_2, chemical stress or molecular oxygen [103]. Similarly, lignin peroxidase (LiP) genes are also expressed under carbon and nitrogen limitation in *P. chrysosporium* [104].

Laccases are multicopper oxidase proteins, and therefore can be induced by copper, although other metals can induce the expression of laccase genes as well, such as manganese and cadmium [105, 106]. Many natural and xenobiotic aromatic compounds, which are often related to lignin or humic substances, were shown to induce genes related to laccases [107]. In general, it has been postulated that laccases are the first enzymes degrading lignin, and possible further degradation products released from the polymer could act as inducers to amplify laccase expression, and subsequently induce other ligninolytic genes [105].

14. Transcription factors involved in the expression of cellulase and hemicellulase-encoding genes

A number of genes encoding plant cell wall degrading enzymes appears to present in their promoter regions regulatory elements for binding of transcriptional activators. The filamentous fungi *A. niger* and *T. reesei* have been the subject of many studies regarding the mechanism of transcriptional regulation of cellulase- and hemicellulase-encoding genes. Also, genes responsible for the expression of transcription factors found in *A. niger* and *T. reesei* have their homologs in other *Aspergilli* species and *Neurospora crassa*. In this section, the main transcription regulators involved in plant-polysaccharide degradation found in fungi are presented.

15. The transcriptional activator XlnR

Complementation by transformation of an *A. niger* mutant lacking xylanolytic activity led to the isolation of *xlnR* gene, a gene encoding the first known transcriptional activator controlling the expression of genes for xylanolytic and cellulolytic enzymes in filamentous fungi [108]. Initially, it was thought that XlnR was able to regulate the expression of the xylanolytic genes encoding two endoxylanases (*xlnB* and *xlnC*) and a β-xylosidase (*xlnD*), and the transcription of genes encoding some accessory enzymes involved in xylan degradation including α-glucuronidase A (*aguA*), acetylxylan esterase A (*axeA*), arabinoxylan arabinofuranohydrolase A (*axhA*), and feruloyl esterase A (*faeA*). Furthermore, XlnR has been found to activate the transcription of two endoglucanase-encoding genes, *eglA* and *eglB*, indicating that transcriptional regulation by XlnR includes cellulase-encoding genes [109]. Currently, it is known that XlnR actually controls the transcription of about 20-30 genes encoding hemicellulases and cellulases, and a gene encoding a D-xylose reductase (*xyrA*), involved in intracellular D-xylose metabolism [110]. These results demonstrate an interconnection of extracellular xylan degradation and intracellular D-xylose metabolism, coupled via transcriptional regulation of *xyrA* gene by XlnR. In this way, the fungus is able to adapt intracellular D-xylose metabolism to extracellular xylan degradation, indicating a high level of metabolic regulation. Based on these findings, a model was proposed to explain the activation of XlnR regulon [111]. Basically, carbon limitation minimize carbon catabolite repression, and this de-repressed condition favours monomeric sugars or their derivatives acting as inducers of cellulolytic/hemicellulolytic system. The nature of the monomeric sugar drives the polysaccharide enzyme system to be induced [110]. Figure 5 shows a schematic model for the regulation through XlnR of genes encoding CWDEs in *A. niger*. Besides *A. niger*, the gene encoding XlnR has also been isolated from *A. oryzae*, where the corresponding protein AoXlnR demonstrated to control the expression of the xylanase-encoding genes *xynF1* and *xynF2* [112].

The XlnR transcriptional activator belongs to the class of zinc binuclear cluster domain proteins (PF00172) [113]. The DNA-binding domain is found in the XlnR at the N-terminal of the protein and, in addition to this domain, a fungal specific transcription factor domain is also present (PF04082) [110]. Functional studies have been demonstrated that a putative coiled-coil domain is important for the XlnR function, as the disruption of the α-helix structure (Leu650Pro mutation) lead to cytoplasmic localization and loss of function of XlnR, due to a loss of transcription of the structural genes of the regulon [114]. As demonstrated by the same study, a C-terminal portion of XlnR appeared to be involved in transcriptional regulation, as a deletion of some amino acids of the C-terminus increased the expression of XlnR target genes, even under D-glucose repression conditions [114]. Efforts have been done in order to evaluate the behavior of XlnR regulon to optimize the expression of target genes. For instance, a modeling study for the observation of XlnR regulon dynamics under D-xylose induction was performed. In this study, it was demonstrated that regulation of the *A. niger* XlnR network system was dictated mainly by transcription and translation degradation rate parameters, and by D-xylose consumption profile [115, 116]. Structural and functional studies of XlnR are pivotal for the development of new strains with improved cellulase/hemicellulase production, given its importance for the transcription of hydrolase-encoding genes.

16. The transcriptional regulators Xyr1, Ace1 and Ace2

In *Aspergillus*, the xylanolytic and cellulolytic enzymatic system is strictly co-regulated via the inducer D-xylose, while enzymes involved in the same systems in *T. reesei* are not mainly acti-

http://dx.doi.org/10.5772/54325

vated by this sugar. As described above, the most potent inducer for cellulolytic system in *T. reesei* is sophorose, whereas hemicellulases appear to be induced during *T. reesei* growth in the presence of cellulose, xylan, sophorose, xylobiose and L-arabitol [6]. Despite of the diversity of inducers, it was demonstrated that the transcriptional regulation of the major hydrolytic enzyme-encoding genes *cbh1*, *cbh2* and *egl1* (cellulases), *xyn1*, *xyn2* (xylanases) is dependent on Xyr1, the XlnR homologue in *T. reesei*. Xyr 1 appears to be an essential activator for all levels of xylanase genes *xyn1* and *xyn2* transcription, receiving or mediating different signals from the inducer molecules [114]. In *xyn1* and *xyn2* promoters, the Xyr1-binding elements resemble the consensus sequences of *T. reesei* transcriptional regulator Ace1 [117]. Ace1 apparently acts as an antagonist of the Xyr1-driven *xyn1* gene transcription. Deletion of *xyr1* gene in *T. reesei* led to increased expression of cellulase and xylanase genes studied in cellulose- and sophorose-induced cultures, suggesting a negative effect of Ace1 on the induced expression of these genes [118]. In contrast, *ace1* activated the *cbh1* promoter in yeast, suggesting that Ace1 could be able to act as an inducer or a repressor, depending on the context, but the mechanisms involved in such regulation still remain to be investigated [6].

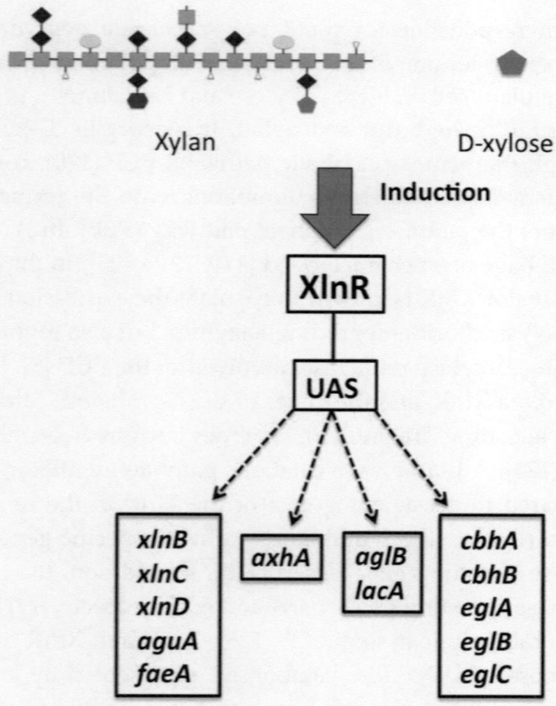

Figure 5. A proposed model for regulation of genes encoding polysaccharide-degrading enzymes in *A. niger*. Sugar monomers or polysaccharides provoke the binding of the transcription factor XlnR to the upstream activating system (UAS) in the promoter region of the genes, inducing the respective enzymatic system. Genes encoding xylanases (*xlnB*, *xlnC*), xylosidase (*xlnD*), glucuronidase (*aguA*) and feruloyl esterase (*faeA*) induce the xylanolytic enzymatic system. Genes encoding cellobiohydrolases (*cbhA*, *cbhB*) and endoglucanases (*eglA*, *eglB*, *eglC*) induce the cellulolytic system. Genes encoding galactosidases (*aglB*, *lacA*) induce the galactolytic system and the gene encoding arabinoxylan arabinofuranohydrolase induces the arabinolytic system.

Ace2 belongs to a zinc-binuclear cluster DNA-binding protein and appears to be an activator of cellulase and hemicellulase genes in cellulose-induced cultures of *T. reesei*. Deletion of the *ace2* gene resulted in decreased expression of the cellulase genes *cbh1, cbh2, egl1, egl2,* and the xylanase gene *xyn2* upon *T. reesei* growth on cellulose as sole carbon source [119]. The same study has been demonstrated that the expression of these genes were not affected after *T. reesei* growing on sophorose, suggesting partially different mechanisms for hydrolase-encoding genes expression upon different carbon sources [119]. Homologs of *ace2* in *A. nidulans, A. niger, N. crassa* and *Magnaporthe grisea* were not found to date, suggesting that different transcriptional regulators are used by these fungi to induce the expression of hydrolytic-encoding genes.

17. Other regulators involved in plant polysaccharide degradation in Aspergilli

The enzymatic system responsible for plant polysaccharide degradation is induced and commonly amplified after releasing of the plant cell wall polymers components. Two important components of the plant cell wall are D-xylose and L-arabinose, present in the polymers arabinan, arabinogalactan, xyloglucan and xylan. In Aspergilli, D-xylose and L-arabinose are catabolized through the pentose catabolic pathway, PCP [120], consisting of a series of reversible reductase/dehydrogenase steps culminating with the formation of D-xylulose-5-phosphate, which enters the pentose phosphate pathway (PPP). In *A. niger,* the most of the genes involved in PCP have been characterized [110, 121 - 123]. In the presence of D-xylose, the transcriptional activator XlnR is known to regulate the expression of genes not only encoding extracellular polysaccharide-degrading enzymes, but also regulates the expression of *xyrA,* the gene encoding D-xylose reductase, involved in the PCP [5]. L-arabinose induction of PCP is not mediated via XlnR, and the genes of the L-arabinose catabolic pathway are co-regulated with genes encoding arabinolytic enzymes such as α-L-arabinofuranosidase and endoarabinanase [93, 124]. A L-arabinose catabolic pathway regulator, AraR, was identified in *A. niger,* and appeared to act as antagonist of the XlnR in the regulation of PCP [125]. AraR demonstrated to regulate the L-arabinose pathway specific genes, for instance, the L-arabitol dehydrogenase encoding gene (*ladA*) [125]. In addition, the L-arabinose reductase and L-xylulose reductase encoding genes (*larA* and *lxrA,* respectively) appeared to be under the regulation of the arabinolytic system [125, 126]. AraR and XlnR both regulate the common steps of L-arabinose and D-xylose catabolism, represented by xylitol dehydrogenase- and xylulose kinase-encoding genes, *xdhA* and *xkiA,* respectively [123, 125]. Comparative analysis have been shown that the regulation of PCP by AraR differs in *A. nidulans* and *A. niger,* whereas the regulation of the PCP by XlnR was similar in both species, suggesting different evolutionary changes in these two species affecting pentose utilization [125]. These authors suggest that the differences in Aspergilli species implies that manipulating regulatory systems to improve the production of polysaccharide degrading enzymes may give different results in different industrial fungi.

http://dx.doi.org/10.5772/54325

Furthermore, it was found that in *Aspergillus aculeatus,* a new XlnR-independent pathway for the regulation of cellulase- and hemicellulase-encoding genes exists [127]. This study suggests that cellobiose from Avicel (crystalline cellulose) degradation stimulates only the XlnR-independent signaling pathway for cellulase/hemicellulase production in *A. aculeatus*. In addition, sequential promoter truncation studies on *cbh1* gene demonstrated that a conserved sequence in the promoter region of *cbh1,* namely CeRe (required for *eglA* induction in *A. nidulans;* [128], plays a pivotal role on the expression regulated by the XlnR-independent signaling pathway triggered by cellulosic compounds in *A. aculeatus* [127].

AmyR is another transcriptional activator found in Aspergilli. AmyR was first described as a transcriptional regulator of genes encoding enzymes involved in starch and maltose hydrolysis [129]. Nowadays, studies have been demonstrated a broader role for AmyR, which appears to regulate another gene expression systems. High levels of both α- and β-glucosidase as well as α- and β-galactosidase in the *amyR* multicopy strain of *A. niger* were found [130]. This study also demonstrated that AmyR-regulated genes in *A. niger* are induced during growth in low levels of D-glucose, as their expression increased during the cultivation. As D-glucose has been shown to act as a repressor through the carbon catabolite repressor protein CreA [75] (see next section), the authors suggested that repression levels of those genes are likely a balance between induction through AmyR and repression through CreA. Similar results were obtained for AmyR in *Aspergillus oryzae* [131].

18. Transcriptional regulators of plant polysaccharide degradation genes in Neurospora crassa

The filamentous ascomycete fungus *Neurospora crassa* has been commonly used as a model laboratory organism [132]. In nature, *N. crassa* can be found on burnt plant material, primarily grasses, including sugarcane and *Miscanthus* [133]. Previous studies have been demonstrated that *N. crassa* is able to express and secrete many plant cell wall degrading enzymes after grown on ground *Miscanthus* stems and crystalline cellulose [134]. Studies conducted with strains containing deletions of predicted transcription factors (TFs) in *N. crassa* demonstrated that a specific TF, named XLR1 (xylan degradation regulator-1), is essential for hemicellulose degradation in *N. crassa* [135]. The *xlr-1* gene is an ortholog to XlnR/Xyr1 found in *Aspergillus* and *Trichoderma* species, respectively. The results presented in this study have been shown that deletion of *xlr-1* in *N. crassa* abolished growth on xylan and xylose, but growth on cellulose and cellulolytic activities were not highly affected. The transcriptional profiling showed that *xlr-1* is required for induction of hemicellulase and xylose metabolism genes, and modulated the expression levels of few cellulase genes, but these genes do not require XLR-1 for induction [135]. These findings suggested that unknown TFs in *N. crassa* could be important for the induced expression of genes encoding cellulases in response to the presence of cellulose.

In fact, studies assessing a near-full genome deletion strain set in *N. crassa,* have been shown two transcription factors, named *clr-1* and *clr-2,* required for degradation of cellulose [136].

Homologs of *clr-1* and *clr-2* are present in the genomes of many filamentous ascomycete species capable of degrading plant-cell wall material, including *A. nidulans*. The *N. crassa* TFs *clr-1* and *clr-2* were able to induce all major cellulase and some hemicellulase genes, and functional CLR-1 was necessary for the expression of *clr-2* and efficient cellobiose utilization by the fungus. Besides, in *A. nidulans*, a deleted strain of the *clr-2* homolog (*clrB*) failed to induce cellulase gene expression and lacked cellulolytic activity on Avicel [136]. These authors reinforced the idea that further manipulation of the transcriptional regulation of cellulase/hemicellulase system may improve yields of cellulases for industrial applications, e.g., for biofuel production.

19. Carbon catabolite repression in Aspergilli

As briefly described above, microorganisms are known to adjust their carbon metabolism in order to minimize energy demands. One of these regulatory mechanisms is the carbon catabolite repression (CCR). Readily metabolizable carbon sources, such as glucose, are preferably catabolized and, in general, suppress the utilization of alternative carbon sources, repressing mainly the enzymatic system required for the catabolism of less favorable carbohydrates. For general carbon catabolite repression in some Aspergilli species, the DNA-binding Cys_2His_2 zinc-finger repressor CreA is absolutely necessary [75]. In general, the negative effect of this regulatory system depends on the concentration of the preferable carbon source (elicitor). For instance, higher concentrations of the elicitor usually induce stronger transcriptional repression [137]. The presence of the repressing elicitors initiates signal transduction pathways to result in transcriptional repression of the catabolism of poor carbon sources. In this context, the molecular mechanisms leading to CCR is well known for the ethanol utilization in *A. nidulans*, and therefore ethanol catabolism in this specie is commonly used as a model for studying CCR gene regulation [138]. In *A. nidulans*, ethanol, ethylamine and L-threonine can be used as sole carbon sources via their conversion into acetaldehyde and acetate [139, 140]. Furthermore, acetate enters the main metabolism in its activated form, acetyl-CoA. Alcohol dehydrogenase I (*alcA* gene) and aldehyde dehydrogenase (*aldA* gene) are the two enzymes involved in the oxidation of ethanol into acetate. The genes *alcA* and *aldA* are activated through the transcriptional regulator AlcR in the presence of a co-inducer compound [141]. AlcR is a positive regulatory protein of the zinc binuclear class, is autoregulated and binds to specific sites on the *alcA* and *alcR* promoter regions.

In this context, CreA appears as a sole transcriptional repressor of the system, exerting its function in the presence of a co-repressor [139, 142 - 143]. It is well known that in the *alc* genes repression system, CreA exerts its repressing function in three main different levels: (i) direct repression of *alcR*; (ii) indirect repression, via *alcR* repression, of the structural genes (*aldA* and *alcM*); and (iii) combined direct and indirect repression of the structural *alcA* and *alcS* genes, by the "double-lock" mechanism. At the molecular level, a competition between AlcR and CreA results in partial repression of the *alcR* gene and a complete repression of *alcA*. This mechanism is important under growth conditions in which poor carbon sources such as ethanol are simultaneously present with high amounts of a preferable carbon source such as glucose, and the fungus is able to fine-tune the regulation in order to adapt to new nutrient. A second mechanism involves complete repression of the *alcR* gene, operating at high glucose concentration. Under these conditions, expression of the *alc* genes does not occur, and the fungus metabolizes only the rich carbon source (reviewed in [141]).

A variety of studies have been demonstrated the mechanisms through which CreA represses some polysaccharide-degrading enzymatic systems in fungi. It was shown that CreA appears to repress *xlnA* transcription by the "double-lock" mechanism in *A. nidulans*, repressing directly the gene through its binding to the consensus *xlnA*.C1 site of the promoter, as well as indirect repression [144]. Studies on *A. nidulans xlnB* gene repression demonstrated that the four CreA target sites located in *xlnB* gene promoter region lack physiological relevance, suggesting that the repression exerted by CreA on *xlnB* is by an indirect mechanism [145]. The latter results suggested that an additional level of CreA repression via the xylanolytic activator is present in *A. nidulans*. The authors suggested that this mechanism of regulation would be analogous to that described above for the *alc* regulon, where certain genes are under a double-lock mechanism of repression by CreA while others are not subject to direct repression, being regulated via CreA repression of the *alcR* regulatory gene. In fact, studies have been shown that the *xlnR* (the xylanolytic transcriptional activator) promoter is repressed by glucose via CreA in *A. nidulans*, and when this repression is eliminated, by promoter exchange, transcription of xylanolytic genes such as *xlnA*, *xlnB* and *xlnD* is derepressed [146]. These results demonstrated that a transcription factor cascade involving CreA and XlnR regulates CCR of *A. nidulans* xylanolytic genes.

F-box proteins are proteins containing at least one F-box domain in their structures. The F-box domain is a protein structural motif of about 50 amino acids that mediates protein-protein interactions [147]. Usually, F-box proteins mediate ubiquitination of proteins targeted for degradation by the proteasome, but these proteins have also been associated with cellular functions such as signal transduction and regulation of cell-cycle [148]. A study that performed a screening of 42 *A. nidulans* F-box deletion mutants grown either on xylose or xylan as a sole carbon source in the presence of 2-deoxy-D-glucose was able to identify mutants with de-regulated xylanase induction [149]. In this study, a null mutant in a gene (*fbxA*) with decreased xylanase activity and reduced *xlnA* and *xlnD* mRNA accumulation was identified. This mutant interacted genetically with *creA* mutants, emphasizing the importance of the CCR and ubiquitination in the *A. nidulans* xylanase induction. In addition, the identification of FbxA protein provides evidence for another level of regulatory network concerning xylanase induction in filamentous fungi [149].

In summary, an intricate and fine-tuned regulation network exists in order to control the expression of plant cell-wall degradation genes in fungi. A variety of transcriptional regulators are able to respond to different nutritional requirements of the fungus, depending on its lifestyle. In general, readily metabolizable carbon sources such as glucose represses the transcription of genes responsible for the poor carbon source catabolism, via different mechanisms. The carbon catabolite repression in fungi is a common mechanism of regulation through which the organism adapts to nutritional availability in their environment. For instance, in *A. nidulans*, the xylanolytic transcriptional activator XlnR is repressed by glucose via CreA, the transcriptional factor responsible for CCR in this specie. The comprehension of such sophisticated regulatory network is essential for genetic engineering of new strains able to produce a wide range of lignocellulolytic enzymes.

20. Improving microbial strains for degradation of lignocellulosic biomass

In order to degrade efficiently plant biomass, a microorganism should possess characteristics that make the process economically viable. For cellulosic ethanol production, for instance, an efficient microorganism should produce high yields of the desired product, must have a broad substrate range and high ethanol tolerance and it has to be tolerant to the inhibitors present in lignocellulosic hydrolysates. Therefore, engineering microbial strains for improvement of effectiveness in industrial applications is not a simple task. Concerning to bioethanol production, the most promising organism for genetic bioengineering is the yeast *Saccharomices cerevisiae*. This microorganism has a great capacity for cell-recycle fermentation, and it is tolerant against various stresses, such as high temperature, low pH and many inhibitors [150]. Moreover, *S. cerevisiae* is a well-characterized model organism, with a diverse array of research tools and resources, which facilitates metabolic engineering [151, 152].

However, the metabolism of *S. cerevisiae* is very complex and punctual genetic modifications in the yeast may lead to unpredicted modifications in the whole metabolism, with undesired effects for industrial applications, such as generation of toxic by-products. In general, metabolic engineering of *S. cerevisiae* is performed on a trial-and error basis, with various modifications being tested at the same time. In this way, the most promising approaches are implemented to increase target production [153]. Due to this challenge, a novel and rational strategy has recently emerged for a system-wide modification of metabolism. The novel approach, termed synthetic bioengineering, is essential for creating effective yeast cell factories [152 - 154]. The first step in the synthetic bioengineering consists of an optimized metabolic pathway designed by computational simulation. Based on this metabolic profile, a list of target genes to be inserted or deleted is determined. Next, a customized microbial cell factory is assembled using advanced gene manipulation techniques. After the detection of metabolic problems concerning the prototype strains by using transcriptomics and metabolomics, further improvement of the assembled strains are performed [153]. Currently, many researchers have been using synthetic bioengineering approaches to improve microbial strains for industrial applications. For instance, efforts have been done in order to produce *S. cerevisiae* strains able to use xylose as carbon source, once this sugar is the second most abundant in lignocellulosic biomass. The yeast is not able to ferment xylose, but some groups already produced *S. cerevisiae* strains with improved capacity of xylose fermentation [30, 155 - 156]. A detailed description of synthetic bioengineering and its applications could be found at [153]. In this review, the authors describe engineered microbial strains producing higher alcohols such as 1-butanol and isobutanol and strains overproducing glutathione, for instance.

It is worth to mention that synthetic bioengineering could be applied for any microorganism, since it is a rational design of metabolic pathways.

While most biological routes being studied for the processing of lignocellulosic biomass focused on the separate production of hydrolytic enzymes, in a process that usually comprises several steps, another approach is suggested to achieve this goal. This approach, termed consolidated bioprocessing (CBP) involves the production of cellulolytic enzymes, hydrolysis of biomass, and fermentation of resulting sugars in a single stage via microorganisms or a consortium [157]. CBP appears to offer very large costs reduction if microorganisms can be developed that possess the required combination of substrate utilization and product formation properties [158]. In a 2006 report in biomass conversion to biofuels, the U.S. Department of Energy endorsed the view that CBP technology is "the ultimate low-cost configuration for cellulose hydrolysis and fermentation" (DOE Joint Task Force, 2006; energy.gov). Currently, CBP technology is developing fast, especially due to partnerships with venture capital investors and researchers. The main challenge of CBP is to generate engineered microorganisms able to produce the saccharolytic enzymes and converting the sugars released by those enzymes into the desired end-products. In addition, CBP microorganisms need to be able to perform these tasks rapidly and efficiently under challenging, industrial processes. A successful microbial platform for production of bioethanol from microalgae is currently available, and demonstrates an application of the CBP [159]. A DNA fragment encoding enzymes for alginate transport and metabolism from *Vibrio splendidus*, abundant and ubiquitous marine bacteria, was introduced in the genome of *Escherichia coli*, a well-characterized microorganism. This microbial platform was able to simultaneously degrade, uptake and metabolize alginate, an abundant polysaccharide present in microalgae. When further engineered for ethanol synthesis, this platform enabled bioethanol production with satisfactory yield directly from microalgae via a consolidated bioprocess [159].

The approach required for generation of CBP microorganisms involves the knowledge of many topics discussed in this chapter, concerning to fundamental principles of microbial cellulose utilization and its regulation. Moreover, the principles of synthetic bioengineering discussed above can be applied to the development of new strains for CBP technology, and therefore the generation of new microbial platforms able to uptake and metabolize completely the lignocellulosic biomass.

21. Conclusions and future perspectives

A large quantity of lignocellulosic residues is accumulating over the world, mainly due to the expansion of industrial processes, but other sources such as wood, grass, agricultural, forestry and urban solid wastes contribute to accumulation of lignocellulosic material. These residues constitute a renewable resource from which many useful biological and chemical products can be derived. The natural ability of fungi and other microorganisms to degrade lignocellulosic biomass, due to highly efficient enzymatic systems, is very attractive for the development of new strategies concerning industrial processes. Paper manufacture, composting, human and animal feeding, economically important chemical compounds and biomass fuel production are among some industrial applications derived from microbial lignocellulosic degradation.

Global climate change and future energy demands initiate a race in order to achieve sustainable fuels derived from biomass residues. Conversion of sugars to ethanol is already currently done at very low cost from sugarcane in Brazil, and from corn, in United States. However, the challenge is how to obtain the biofuel from the wastes generated from the mills producing ethanol. Residues such as sugarcane bagasse and corncobs contain large amounts of lignocellulosic material and therefore can be transformed into biofuels. A major advantage of using residues to produce biofuels is to reduce the competition between fuels and food. In this context, hydrolytic enzymes such as cellulases contribute for the large cost of cellulosic ethanol nowadays. The great bottleneck to achieve cellulosic biofuels is the plant biomass recalcitrance, and overcome such barrier is the key for the development of feasible industrial processes for biofuels production. For instance, it was recently demonstrated that *Aspergillus niger* growing on steam-exploded sugarcane bagasse was able to produce and secrete a high number of (hemi)cellulases [98]. The challenge is to adapt the process of steam explosion of a waste residue, such as sugarcane bagasse, with its hydrolysis by the fungi. The steam-explosion is a pretreatment that can decrease biomass recalcitrance, allowing the fungi to penetrate deeply within the biomass.

The comprehension of the machinery behind the enzymatic systems of fungi able to degrade plant cell wall polysaccharides favors the use of the microorganisms in industrial applications. Currently, through advanced molecular techniques, it is possible to engineer new microbial strains by insertion or deletion of genes involved in important metabolic pathways responsible for biomass degradation. The useful host cells to develop the synthetic bioengineering should have versatile genetic tools, resources and suitability for bio-refinery processes, such as stress tolerance. Therefore, a strain development in future requires insertion, deletion and expression controls of multiple genes and it is a difficult task to achieve. However, integrated advanced techniques could be able to overcome these challenges, including computational simulation of metabolic pathways, genome synthesis, directed evolution and minimum genome factory. The synthesis of the whole genome has already been done [160, 162] and, as discussed in reference [153], in a near future the synthesis of very large fragments of DNA will make it possible to design a whole yeast artificial chromosome (YAC) encoding a number of genes. According to these authors, the *de novo* synthesis of YAC should be a breakthrough methodology for the future synthetic bioengineering, and cloning of individual genes and a construction of plasmid vectors would be obsolete. A rational design of metabolic pathways along with customized design of genes with optimized expression may be obtained, making it possible to produce a whole sequence of the artificial chromosome [153, 163].

As said by Lee Lynd, a pioneering researcher in the field of biomass: "the first step toward realizing currently improbable futures is to show that they are possible". These technologies described above are currently available for scientific community and, along with advances in industrial processes, endorse the possibility to take energy from plant biomass using microorganisms. Thus, the Humanity has never been so close to use new and sustainable ways of energy.

Author details

Wagner Rodrigo de Souza*

Address all correspondence to: wagnerusp@gmail.com

Institute of Biology, State University of Campinas (UNICAMP), Brasil

References

[1] Goldemberg J, Coelho ST. Renewable energy - traditional biomass vs. modern biomass. Energ Policy. 2004 Mar;32(6):711-4. PubMed PMID: ISI:000187896900001. English.

[2] Himmel ME. Biomass recalcitrance: engineering plants and enzymes for biofuels production (vol 315, pg 804, 2007). Science. 2007 May 18;316(5827):982-. PubMed PMID: ISI:000246554000023. English.

[3] de Vries R, Visser, J. . Aspergillus enzymes involved in degradation of plant cell wall polysaccharides. . Microbiology and Molecular Biology Reviews. 2001;65(4):497-522.

[4] Aro N, Pakula, T, Penttila, M. . Transcriptional regulation of plat cell wall degradation by filamentous fungi. FEMS Microbiology Reviews. 2005;29:719-39.

[5] de Vries RP. Regulation of Aspergillus genes encoding plant cell wall polysaccharide-degrading enzymes; relevance for industrial production. Appl Microbiol Biotechnol. 2003 Mar;61(1):10-20. PubMed PMID: 12658510. Epub 2003/03/27. eng.

[6] Aro N, Pakula T, Penttila M. Transcriptional regulation of plant cell wall degradation by filamentous fungi. FEMS microbiology reviews. 2005 Sep;29(4):719-39. PubMed PMID: 16102600.

[7] Lynd LR WP, van Zyl WH, Pretorius IS. Microbial cellulose utilization: Fundamentals and biotechnology. Microbiology and Molecular Biology Reviews. 2002;66(3): 506-77.

[8] Lynd LR, Weimer PJ, van Zyl WH, Pretorius IS. Microbial cellulose utilization: Fundamentals and biotechnology (vol 66, pg 506, 2002). Microbiology and Molecular Biology Reviews. 2002 Dec;66(4):739-. PubMed PMID: ISI:000179683700007. English.

[9] Lee S, Ha, JK, Kang, HS, McAllister T, Cheng K-J. . Overview of energy metabolism, substrate utilization and fermentation characteristics of ruminal anaerobic fungi. Korean Journal of Animal and Nutritional Feedstuffs. 1997;21:295–314.

[10] Lobarzewski J. The Characteristics and Functions of the Peroxidases from Trametes-Versicolor in Lignin Biotransformation. J Biotechnol. 1990 Feb;13(2-3):111-7. PubMed PMID: ISI:A1990CU51000003. English.

[11] Galliano H, G. Gas, J.L. Sevis and A.M. Boudet: Lignin degradation by Rigidoporus lignosus involves synergistic action of two oxidizing enzymes: Mn peroxidase and laccase. . Enzyme Microb Technol. 1991;13:478-82.

[12] Stricker SH, Steenpass L, Pauler FM, Santoro F, Latos PA, Huang R, et al. Silencing and transcriptional properties of the imprinted Airn ncRNA are independent of the endogenous promoter. Embo Journal. 2008 Dec 3;27(23):3116-28. PubMed PMID: ISI: 000261330900004. English.

[13] Stricker AR, Mach RL, de Graaff LH. Regulation of transcription of cellulases- and hemicellulases-encoding genes in Aspergillus niger and Hypocrea jecorina (Trichoderma reesei). Appl Microbiol Biot. 2008 Feb;78(2):211-20. PubMed PMID: ISI: 000252614200003. English.

[14] Bayer EA, Belaich JP, Shoham Y, Lamed R. The cellulosomes: multienzyme machines for degradation of plant cell wall polysaccharides. Annu Rev Microbiol. 2004;58:521-54. PubMed PMID: 15487947.

[15] Fontes CM, Gilbert HJ. Cellulosomes: highly efficient nanomachines designed to deconstruct plant cell wall complex carbohydrates. Annual review of biochemistry. 2010;79:655-81. PubMed PMID: 20373916.

[16] Kosugi A, Murashima K, Doi RH. Xylanase and acetyl xylan esterase activities of XynA, a key subunit of the Clostridium cellulovorans cellulosome for xylan degradation. Appl Environ Microbiol. 2002 Dec;68(12):6399-402. PubMed PMID: 12450866. Pubmed Central PMCID: 134393.

[17] Morag E, Bayer EA, Lamed R. Relationship of cellulosomal and noncellulosomal xylanases of Clostridium thermocellum to cellulose-degrading enzymes. J Bacteriol. 1990 Oct;172(10):6098-105. PubMed PMID: 2211528. Pubmed Central PMCID: 526935.

[18] Tamaru Y, Doi RH. Pectate lyase A, an enzymatic subunit of the Clostridium cellulovorans cellulosome. Proc Natl Acad Sci U S A. 2001 Mar 27;98(7):4125-9. PubMed PMID: 11259664. Pubmed Central PMCID: 31190.

[19] Salamitou S, Raynaud O, Lemaire M, Coughlan M, Beguin P, Aubert JP. Recognition specificity of the duplicated segments present in Clostridium thermocellum endoglucanase CelD and in the cellulosome-integrating protein CipA. J Bacteriol. 1994 May; 176(10):2822-7. PubMed PMID: 8188583. Pubmed Central PMCID: 205435.

[20] Tokatlidis K, Salamitou S, Beguin P, Dhurjati P, Aubert JP. Interaction of the duplicated segment carried by Clostridium thermocellum cellulases with cellulosome components. FEBS Lett. 1991 Oct 21;291(2):185-8. PubMed PMID: 1936262.

[21] Poole DM, Morag E, Lamed R, Bayer EA, Hazlewood GP, Gilbert HJ. Identification of the cellulose-binding domain of the cellulosome subunit S1 from Clostridium thermocellum YS. FEMS Microbiol Lett. 1992 Dec 1;78(2-3):181-6. PubMed PMID: 1490597.

[22] Mingardon F, Chanal A, Lopez-Contreras AM, Dray C, Bayer EA, Fierobe HP. Incorporation of fungal cellulases in bacterial minicellulosomes yields viable, synergistically acting cellulolytic complexes. Appl Environ Microbiol. 2007 Jun;73(12):3822-32. PubMed PMID: 17468286. Pubmed Central PMCID: 1932714.

[23] Mingardon F, Perret S, Belaich A, Tardif C, Belaich JP, Fierobe HP. Heterologous production, assembly, and secretion of a minicellulosome by Clostridium acetobutylicum ATCC 824. Appl Environ Microbiol. 2005 Mar;71(3):1215-22. PubMed PMID: 15746321. Pubmed Central PMCID: 1065181.

[24] McClendon SD, Mao Z, Shin HD, Wagschal K, Chen RR. Designer xylanosomes: protein nanostructures for enhanced xylan hydrolysis. Appl Biochem Biotechnol. 2012 Jun;167(3):395-411. PubMed PMID: 22555497.

[25] Lee J. Biological conversion of lignocellulosic biomass to ethanol. J Biotechnol. 1997 Jul 23;56(1):1-24. PubMed PMID: 9246788.

[26] Galazka JM, Tian CG, Beeson WT, Martinez B, Glass NL, Cate JHD. Cellodextrin Transport in Yeast for Improved Biofuel Production. Science. 2010 Oct 1;330(6000): 84-6. PubMed PMID: ISI:000282334500039. English.

[27] Hahn-Hagerdal B, Karhumaa K, Fonseca C, Spencer-Martins I, Gorwa-Grauslund MF. Towards industrial pentose-fermenting yeast strains. Appl Microbiol Biotechnol. 2007 Apr;74(5):937-53. PubMed PMID: 17294186. Epub 2007/02/13. eng.

[28] Chu BC, Lee H. Genetic improvement of Saccharomyces cerevisiae for xylose fermentation. Biotechnol Adv. 2007 Sep-Oct;25(5):425-41. PubMed PMID: 17524590.

[29] Jeffries TW. Engineering yeasts for xylose metabolism. Curr Opin Biotechnol. 2006 Jun;17(3):320-6. PubMed PMID: 16713243.

[30] Hahn-Hagerdal B, Karhumaa K, Jeppsson M, Gorwa-Grauslund MF. Metabolic engineering for pentose utilization in Saccharomyces cerevisiae. Adv Biochem Eng Biotechnol. 2007;108:147-77. PubMed PMID: 17846723. Epub 2007/09/12. eng.

[31] Tamakawa H, Ikushima S, Yoshida S. Ethanol production from xylose by a recombinant Candida utilis strain expressing protein-engineered xylose reductase and xylitol dehydrogenase. Biosci Biotechnol Biochem. 2011;75(10):1994-2000. PubMed PMID: 21979076.

[32] Cantarel B, Rancurel, C, Bernard, T, Lombard, V, Coutinho, PM, Henrissat, B. The Carbohydrate-Active EnZymes database: an expert resource for Glycogenomics. Nucleic Acids Research. 2009;37:D233-D8.

[33] van den Brink J, de Vries RP. Fungal enzyme sets for plant polysaccharide degradation. Appl Microbiol Biot. 2011 Sep;91(6):1477-92. PubMed PMID: ISI: 000294214900002. English.

[34] Dias FM, Vincent F, Pell G, Prates JA, Centeno MS, Tailford LE, et al. Insights into the molecular determinants of substrate specificity in glycoside hydrolase family 5 re-

vealed by the crystal structure and kinetics of Cellvibrio mixtus mannosidase 5A. J Biol Chem. 2004 Jun 11;279(24):25517-26. PubMed PMID: 15014076. Epub 2004/03/12. eng.

[35] Beguin P. Molecular-Biology of Cellulose Degradation. Annu Rev Microbiol. 1990;44:219-48. PubMed PMID: ISI:A1990EA66200010. English.

[36] Ward M, Wu, S., Dauberman, J., Weiss, G., Larenas, E., Bower, B., Rey, M., Clarkson, K. and Bott, R. Cloning, sequnece and preliminary structural analysis of a small, high pI endoglucanase (EGIII) form Trichoderma reesei (Suominen, P. and Reinikainen, T., Eds.), Trichoderma reesei cellulases and other hydrolases (TRICEL 93). Kirk-konummi, Finland.; 1993.

[37] Reczey K, Brumbauer A, Bollok M, Szengyel Z, Zacchi G. Use of hemicellulose hydrolysate for beta-glucosidase fermentation. Appl Biochem Biotechnol. 1998 Spring; 70-72:225-35. PubMed PMID: 18575992. Epub 2008/06/26. eng.

[38] Kubicek CP, Herrera-Estrella A, Seidl-Seiboth V, Martinez DA, Druzhinina IS, Thon M, et al. Comparative genome sequence analysis underscores mycoparasitism as the ancestral life style of Trichoderma. Genome Biol. 2011;12(4):R40. PubMed PMID: 21501500. Pubmed Central PMCID: 3218866. Epub 2011/04/20. eng.

[39] Chen H, Hayn M, Esterbauer H. Purification and characterization of two extracellular beta-glucosidases from Trichoderma reesei. Biochim Biophys Acta. 1992 May 22;1121(1-2):54-60. PubMed PMID: 1599951. Epub 1992/05/22. eng.

[40] Polizeli ML, Rizzatti AC, Monti R, Terenzi HF, Jorge JA, Amorim DS. Xylanases from fungi: properties and industrial applications. Appl Microbiol Biotechnol. 2005 Jun; 67(5):577-91. PubMed PMID: 15944805. Epub 2005/06/10. eng.

[41] Biely P, Vrsanska M, Tenkanen M, Kluepfel D. Endo-beta-1,4-xylanase families: differences in catalytic properties. J Biotechnol. 1997 Sep 16;57(1-3):151-66. PubMed PMID: 9335171. Epub 1997/10/23. eng.

[42] Pollet A, Delcour JA, Courtin CM. Structural determinants of the substrate specificities of xylanases from different glycoside hydrolase families. Crit Rev Biotechnol. 2010 Sep;30(3):176-91. PubMed PMID: 20225927. Epub 2010/03/17. eng.

[43] Shinoyama H, A. Ando, T. Fujii, and T. Yasui. The possibility of enzymatic synthesis of a variety of -xylosides using the transfer reaction of Aspergillus niger xylosidase. Agric Biol Chem. 1991;55:849–50.

[44] Sulistyo J, Kamiyama Y, Yasui T. Purification and Some Properties of Aspergillus-Pulverulentus Beta-Xylosidase with Transxylosylation Capacity. J Ferment Bioeng. 1995;79(1):17-22. PubMed PMID: ISI:A1995QF31200004. English.

[45] Pauly M, Andersen LN, Kauppinen S, Kofod LV, York WS, Albersheim P, et al. A xyloglucan-specific endo-beta-1,4-glucanase from Aspergillus aculeatus: expression cloning in yeast, purification and characterization of the recombinant enzyme. Glycobiology. 1999 Jan;9(1):93-100. PubMed PMID: 9884411. Epub 1999/01/13. eng.

[46] Master ER, Zheng Y, Storms R, Tsang A, Powlowski J. A xyloglucan-specific family 12 glycosyl hydrolase from Aspergillus niger: recombinant expression, purification and characterization. Biochem J. 2008 Apr 1;411(1):161-70. PubMed PMID: 18072936. Epub 2007/12/13. eng.

[47] Desmet T, Cantaert T, Gualfetti P, Nerinckx W, Gross L, Mitchinson C, et al. An investigation of the substrate specificity of the xyloglucanase Cel74A from Hypocrea jecorina. FEBS J. 2007 Jan;274(2):356-63. PubMed PMID: 17229143. Epub 2007/01/19. eng.

[48] de Vries R V, J. Aspergillus enzymes involved in degradation of plant cell wall polysaccharides. Microbiology and Molecular Biology Reviews. 2001;65(4):497-522.

[49] Mccleary BV. Comparison of Endolytic Hydrolases That Depolymerize 1,4-Beta-D-Mannan, 1,5-Alpha-L-Arabinan, and 1,4-Beta-D-Galactan. Acs Sym Ser. 1991;460:437-49. PubMed PMID: ISI:A1991BT27V00034. English.

[50] Civas A, Eberhard R, Ledizet P, Petek F. Glycosidases Induced in Aspergillus-Tamarii - Secreted Alpha-D-Galactosidase and Beta-D-Mannanase. Biochem J. 1984;219(3): 857-63. PubMed PMID: ISI:A1984SQ61800022. English.

[51] Mccleary BV, Matheson NK. Action Patterns and Substrate-Binding Requirements of Beta-Deuterium-Mannanase with Mannosaccharides and Mannan-Type Polysaccharides. Carbohyd Res. 1983;119(Aug):191-219. PubMed PMID: ISI:A1983RE59300017. English.

[52] Ademark P, Lundqvist J, Hagglund P, Tenkanen M, Torto N, Tjerneld F, et al. Hydrolytic properties of a beta-mannosidase purified from Aspergillus niger. Journal of Biotechnology. 1999 Oct 8;75(2-3):281-9. PubMed PMID: ISI:000083683900019. English.

[53] Verbruggen MA, Spronk BA, Schols HA, Beldman G, Voragen AGJ, Thomas JR, et al. Structures of enzymically derived oligosaccharides from sorghum glucuronoarabinoxylan. Carbohyd Res. 1998 Jan;306(1-2):265-74. PubMed PMID: ISI: 000074837200026. English.

[54] Sakamoto T, Ogura A, Inui M, Tokuda S, Hosokawa S, Ihara H, et al. Identification of a GH62 alpha-l-arabinofuranosidase specific for arabinoxylan produced by Penicillium chrysogenum. Appl Microbiol Biot. 2011 Apr;90(1):137-46. PubMed PMID: ISI: 000288252000013. English.

[55] Yoshikawa K, Yamamoto K, Okada S. Isolation of Aspergillus flavus MO-5 producing two types of intracellular alpha-D-xylosidases: purification and characterization of alpha-D-xylosidase I. Biosci Biotechnol Biochem. 1993 Aug;57(8):1275-80. PubMed PMID: 7764013. Epub 1993/08/01. eng.

[56] Yoshikawa K, Yamamoto K, Okada S. Classification of some alpha-glucosidases and alpha-xylosidases on the basis of substrate specificity. Biosci Biotechnol Biochem. 1994 Aug;58(8):1392-8. PubMed PMID: 7765271. Epub 1994/08/01. eng.

[57] Suykerbuyk ME, Schaap PJ, Stam H, Musters W, Visser J. Cloning, sequence and expression of the gene coding for rhamnogalacturonase of Aspergillus aculeatus; a novel pectinolytic enzyme. Appl Microbiol Biotechnol. 1995 Oct;43(5):861-70. PubMed PMID: 7576553. Epub 1995/10/01. eng.

[58] van der Vlugt-Bergmans CJ, Meeuwsen PJ, Voragen AG, van Ooyen AJ. Endo-xylogalacturonan hydrolase, a novel pectinolytic enzyme. Appl Environ Microbiol. 2000 Jan;66(1):36-41. PubMed PMID: 10618200. Pubmed Central PMCID: 91782. Epub 2000/01/05. eng.

[59] Lombard V, Bernard T, Rancurel C, Brumer H, Coutinho PM, Henrissat B. A hierarchical classification of polysaccharide lyases for glycogenomics. Biochem J. 2010 Dec 15;432(3):437-44. PubMed PMID: 20925655. Epub 2010/10/12. eng.

[60] Jurnak F, Kita N, Garrett M, Heffron SE, Scavetta R, Boyd C, et al. Functional implications of the three-dimensional structures of pectate lyases. Progr Biotechnol. 1996;14:295-308. PubMed PMID: ISI:A1996BJ39G00022. English.

[61] Isroi, Millati R, Syamsiah S, Niklasson C, Cahyanto MN, Lundquist K, et al. Biological Pretreatment of Lignocelluloses with White-Rot Fungi and Its Applications: A Review. Bioresources. 2011;6(4):5224-59. PubMed PMID: ISI:000298119500121. English.

[62] Hattaka A. Biodegradation of lignin. (eds.) MHaASc, editor: Wiley-WCH; 2001.

[63] Hammel KE, Cullen D. Role of fungal peroxidases in biological ligninolysis. Curr Opin Plant Biol. 2008 Jun;11(3):349-55. PubMed PMID: 18359268. Epub 2008/03/25. eng.

[64] Wong DW. Structure and action mechanism of ligninolytic enzymes. Appl Biochem Biotechnol. 2009 May;157(2):174-209. PubMed PMID: 18581264. Epub 2008/06/27. eng.

[65] Lundquist K, Kirk, T. K., and Connors, W. J. Fungal degradation of kraft lignin and lignin sulfonates prepared from synthetic 14C-lignins. Arch Microbiol. 1977;112:291-6.

[66] Hatakka AI. Pretreatment of wheat straw by white-rot fungi for enzymatic saccharification of cellulose. Eur J Appl Microbiol Biotechnol. 1983;18:350-7..

[67] Kirk TK, and Chang, H.-M. Potential applications of bio-ligninolytic systems. Enzyme and Microbial Technology. 1981;3:189-96..

[68] Keller FA, Hamilton JE, Nguyen QA. Microbial pretreatment of biomass: potential for reducing severity of thermochemical biomass pretreatment. Appl Biochem Biotechnol. 2003 Spring;105 -108:27-41. PubMed PMID: 12721473. Epub 2003/05/02. eng.

[69] Lomascolo A, Uzan-Boukhris E, Herpoel-Gimbert I, Sigoillot JC, Lesage-Meessen L. Peculiarities of Pycnoporus species for applications in biotechnology. Appl Microbiol Biotechnol. 2011 Dec;92(6):1129-49. PubMed PMID: 22038244.

[70] Eggert C, Temp U, Eriksson KE. The ligninolytic system of the white rot fungus Pycnoporus cinnabarinus: purification and characterization of the laccase. Appl Environ Microbiol. 1996 Apr;62(4):1151-8. PubMed PMID: 8919775. Pubmed Central PMCID: 167880.

[71] Record E, Punt PJ, Chamkha M, Labat M, van Den Hondel CA, Asther M. Expression of the Pycnoporus cinnabarinus laccase gene in Aspergillus niger and characterization of the recombinant enzyme. Eur J Biochem. 2002 Jan;269(2):602-9. PubMed PMID: 11856319.

[72] Camarero S, Pardo I, Canas AI, Molina P, Record E, Martinez AT, et al. Engineering platforms for directed evolution of Laccase from Pycnoporus cinnabarinus. Appl Environ Microbiol. 2012 Mar;78(5):1370-84. PubMed PMID: 22210206. Pubmed Central PMCID: 3294479.

[73] Abbott TP, Wicklow DT. Degradation of lignin by cyathus species. Appl Environ Microbiol. 1984 Mar;47(3):585-7. PubMed PMID: 16346497. Pubmed Central PMCID: 239724.

[74] Salony, Mishra S, Bisaria VS. Production and characterization of laccase from Cyathus bulleri and its use in decolourization of recalcitrant textile dyes. Appl Microbiol Biotechnol. 2006 Aug;71(5):646-53. PubMed PMID: 16261367.

[75] Ruijter GJG, J. Visser. Carbon repression in aspergilli. FEMS Microbiology Letters. 1997;151:103-14.

[76] Gritzali MaB, R.D.J. The cellulase system of Trichoderma: relationship betweeen purified extracellular enzymes from induced or cellulose-grown cells. Adv Chem Ser. 1979;181:237–60.

[77] Vaheri M, Leisola, M. and Kauppinen, V. Transgly- cosylation products of cellulase system of Trichoderma reesei. Biotechnol Tech. 1979;1:696–9.

[78] Fowler T, Brown RD, Jr. The bgl1 gene encoding extracellular beta-glucosidase from Trichoderma reesei is required for rapid induction of the cellulase complex. Mol Microbiol. 1992 Nov;6(21):3225-35. PubMed PMID: 1453960. Epub 1992/11/11. eng.

[79] Bisaria VS, Mishra S. Regulatory aspects of cellulase biosynthesis and secretion. Crit Rev Biotechnol. 1989;9(2):61-103. PubMed PMID: 2509081. Epub 1989/01/01. eng.

[80] Hrmova M, Petrakova E, Biely P. Induction of cellulose- and xylan-degrading enzyme systems in Aspergillus terreus by homo- and heterodisaccharides composed of glucose and xylose. J Gen Microbiol. 1991 Mar;137(3):541-7. PubMed PMID: 2033377. Epub 1991/03/01. eng.

[81] Ilmen M, Saloheimo A, Onnela ML, Penttila ME. Regulation of cellulase gene expression in the filamentous fungus Trichoderma reesei. Appl Environ Microbiol. 1997 Apr;63(4):1298-306. PubMed PMID: 9097427. Pubmed Central PMCID: 168424. Epub 1997/04/01. eng.

[82] Ding SJ, Ge W, Buswell JA. Endoglucanase I from the edible straw mushroom, Volvariella volvacea. Purification, characterization, cloning and expression. Eur J Biochem. 2001 Nov;268(22):5687-95. PubMed PMID: 11722552. Epub 2001/11/28. eng.

[83] Mernitz G, Koch A, Henrissat B, Schulz G. Endoglucanase II (EGII) of Penicillium janthinellum: cDNA sequence, heterologous expression and promotor analysis. Curr Genet. 1996 Apr;29(5):490-5. PubMed PMID: 8625430. Epub 1996/04/01. eng.

[84] Chikamatsu G, Shirai K, Kato M, Kobayashi T, Tsukagoshi N. Structure and expression properties of the endo-beta-1,4-glucanase A gene from the filamentous fungus Aspergillus nidulans. FEMS Microbiol Lett. 1999 Jun 15;175(2):239-45. PubMed PMID: 10386374. Epub 1999/07/01. eng.

[85] Fekete E, Padra J, Szentirmai A, Karaffa L. Lactose and D-galactose catabolism in the filamentous fungus Aspergillus nidulans. Acta Microbiol Imm H. 2008 Jun;55(2): 119-24. PubMed PMID: ISI:000256797900004. English.

[86] Seiboth B, Gamauf C, Pail M, Hartl L, Kubicek CP. The D-xylose reductase of Hypocrea jecorina is the major aldose reductase in pentose and D-galactose catabolism and necessary for beta-galactosidase and cellulase induction by lactose. Mol Microbiol. 2007 Nov;66(4):890-900. PubMed PMID: 17924946. Epub 2007/10/11. eng.

[87] Martinez D, Berka RM, Henrissat B, Saloheimo M, Arvas M, Baker SE, et al. Genome sequencing and analysis of the biomass-degrading fungus Trichoderma reesei (syn. Hypocrea jecorina). Nat Biotechnol. 2008 May;26(5):553-60. PubMed PMID: 18454138. Epub 2008/05/06. eng.

[88] Pel HJ, de Winde JH, Archer DB, Dyer PS, Hofmann G, Schaap PJ, et al. Genome sequencing and analysis of the versatile cell factory Aspergillus niger CBS 513.88. Nat Biotechnol. 2007 Feb;25(2):221-31. PubMed PMID: 17259976. Epub 2007/01/30. eng.

[89] de Vries RP, Visser J. Regulation of the feruloyl esterase (faeA) gene from Aspergillus niger. Appl Environ Microbiol. 1999 Dec;65(12):5500-3. PubMed PMID: 10584009. Pubmed Central PMCID: 91749. Epub 1999/12/03. eng.

[90] de Vries RP, van de Vondervoort PJ, Hendriks L, van de Belt M, Visser J. Regulation of the alpha-glucuronidase-encoding gene (aguA) from Aspergillus niger. Mol Genet Genomics. 2002 Sep;268(1):96-102. PubMed PMID: 12242504. Epub 2002/09/21. eng.

[91] de Vries RP, Visser J, de Graaff LH. CreA modulates the XlnR-induced expression on xylose of Aspergillus niger genes involved in xylan degradation. Res Microbiol. 1999 May;150(4):281-5. PubMed PMID: 10376490. Epub 1999/06/22. eng.

[92] Mach-Aigner AR, Omony J, Jovanovic B, van Boxtel AJ, de Graaff LH. d-Xylose concentration-dependent hydrolase expression profiles and the function of CreA and XlnR in Aspergillus niger. Appl Environ Microbiol. 2012 May;78(9):3145-55. PubMed PMID: 22344641. Pubmed Central PMCID: 3346484. Epub 2012/02/22. eng.

[93] de Vries RP, Flipphi MJ, Witteveen CF, Visser J. Characterization of an Aspergillus nidulans L-arabitol dehydrogenase mutant. FEMS Microbiol Lett. 1994 Oct 15;123(1-2):83-90. PubMed PMID: 7988903. Epub 1994/10/15. eng.

[94] de Groot MJ, van de Vondervoort PJ, de Vries RP, vanKuyk PA, Ruijter GJ, Visser J. Isolation and characterization of two specific regulatory Aspergillus niger mutants shows antagonistic regulation of arabinan and xylan metabolism. Microbiology. 2003 May;149(Pt 5):1183-91. PubMed PMID: 12724380. Epub 2003/05/02. eng.

[95] Mach RL, Strauss J, Zeilinger S, Schindler M, Kubicek CP. Carbon catabolite repression of xylanase I (xyn1) gene expression in Trichoderma reesei. Mol Microbiol. 1996 Sep;21(6):1273-81. PubMed PMID: 8898395. Epub 1996/09/01. eng.

[96] Zeilinger S, Mach RL, Schindler M, Herzog P, Kubicek CP. Different inducibility of expression of the two xylanase genes xyn1 and xyn2 in Trichoderma reesei. J Biol Chem. 1996 Oct 11;271(41):25624-9. PubMed PMID: 8810338. Epub 1996/10/11. eng.

[97] Margolles-Clark M, Ilme´n, M. and Penttila, M. Expression patterns of 10 hemicellulase genes from filamentous fungus Trichoderma reesei on various carbon sources. J Biotechnol. 1997;57:167–79..

[98] de Souza WR, de Gouvea PF, Savoldi M, Malavazi I, Bernardes LAD, Goldman MHS, et al. Transcriptome analysis of Aspergillus niger grown on sugarcane bagasse. Biotechnol Biofuels. 2011 Oct 18;4. PubMed PMID: ISI:000297110100001. English.

[99] Gold MH, Alic M. Molecular-Biology of the Lignin-Degrading Basidiomycete Phanerochaete-Chrysosporium. Microbiol Rev. 1993 Sep;57(3):605-22. PubMed PMID: ISI:A1993LW44100005. English.

[100] Gettemy JM, Ma B, Alic M, Gold MH. Reverse transcription-PCR analysis of the regulation of the manganese peroxidase gene family. Appl Environ Microbiol. 1998 Feb; 64(2):569-74. PubMed PMID: 9464395. Pubmed Central PMCID: 106084. Epub 1998/02/17. eng.

[101] Brown JA, Alic M, Gold MH. Manganese peroxidase gene transcription in Phanerochaete chrysosporium: activation by manganese. J Bacteriol. 1991 Jul;173(13):4101-6. PubMed PMID: 2061289. Pubmed Central PMCID: 208059. Epub 1991/07/01. eng.

[102] Mayfield MB, Godfrey BJ, Gold MH. Characterization of the mnp2 gene encoding manganese peroxidase isozyme 2 from the basidiomycete Phanerochaete chrysosporium. Gene. 1994 May 16;142(2):231-5. PubMed PMID: 8194756. Epub 1994/05/16. eng.

[103] Li D, Alic M, Brown JA, Gold MH. Regulation of manganese peroxidase gene transcription by hydrogen peroxide, chemical stress, and molecular oxygen. Appl Environ Microbiol. 1995 Jan;61(1):341-5. PubMed PMID: 7887613. Pubmed Central PMCID: 167287. Epub 1995/01/01. eng.

[104] Stewart P, Kersten P, Vanden Wymelenberg A, Gaskell J, Cullen D. Lignin peroxidase gene family of Phanerochaete chrysosporium: complex regulation by carbon

and nitrogen limitation and identification of a second dimorphic chromosome. J Bacteriol. 1992 Aug;174(15):5036-42. PubMed PMID: 1629160. Pubmed Central PMCID: 206318. Epub 1992/08/01. eng.

[105] Giardina P, Faraco V, Pezzella C, Piscitelli A, Vanhulle S, Sannia G. Laccases: a never-ending story. Cell Mol Life Sci. 2010 Feb;67(3):369-85. PubMed PMID: 19844659. Epub 2009/10/22. eng.

[106] Soden DM, Dobson ADW. Differential regulation of laccase gene expression in Pleurotus sajor-caju. Microbiol-Sgm. 2001 Jul;147:1755-63. PubMed PMID: ISI: 000169732100005. English.

[107] Scheel T, Hofer M, Ludwig S, Holker U. Differential expression of manganese peroxidase and laccase in white-rot fungi in the presence of manganese or aromatic compounds. Appl Microbiol Biotechnol. 2000 Nov;54(5):686-91. PubMed PMID: 11131396. Epub 2000/12/29. eng.

[108] van Peij NN, Visser J, de Graaff LH. Isolation and analysis of xlnR, encoding a transcriptional activator co-ordinating xylanolytic expression in Aspergillus niger. Mol Microbiol. 1998 Jan;27(1):131-42. PubMed PMID: 9466262. Epub 1998/02/18. eng.

[109] van Peij NN, Gielkens MM, de Vries RP, Visser J, de Graaff LH. The transcriptional activator XlnR regulates both xylanolytic and endoglucanase gene expression in Aspergillus niger. Appl Environ Microbiol. 1998 Oct;64(10):3615-9. PubMed PMID: 9758775. Pubmed Central PMCID: 106473. Epub 1998/10/06. eng.

[110] Hasper AA, Visser J, de Graaff LH. The Aspergillus niger transcriptional activator XlnR, which is involved in the degradation of the polysaccharides xylan and cellulose, also regulates D-xylose reductase gene expression. Mol Microbiol. 2000 Apr; 36(1):193-200. PubMed PMID: 10760176. Epub 2000/04/12. eng.

[111] Gielkens MM, Dekkers E, Visser J, de Graaff LH. Two cellobiohydrolase-encoding genes from Aspergillus niger require D-xylose and the xylanolytic transcriptional activator XlnR for their expression. Appl Environ Microbiol. 1999 Oct;65(10):4340-5. PubMed PMID: 10508057. Pubmed Central PMCID: 91575. Epub 1999/10/03. eng.

[112] Marui J, Tanaka A, Mimura S, de Graaff LH, Visser J, Kitamoto N, et al. A transcriptional activator, AoXlnR, controls the expression of genes encoding xylanolytic enzymes in Aspergillus oryzae. Fungal Genet Biol. 2002 Mar;35(2):157-69. PubMed PMID: 11848678. Epub 2002/02/19. eng.

[113] Finn RD, Mistry J, Schuster-Bockler B, Griffiths-Jones S, Hollich V, Lassmann T, et al. Pfam: clans, web tools and services. Nucleic Acids Res. 2006 Jan 1;34(Database issue):D247-51. PubMed PMID: 16381856. Pubmed Central PMCID: 1347511. Epub 2005/12/31. eng.

[114] Hasper AA, Trindade LM, van der Veen D, van Ooyen AJ, de Graaff LH. Functional analysis of the transcriptional activator XlnR from Aspergillus niger. Microbiology. 2004 May;150(Pt 5):1367-75. PubMed PMID: 15133098. Epub 2004/05/11. eng.

[115] Omony J, de Graaff LH, van Straten G, van Boxtel AJ. Modeling and analysis of the dynamic behavior of the XlnR regulon in Aspergillus niger. BMC Syst Biol. 2011;5 Suppl 1:S14. PubMed PMID: 21689473. Pubmed Central PMCID: 3121114. Epub 2011/06/28. eng.

[116] Omony J, Mach-Aigner AR, de Graaff LH, van Straten G, van Boxtel AJ. Evaluation of Design Strategies for Time Course Experiments in Genetic Networks: Case Study of the XlnR Regulon in Aspergillus niger. IEEE/ACM Trans Comput Biol Bioinform. 2012 Apr 16. PubMed PMID: 22529332. Epub 2012/04/25. Eng.

[117] Saloheimo A, Aro N, Ilmen M, Penttila M. Isolation of the ace1 gene encoding a Cys(2)-His(2) transcription factor involved in regulation of activity of the cellulase promoter cbh1 of Trichoderma reesei. J Biol Chem. 2000 Feb 25;275(8):5817-25. PubMed PMID: 10681571. Epub 2000/02/22. eng.

[118] Aro N, Ilme´n, M., Saloheimo, A. and Penttila , M. ACEI is a repressor of cellulase and xylanase genes in Trichoderma reesei. Appl Environ Microbiol. 2002;69:56–65..

[119] Aro N, Saloheimo A, Ilmen M, Penttila M. ACEII, a novel transcriptional activator involved in regulation of cellulase and xylanase genes of Trichoderma reesei. Journal of Biological Chemistry. 2001 Jun 29;276(26):24309-14. PubMed PMID: ISI: 000169531100142. English.

[120] Witteveen CFB, Busink R, Vandevondervoort P, Dijkema C, Swart K, Visser J. L-Arabinose and D-Xylose Catabolism in Aspergillus-Niger. Journal of General Microbiology. 1989 Aug;135:2163-71. PubMed PMID: ISI:A1989AL64700004. English.

[121] vanKuyk PA, de Groot MJ, Ruijter GJ, de Vries RP, Visser J. The Aspergillus niger D-xylulose kinase gene is co-expressed with genes encoding arabinan degrading enzymes, and is essential for growth on D-xylose and L-arabinose. Eur J Biochem. 2001 Oct;268(20):5414-23. PubMed PMID: 11606204. Epub 2001/10/19. eng.

[122] Seiboth B, Hartl L, Pail M, Kubicek CP. D-xylose metabolism in Hypocrea jecorina: loss of the xylitol dehydrogenase step can be partially compensated for by lad1-encoded L-arabinitol-4-dehydrogenase. Eukaryot Cell. 2003 Oct;2(5):867-75. PubMed PMID: 14555469. Pubmed Central PMCID: 219359. Epub 2003/10/14. eng.

[123] de Groot MJ vdDC, Wosten HAB, Levisson M, vanKuyk PA, Ruijter GJG, de Vries RP. Regulation of pentose catabolic pathway genes of Aspergillus niger. Food Technol Biotechnol 2007;45:134–8.

[124] Flipphi MJ, Visser J, van der Veen P, de Graaff LH. Arabinase gene expression in Aspergillus niger: indications for coordinated regulation. Microbiology. 1994 Oct;140 (Pt 10):2673-82. PubMed PMID: 8000538. Epub 1994/10/01. eng.

[125] Battaglia E, Hansen SF, Leendertse A, Madrid S, Mulder H, Nikolaev I, et al. Regulation of pentose utilisation by AraR, but not XlnR, differs in Aspergillus nidulans and Aspergillus niger. Appl Microbiol Biotechnol. 2011 Jul;91(2):387-97. PubMed PMID: 21484208. Pubmed Central PMCID: 3125510. Epub 2011/04/13. eng.

[126] de Groot MJ, Prathumpai W, Visser J, Ruijter GJ. Metabolic control analysis of Aspergillus niger L-arabinose catabolism. Biotechnol Prog. 2005 Nov-Dec;21(6):1610-6. PubMed PMID: 16321042. Epub 2005/12/03. eng.

[127] Tani S, Kanamasa S, Sumitani J, Arai M, Kawaguchi T. XlnR-independent signaling pathway regulates both cellulase and xylanase genes in response to cellobiose in Aspergillus aculeatus. Curr Genet. 2012 Apr;58(2):93-104. PubMed PMID: 22371227. Epub 2012/03/01. eng.

[128] Endo Y, Yokoyama M, Morimoto M, Shirai K, Chikamatsu G, Kato N, et al. Novel promoter sequence required for inductive expression of the Aspergillus nidulans endoglucanase gene eglA. Biosci Biotechnol Biochem. 2008 Feb;72(2):312-20. PubMed PMID: 18256482. Epub 2008/02/08. eng.

[129] Petersen KL, Lehmbeck J, Christensen T. A new transcriptional activator for amylase genes in Aspergillus. Mol Gen Genet. 1999 Dec;262(4-5):668-76. PubMed PMID: 10628849. Epub 2000/01/11. eng.

[130] vanKuyk PA, Benen JA, Wosten HA, Visser J, de Vries RP. A broader role for AmyR in Aspergillus niger: regulation of the utilisation of D-glucose or D-galactose containing oligo- and polysaccharides. Appl Microbiol Biotechnol. 2012 Jan;93(1):285-93. PubMed PMID: 21874276. Pubmed Central PMCID: 3251782. Epub 2011/08/30. eng.

[131] Carlsen M, Nielsen J. Influence of carbon source on alpha-amylase production by Aspergillus oryzae. Appl Microbiol Biotechnol. 2001 Oct;57(3):346-9. PubMed PMID: 11759683. Epub 2002/01/05. eng.

[132] Davis RH, Perkins DD. Timeline: Neurospora: a model of model microbes. Nat Rev Genet. 2002 May;3(5):397-403. PubMed PMID: 11988765. Epub 2002/05/04. eng.

[133] Turner BC, Perkins DD, Fairfield A. Neurospora from natural populations: a global study. Fungal Genet Biol. 2001 Mar;32(2):67-92. PubMed PMID: 11352529. Epub 2001/05/16. eng.

[134] Tian C, Beeson WT, Iavarone AT, Sun J, Marletta MA, Cate JH, et al. Systems analysis of plant cell wall degradation by the model filamentous fungus Neurospora crassa. Proc Natl Acad Sci U S A. 2009 Dec 29;106(52):22157-62. PubMed PMID: 20018766. Pubmed Central PMCID: 2794032. Epub 2009/12/19. eng.

[135] Sun J, Tian C, Diamond S, Glass NL. Deciphering transcriptional regulatory mechanisms associated with hemicellulose degradation in Neurospora crassa. Eukaryot Cell. 2012 Apr;11(4):482-93. PubMed PMID: 22345350. Pubmed Central PMCID: 3318299. Epub 2012/02/22. eng.

[136] Coradetti ST, Craig JP, Xiong Y, Shock T, Tian C, Glass NL. Conserved and essential transcription factors for cellulase gene expression in ascomycete fungi. Proc Natl Acad Sci U S A. 2012 May 8;109(19):7397-402. PubMed PMID: 22532664. Pubmed Central PMCID: 3358856. Epub 2012/04/26. eng.

[137] Bailey CaA, H.N. Jr. Carbon catabolite repression in Aspergillus nidulans. Eur J Biochem. 1975;21:573– 7.

[138] Felenbock B, Flipphi, M., Nikolaev, I. Ethanol catabolism in Aspergillus nidulans: a model system for studying gene regulation. Prog Nucleic Acid Res Mol Biol. 2001;69:149-204.

[139] Flipphi M, Mathieu, M., Cirpus, I., Panozzo, C., Felenbock, B. Regulation of the aldehyde dehydrogenase gene (aldA) and its role in the control of the coinducer level necessary for the induction of the ethanol utilization pathway in Aspergillus nidulans. Journal of Biological Chemistry. 2001;276:6950-8.

[140] Flipphi M, Kocialkowska J, Felenbok B. Characteristics of physiological inducers of the ethanol utilization (alc) pathway in Aspergillus nidulans. Biochem J. 2002 May 15;364(Pt 1):25-31. PubMed PMID: 11988072. Pubmed Central PMCID: 1222541. Epub 2002/05/04. eng.

[141] Flipphi MaF, B. The onset of carbon catabolic repression and interplay between specific induction and carbon catabolite repression in Aspergillus nidulans. The Mycota III. 2nd ed. Berlin: Springer-Verlag; 2004. p. 403-20.

[142] Panozzo C, Cornillot E, Felenbok B. The CreA repressor is the sole DNA-binding protein responsible for carbon catabolite repression of the alcA gene in Aspergillus nidulans via its binding to a couple of specific sites. J Biol Chem. 1998 Mar 13;273(11): 6367-72. PubMed PMID: 9497366. Epub 1998/04/16. eng.

[143] Mathieu M, Fillinger S, Felenbok B. In vivo studies of upstream regulatory cis-acting elements of the alcR gene encoding the transactivator of the ethanol regulon in Aspergillus nidulans. Mol Microbiol. 2000 Apr;36(1):123-31. PubMed PMID: 10760169. Epub 2000/04/12. eng.

[144] Orejas M, MacCabe AP, Perez Gonzalez JA, Kumar S, Ramon D. Carbon catabolite repression of the Aspergillus nidulans xlnA gene. Mol Microbiol. 1999 Jan;31(1): 177-84. PubMed PMID: 9987120. Epub 1999/02/13. eng.

[145] Orejas M, MacCabe AP, Perez-Gonzalez JA, Kumar S, Ramon D. The wide-domain carbon catabolite repressor CreA indirectly controls expression of the Aspergillus nidulans xlnB gene, encoding the acidic endo-beta-(1,4)-xylanase X(24). J Bacteriol. 2001 Mar;183(5):1517-23. PubMed PMID: 11160081. Pubmed Central PMCID: 95035. Epub 2001/02/13. eng.

[146] Tamayo EN, Villanueva A, Hasper AA, de Graaff LH, Ramon D, Orejas M. CreA mediates repression of the regulatory gene xlnR which controls the production of xylanolytic enzymes in Aspergillus nidulans. Fungal Genet Biol. 2008 Jun;45(6):984-93. PubMed PMID: 18420433. Epub 2008/04/19. eng.

[147] Bai C, Sen P, Hofmann K, Ma L, Goebl M, Harper JW, et al. SKP1 connects cell cycle regulators to the ubiquitin proteolysis machinery through a novel motif, the F-box. Cell. 1996 Jul 26;86(2):263-74. PubMed PMID: 8706131. Epub 1996/07/26. eng.

[148] Craig KL, Tyers M. The F-box: a new motif for ubiquitin dependent proteolysis in cell cycle regulation and signal transduction. Prog Biophys Mol Biol. 1999;72(3): 299-328. PubMed PMID: 10581972. Epub 1999/12/03. eng.

[149] Colabardini AC, Humanes AC, Gouvea PF, Savoldi M, Goldman MH, Kress MR, et al. Molecular characterization of the Aspergillus nidulans fbxA encoding an F-box protein involved in xylanase induction. Fungal Genet Biol. 2012 Feb;49(2):130-40. PubMed PMID: 22142781. Epub 2011/12/07. eng.

[150] Hasunuma T, Kondo A. Development of yeast cell factories for consolidated bioprocessing of lignocellulose to bioethanol through cell surface engineering. Biotechnol Adv. 2011 Nov 4. PubMed PMID: 22085593. Epub 2011/11/17. Eng.

[151] Cherry JM, Hong EL, Amundsen C, Balakrishnan R, Binkley G, Chan ET, et al. Saccharomyces Genome Database: the genomics resource of budding yeast. Nucleic Acids Res. 2012 Jan;40(Database issue):D700-5. PubMed PMID: 22110037. Pubmed Central PMCID: 3245034. Epub 2011/11/24. eng.

[152] Krivoruchko A, Siewers V, Nielsen J. Opportunities for yeast metabolic engineering: Lessons from synthetic biology. Biotechnol J. 2011 Mar;6(3):262-76. PubMed PMID: 21328545. Epub 2011/02/18. eng.

[153] Kondo A, Ishii J, Hara KY, Hasunuma T, Matsuda F. Development of microbial cell factories for bio-refinery through synthetic bioengineering. J Biotechnol. 2012 Jun 19. PubMed PMID: 22728424. Epub 2012/06/26. Eng.

[154] Nielsen J, Jewett MC. Impact of systems biology on metabolic engineering of Saccharomyces cerevisiae. FEMS Yeast Res. 2008 Feb;8(1):122-31. PubMed PMID: 17727659. Epub 2007/08/31. eng.

[155] Matsushika A, Inoue H, Kodaki T, Sawayama S. Ethanol production from xylose in engineered Saccharomyces cerevisiae strains: current state and perspectives. Appl Microbiol Biotechnol. 2009 Aug;84(1):37-53. PubMed PMID: 19572128. Epub 2009/07/03. eng.

[156] Van Vleet JH, Jeffries TW. Yeast metabolic engineering for hemicellulosic ethanol production. Curr Opin Biotechnol. 2009 Jun;20(3):300-6. PubMed PMID: 19545992. Epub 2009/06/24. eng.

[157] Lynd LR, van Zyl WH, McBride JE, Laser M. Consolidated bioprocessing of cellulosic biomass: an update. Curr Opin Biotechnol. 2005 Oct;16(5):577-83. PubMed PMID: 16154338. Epub 2005/09/13. eng.

[158] Lynd LR, Elander RT, Wyman CE. Likely features and costs of mature biomass ethanol technology. Appl Biochem Biotech. 1996 Spr;57-8:741-61. PubMed PMID: ISI:A1996UL85700075. English.

[159] Wargacki AJ, Leonard E, Win MN, Regitsky DD, Santos CNS, Kim PB, et al. An Engineered Microbial Platform for Direct Biofuel Production from Brown Macroalgae. Science. 2012 Jan 20;335(6066):308-13. PubMed PMID: ISI:000299273400040. English.

[160] Gibson DG, Glass JI, Lartigue C, Noskov VN, Chuang RY, Algire MA, et al. Creation of a Bacterial Cell Controlled by a Chemically Synthesized Genome. Science. 2010 Jul 2;329(5987):52-6. PubMed PMID: ISI:000279402700028. English.

[161] Itaya M, Tsuge K. Construction and manipulation of giant DNA by a genome vector. Methods Enzymol. 2011;498:427-47. PubMed PMID: 21601689. Epub 2011/05/24. eng.

[162] Kaneko S, Itaya M. Designed horizontal transfer of stable giant DNA released from Escherichia coli. J Biochem. 2010 Jun;147(6):819-22. PubMed PMID: 20145021. Epub 2010/02/11. eng.

[163] Dymond JS, Richardson SM, Coombes CE, Babatz T, Muller H, Annaluru N, et al. Synthetic chromosome arms function in yeast and generate phenotypic diversity by design. Nature. 2011 Sep 22;477(7365):471-6. PubMed PMID: 21918511. Epub 2011/09/16. eng.

Section 4

Sustainable Products from Lignocellulosics

Chapter 10

A Review of Xylanase Production by the Fermentation of Xylan: Classification, Characterization and Applications

F. L. Motta, C. C. P. Andrade and M. H. A. Santana

Additional information is available at the end of the chapter

http://dx.doi.org/10.5772/53544

1. Introduction

The enzymatic hydrolysis of xylan, which is the second most abundant natural polysaccharide, is one of the most important industrial applications of this polysaccharide [1, 2]. The primary chain of xylan is composed of β-xylopyranose residues, and its complete hydrolysis requires the action of several enzymes, including endo-1,4-β-D-xylanase (EC3.2.1.8), which is crucial for xylan depolymerization [2]. Due to the diversity in the chemical structures of xylans derived from the cell walls of wood, cereal or other plant materials, a large variety of xylanases with various hydrolytic activities, physicochemical properties and structures are known. Moreover, xylan derivatives are frequently used to induce the production of xylanases [3] by microorganisms [4], using either solid-state or submerged fermentation [5].

Xylanases and the microorganisms that produce them are currently used in the management of waste, to degrade xylan to renewable fuels and chemicals, in addition to their use in food, agro-fiber, and the paper and pulp industries, where the enzymes help to reduce their environmental impact [6]. Oligosaccharides produced by the action of xylanases are further used as functional food additives or alternative sweeteners with beneficial properties [7].

To meet the needs of industry, more attention has been focused on the enzyme stability under different processing conditions, such as pH, temperature and inhibitory irons, in addition to its ability to hydrolyze soluble or insoluble xylans. Although many wild-type xylanases contain certain desired characteristics, such as thermostability, pH stability or high activity, no individual xylanase is capable of meeting all of the requirements of the feed and food industries. Moreover, as industrial applications require cheaper enzymes, the elevation of expression levels and the efficient secretion of xylanases are crucial to ensure the viability of the process; therefore, genetic engineering and recombinant DNA technology have an important role in the large-scale expression of xylanases in homologous or heterologous protein-expression hosts.

Considering the future prospects of xylanases in biotechnological applications, the goal of this review chapter is to present an overview of xylanase production via fermentation and to describe some of the characteristics of these enzymes and their primary substrate, xylan. Moreover, this review will discuss the fermentation processes as well as the genetic techniques applied to improve xylanase yields.

2. Xylan

The three main components that constitute lignocellulosic substrates are cellulose, hemicellulose and lignin [8]. Schulze [9] first introduced the term 'hemicellulose' to represent the fractions isolated or extracted from plant materials using a dilute alkali. Hemicelluloses are composed of complex mixtures of xylan, xyloglucan, glucomannan, galactoglucomannan, arabinogalactan or other heteropolymers [8].

The substrate of xylanase, xylan, is the second most-abundant polysaccharide in nature, accounting for approximately one-third of the renewable organic carbon on Earth [10], and it constitutes the major component of hemicellulose, a complex of polymeric carbohydrates, including xylan, xyloglucan (heteropolymer of D-xylose and D-glucose), glucomannan (heteropolymer of D-glucose and D-mannose), galactoglucomannan (heteropolymer of D-galactose, D-glucose and D-mannose) and arabinogalactan (heteropolymer of D-galactose and arabinose) [11]. Xylan is primarily present in the secondary cell wall and together with cellulose (1,4-β-glucan) and lignin (a complex polyphenolic compound) make up the major polymeric constituents of plant cell walls [12]. Within the cell wall structure, all three constituents interact via covalent and non-covalent linkages, with xylan being found at the interface between lignin and cellulose, where it is believed to be important for fiber cohesion and plant cell wall integrity [1].

2.1. Structure and distribution

A complex, highly branched heteropolysaccharide, xylan varies in structure between different plant species, and the homopolymeric backbone chain of 1,4-linked β-D-xylopyranosyl units can be substituted to varying degrees with glucuronopyranosyl, 4-O-methyl-D-glucuronopyranosyl, α-L-arabinofuranosyl, acetyl, feruloyl or p-coumaroyl side-chain groups [12,13] (Figure 1).

Xylan is distributed in several types of tissues and cells and is present in a variety of plant species [12], being found in large quantities in hardwoods from angiosperms (15–30% of the cell wall content) and softwoods from gymnosperms (7–10%), as well as in annual plants (<30%) [14]. Wood xylan exists as O-acetyl-4-O-methylglucuronoxylan in hardwoods and as arabino-4-O-methylglucuronoxylan in softwoods, while xylans in grasses and annual plants are typically arabinoxylans [12]. Linear unsubstituted xylan has also been reported in esparto grass [15], tobacco [16] and certain marine algae [17,18], with the latter containing xylopyranosyl residues linked by both 1,3-β and 1,4-β linkages [17,19].

Similar to other polysaccharides of plant origin, xylan has a large polydiversity and polymolecularity [20]. The degree of polymerization in xylans is also variable, with, for example, hardwood and softwood xylans generally consisting of 150-200 and 70-130 β-xylopyranose residues, respectively [12].

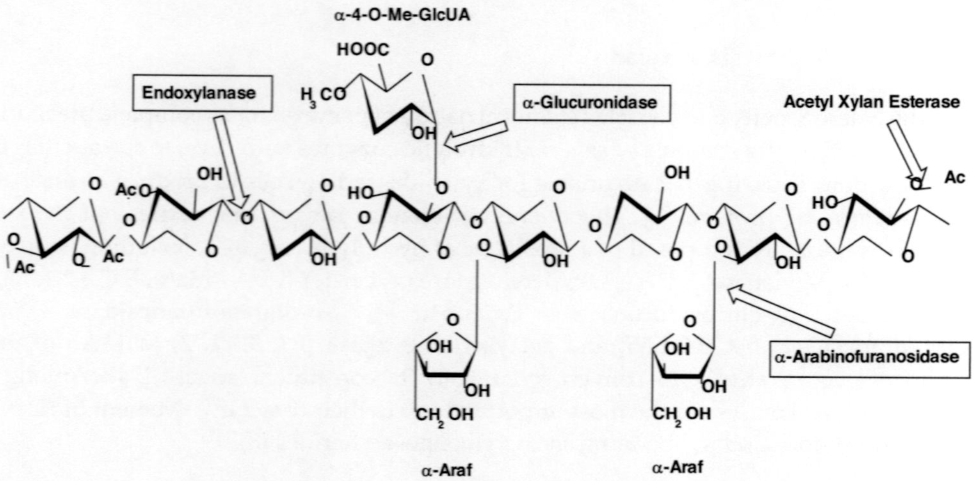

Figure 1. Structure of xylan and the xylanolytic enzymes involved in its degradation. Ac: Acetyl group; α-Araf: α-arabinofuranose; α-4-O-Me-GlcA: α-4-O-methylglucuronic acid. Source: Sunna and Antranikian [20].

Based on the common substituents found on the backbone, xylans are categorized as linear homoxylan, arabinoxylan, glucuronoxylan or glucuronoarabinoxylan. Homoxylans consisting exclusively of xylosyl residues are not widespread in nature; they have been isolated from limited sources, such as esparto grass, tobacco stalks and guar seed husks [20]. However, based on the nature of its substituents, a broad distinction may therefore be made among xylans, in which the complexity increases from linear to highly substituted xylans. Four main families of xylans can be considered [21]:

i. Arabinoxylans, having only side chains of single terminal units of α-L-arabinofuranosyl substituents. In the particular case of cereals, arabinoxylans vary in the degree of arabinosyl substitution, with either 2-O- and 3-O-mono-substituted or double (2-O-, 3-O-) substituted xylosyl residues.

ii. Glucuronoxylans, in which α-D-glucuronic acid or its 4-O-methyl ether derivative represents the only substituent.

iii. Glucuronoarabinoxylan, in which α-D-glucuronic (and 4-O-methyl-α-D-glucuronic) acid and α-L-arabinose are both present.

iv. Galactoglucuronoarabinoxylans, which are characterized by the presence of terminal β-D-galactopyranosyl residues on complex oligosaccharide side chains of xylans and are typically found in perennial plants.

In each category there exists microheterogeneity with respect to the degree and nature of branching. The side chains determine the solubility, physical conformation and reactivity of the xylan molecule with the other hemicellulosic components and hence greatly influence the mode and extent of enzymatic cleavage [12]. Endospermic arabinoxylans of annual plants, also called pentosans, are more soluble in water and alkaline solutions than xylans of lignocellulosic materials because of their branched structures [22].

2.2. Enzymatic hydrolysis of xylan

Due to the heterogeneity and complex chemical nature of plant xylan, its complete breakdown requires the action of a complex of several hydrolytic enzymes with diverse specificities and modes of action. Thus, it is not surprising for xylan-degrading cells to produce an arsenal of polymer-degrading proteins [1]. The xylanolytic enzyme system that carries out the xylan hydrolysis is normally composed of a repertoire of hydrolytic enzymes, including endoxylanase (endo-1,4-β-xylanase, E.C.3.2.1.8), β-xylosidase (xylan-1,4-β-xylosidase, E.C.3.2.1.37), α-glucuronidase (α-glucosiduronase, E.C.3.2.1.139), α-arabinofuranosidase (α-L-arabinofuranosidase, E.C.3.2.1.55) and acetylxylan esterase (E.C.3.1.1.72) [23]. All of these enzymes act cooperatively to convert xylan into its constituent sugars [24]. Among all xylanases, endoxylanases are the most important due to their direct involvement in cleaving the glycosidic bonds and in liberating short xylooligosaccharides [8].

Xylan, being a high molecular mass polymer, cannot penetrate the cell wall. The low molecular mass fragments of xylan play a key role in the regulation of xylanase biosynthesis. These fragments include xylose, xylobiose, xylooligosaccharides, heterodisaccharides of xylose and glucose and their positional isomers. These molecules are liberated from xylan through the action of small amounts of constitutively produced enzymes [12]. Xylanase catalyzes the random hydrolysis of xylan to xylooligosaccharides, while β-xylosidase releases xylose residues from the nonreducing ends of xylooligosaccharides. However, a complete degradation requires the synergistic action of acetyl esterase to remove the acetyl substituents from the β-1,4-linked D-xylose backbone of xylan [25,26].

3. Xylanases

3.1. Classification and mode of action

Xylanases, as glycoside hydrolase members, are able to catalyze the hydrolysis of the glycosidic linkage (β-1,4) of xylosides, leading to the formation of a sugar hemiacetal and the corresponding free aglycone (nonsugar compound remaining after replacement of the glycoside by a hydrogen atom [27]). Xylanases have been classified in at least three ways: based on the molecular weight and isoelectric point (pI) [28], the crystal structure [29] and kinetic properties,

http://dx.doi.org/10.5772/53544

or the substrate specificity and product profile. As the first classification is not sufficient to describe all xylanases, several exceptions have been identified [10] because not all xylanases have a high molecular mass (above 30 kDa) and low pI or a low molecular mass (less than 30 kDa) and high pI [6]. Therefore, a more complete system, based on the primary structure and comparison of the catalytic domains, was introduced [10,30], analyzing both the structural and mechanistic features [10].

Updated information on the characteristics and classification of enzymes may be found in the Carbohydrate-Active Enzyme (CAZy) database. This is a knowledge-based resource specializing on enzymes that build and breakdown complex carbohydrates and glycoconjugates. This database contains information from sequence annotations found in publicly available sources (such as the National Center for Biotechnology Information, NCBI), family classifications and known functional information [31]. According to the CAZy database (http://www.cazy.org), xylanases (EC3.2.1.8) are related to glycoside hydrolase (GH) families 5, 7, 8, 9, 10, 11, 12, 16, 26, 30, 43, 44, 51 and 62. However, the sequences classified in families 16, 51 and 62 appear to be bifunctional enzymes containing two catalytic domains, unlike families 5, 7, 8, 10, 11 and 43, which have a truly distinct catalytic domain with endo-1,4-β-xylanase activity [10]. Using the same analysis, families 9, 12, 26, 30 and 44 may have residual or secondary xylanase activity.

Xylanases have been primarily classified as GH 10 and 11 based on the hydrophobic cluster analysis of the catalytic domains and similarities in the amino acid sequences [8]. Although members of these two families have been thoroughly studied, the catalytic properties of the members of the remaining families (5, 7, 8 and 43) are recent and remain very limited [32].

Members of GH families 5, 7, 8, 10, 11 and 43 differ in their physicochemical properties, structure, mode of action and substrate specificities [10]. Several models have been proposed to explain the mechanism of xylanase action. Xylanase activity leads to the hydrolysis of xylan. Generally, this hydrolysis may result either in the retention (GH families 5, 7, 10 and 11) or the inversion (GH families 8 and 43) of the anomeric center of the reducing sugar monomer of the carbohydrate [33,34].

Families 5, 7, 10 and 11 contain enzymes that catalyze the hydrolysis with the retention of the anomeric configuration, with two glutamate residues being implicated in the catalytic mechanism. This indicates a double-displacement mechanism, in which a covalent glycosyl-enzyme intermediate is formed and subsequently hydrolyzed, and two carboxylic acid residues, suitably located in the active site, are involved in the formation of the intermediate; one acts as a general acid catalyst by protonating the substrate, while the second performs a nucleophilic attack, which results in the departure of the leaving group and the formation of the α-glycosyl enzyme intermediate (β to α inversion). In the second step, the first carboxylate group instead functions as a general base, abstracting a proton from a nucleophilic water molecule, which attacks the anomeric carbon. This leads to a second substitution, in which the anomeric carbon again passes via a transition state to give rise to a product with the β configuration (α to β inversion) [10,34].

In contrast to the mechanism mentioned above, the enzymes in families 8 and 43 generally act via an inversion of the anomeric center, and glutamate and aspartate may be the catalytic residues. Inverting enzymes work via a single displacement reaction, in which one carboxylate provides for a general acid-catalyzed leaving group departure. The second function of these enzymes, acting as general base, activates a nucleophilic water molecule to attack the anomeric carbon, thereby cleaving the glycosidic bond and leading to an inversion of the configuration at the anomeric carbon. Generally, the distance between the two residues allows for the accommodation of the water molecule between the anomeric carbon and the general base [10,34].

3.1.1. GH families 10 and 11

Xylanase from the GH10 family (or family G) have a low molecular mass with a pI between 8–9.5, while those from the GH11 family (or family F) have a high molecular mass and lower pI values [35,36].

Glycoside hydrolase family 10 is composed of endo-1,4-β-xylanases and endo-1,3-β-xylanases (EC 3.2.1.32) [34]. Members of this family are also capable of hydrolyzing the aryl β-glycosides of xylobiose and xylotriose at the aglyconic bond. Furthermore, these enzymes are highly active on short xylooligosaccharides, thereby indicating small substrate-binding sites. Crystal structure analyses, kinetic analyses of the activity on xylooligosaccharides of various sizes and end product analyses have indicated that family 10 xylanases typically have four to five substrate-binding sites [37]. Members of this family also typically have a high molecular mass, a low pI and display an (α/β)8-barrel fold [10,34,38].

Compared to other xylanases, GH11 members display several interesting properties, such as high substrate selectivity and high catalytic efficiency, a small size, and a variety of optimum pH and temperature values, making them suitable in various conditions and in many applications [39]. Family 11 is composed only of xylanases (EC3.2.1.8), leading to their consideration as "true xylanases," as they are exclusively active on D-xylose-containing substrates. GH11 enzymes are generally characterized by a high pI, a low molecular weight, a double-displacement catalytic mechanism, two glutamates that act as the catalytic residues and a β-jelly roll fold structure. Additionally, the products of their action can be further hydrolyzed by the family 10 enzymes [37]. Similar to family 10 xylanases, these enzymes can hydrolyze the aryl β-glycosides of xylobiose and xylotriose at the aglyconic bond, but they are inactive on aryl cellobiosides. Furthermore, in contrast to the family 10 xylanases, but similar to the family 8 cold-adapted xylanases, these enzymes are most active on long-chain xylooligosaccharides, and it has been found that they have larger substrate-binding clefts, containing at least seven subsites [10].

Xylanases belonging to GH10 exhibit greater catalytic versatility and lower substrate specificity than those belonging to GH11 [37,40]. According to Davies *et al.* [41], the binding sites for xylose residues in xylanases are termed subsites, with bond cleavage occurring between the sugar residues at the -1 (non-reducing) and the +1 (reducing) ends of the polysaccharide substrate. As observed in assays using arabinoxylan as the substrate, GH10 products have arabinose residues substituted on xylose at the +1 subsite, whereas GH 11 products have arabinose residues substituted at the +2 subsite [42]. These results suggest that GH 10 enzymes are able to hydrolyze xylose linkages closer to the side-chain residues [43]. Therefore, xylanases from family 11 preferentially cleave the unsubstituted regions of the arabinoxylan backbone, whereas GH10 enzymes cleave the decorated regions, being less hampered by the presence of substituents along the xylan backbone [37]. The xylan side-chain decorations are recognized by xylanases, and the degree of substitution in xylan will influence the hydrolytic products; this difference in substrate specificity has important implications in the deconstruction of xylan [43].

3.1.2. *GH families 5, 7, 8 and 43*

GH family 5 (or family A) is the largest glycoside hydrolase family, and only seven amino acid residues, including the nucleophile and the general acid/base residue, are strictly conserved among all members [10]. Structural alignment among the members of family 5 and 10 showed that these enzymes are as structurally different within family 5 as they are to the family 10 enzymes, therefore both families are classified into clan GH-A. The concept of clan or superfamily demonstrates a broader relationship between GH families, suggesting a more distant common evolutionary ancestor [44,45]. Furthermore, the activity of these enzymes is affected by substituents on the xylan main chain, and it is unable to cleave linkages adjacent to substituted residues. Hydrolysis studies have shown that the shortest substituted fragments formed from glucuronoxylan and arabinoxylan are substituted xylotrioses, with the substitution being found on the internal xylose residue. Therefore, the products produced by family 5 are shorter than those produced by family 7 [10].

GH family 8 (or family D) is composed of cellulases (EC 3.2.1.4), and also contains chitosanases (EC 3.2.1.132), lichenases (EC 3.2.1.73) and endo-1,4-β-xylanases (EC 3.2.1.8). This family of cold-adapted xylanases was found to hydrolyze xylan to xylotriose and xylotetraose and was most active on long-chain xylooligsaccharides. Similar to family 11 xylanases, a large substrate binding cleft containing at least six xylose-binding residues, with the catalytic site in the middle, was proposed [6]. However, unlike family 10 and 11 xylanases, enzymes from family 8 were found to catalyze hydrolysis with the inversion of the anomeric configuration and, under the conditions used, were found to be inactive on aryl β-glycosides of xylose, xylobiose and xylotriose [10,31].

GH families 7 and 43 contain only a few enzymes exhibiting xylanase activity that have been identified and studied. Family 7 has characteristics in common with both family 10 and 11 xylanases. Similar to the former family, those in family 7 have a high molecular weight and low pI, as well as a small substrate-binding site, containing approximately four subsites, with the catalytic site in the middle [10]. The members of family 43 have not been as thoroughly studied, and the structure of only one member has been determined, indicating that members of this family may display a five-blade β-propeller fold. Furthermore, a glutamate and aspartate in the center of a long V-shaped surface groove formed across the face of the propeller have been suggested as the catalytic residues. Family 43 is grouped with family 62 in clan GH-F, and, as also demonstrated in the family 8 enzymes, its members are believed to catalyze hydrolysis via a single displacement mechanism [10,31].

3.2. Properties and applications

The heterogeneity and complexity of xylan have resulted in a diverse range of xylanases, which differ in their physicochemical properties, structure, mode of action and substrate specificities [10]. As the xylosidic linkages in lignocellulose are neither equivalent nor equally accessible, the production of an enzymatic system with specialized functions is a strategy to achieve superior xylan hydrolysis [28]. Together with the heterogeneous nature of xylan, the multiplicity of xylanases in microorganisms may be caused by a redundancy in gene expression. Generally, a single xylanase gene encodes multiple xylanases, and xylanase multiplicity may arise from posttranslational modifications, such as differential glycosylation, proteolysis or both [23].

The potential applications of xylanases also include the bioconversion of lignocellulosic material and agro-wastes into fermentative products, the clarification of juices, the improvement of the consistency of beer and the digestibility of animal feedstocks [28]. One of the most important biotechnological applications of xylanase is its use in pulp bleaching [46]. Xylanases may also be applicable to the production of rayon, cellophane and several chemicals such as cellulose esters (acetates, nitrates, propionates and butyrates) and cellulose ethers (carboxymethyl cellulose and methyl and ethyl cellulose), which are all produced by dissolving pulp and purifying fibers from other carbohydrates [33].

3.2.1. The paper and pulp industries

During the past several years, the use of enzymes in paper and pulp bleaching has caught the attention of researchers and industries all over the world. Xylanase enzymes have proven to be a cost-effective means for mills to take advantage of a variety of bleaching benefits [47]. Xylanases and other side-cleaving enzymes have been used in pulp bleaching primarily to reduce lignin and increase the brightness of the pulp [20,46]. The importance of xylanase in the pulp and paper industries is related to the hydrolysis of xylan, which facilitates the release of lignin from paper pulp and, consequently, reduces the usage of chlorine as the bleaching agent [33].

Bleaching is the process of lignin removal from chemical pulps to produce bright or completely white finished pulp [1]. Thus, the bleaching of pulp using enzymes or ligninolytic microorganisms is called biobleaching [48]. This process is necessary due to the presence of residual lignin and its derivatives in the pulping process, which causes the resultant pulp to gain a characteristic brown color. The intensity of this pulp color is related to the amount and chemical state of the remaining lignin [33].

The bleaching of pulp involves the destruction, alteration or solubilization of the lignin, colored organic matter and other undesirable residues on the fibers [33]. Bleaching of kraft pulp usually requires large amounts of chlorine-based chemicals and sodium hydrosulfite, which cause several effluent-based problems in the pulp and paper industries. The use of these chemicals generates chlorinated organic substances, some of which are toxic, mutagenic, persistent, and highly resistant to biodegradation, in addition to causing numerous harmful disturbances in biological systems and forming one of the major sources of environmental pollution [1,33,49].

As hemicellulose is easier to depolymerize than lignin, biobleaching of pulp appears to be more effective with the use of xylanases than with lignin-degrading enzymes. This is due to the fact of the removal of even a small portion of the hemicellulose could be sufficient to open up the polymer, which facilitates removal of the residual lignin by mild oxidants [33,50].

The use of xylanase in bleaching pulp requires the use of enzymes with special characteristics. A key requirement is to be cellulose-free, to avoid damaging the pulp fibers [3], as cellulose is the primary product in the paper industry [33]. Other desirable characteristics are stability at high temperatures [51] and an alkaline optimal pH [48].

Madlala *et al.* [52] used different preparations of commercial Xylanase P and crude xylanase from *Thermomyces lanuginosus* to evaluate the bleaching process of paper pulp. It was demonstrated that the use of enzymes could increase the pulp brightness (over 5 brightness points over the control) and reduce the amount of bleaching chemicals used (up to 30% for chlorine dioxide). Chipeta *et al.* [53] evaluated crude xylanase preparations from *Aspergillus oryzae* NRRL 3485 and *Aspergillus phoenicis* ATCC 13157 and found that at a charge of 10 U per gram of pulp it was possible to reduce the usage of chlorine dioxide up to 30% without compromising the pulp brightness.

3.2.2. Bioconversion of lignocellulose in biofuels

Currently, second-generation biofuels are the primary products of the bioconversion of lignocellulosic materials. According to Taherzadech and Karimi [54], ethanol is the most important renewable fuel in terms of volume and market value, and following the fossil fuel crisis, it has been identified as an alternative fuel [48]. Despite the primarily first-generation production of ethanol, from sugar and starch, the second-generation production of ethanol has only begun to be tested in pilot plants [55]. And, unlike first-generation biofuels, second-generation biofuels do not compete with food production and can provide environmental, economic, and strategic benefits for the production of fuels [56].

Xylanase, together with other hydrolytic enzymes, can be used for the generation of biological fuels, such as ethanol, from lignocellulosic biomass [1,57]. However, enzymatic hydrolysis is still a major cost factor in the conversion of lignocellulosic raw materials to ethanol [56]. In bioethanol fuel production, the first step is the delignification of lignocellulose, to liberate cellulose and hemicellulose from their complex with lignin. The second step is a depolymerization of the carbohydrate polymers to produce free sugars, followed by the fermentation of mixed pentose and hexose sugars to produce ethanol [1,58]. Simultaneous saccharification and fermentation is an alternative process, in which both hydrolytic enzymes and fermentative microorganism are present in the reaction [48,59].

3.2.3. The pharmaceutical, food and feed industries

Xylanase, together with pectinase, carboxymethylcellulase and amylase, can be used for the clarification of juices because the turbidity observed is due to both pectic materials and other materials suspended in a stable colloidal system [60]. Xylanase may also improve the extraction of coffee, plant oils, and starch [25]. The xylose resulting from xylan depolymerization may also be converted to xylitol, a valuable sweetener that has applications in both the pharmaceutical and food industries [61-63].

In the bakery industry, xylanase may improve the quality of bread, by increasing the bread's specific volume. In combination with amylases, this characteristic was enhanced, as observed upon the introduction of *Aspergillus niger* var. *awamori* [64]. According to Collins *et al.* [65], psychrophilic enzymes may be suitable for use in the baking industry as they are generally optimally active at the temperatures most frequently used for dough preparation (at or below 35 °C). These enzymes could also be used as more efficient baking additives than the currently used commercial mesophilic enzymes, which are optimally active at higher temperatures.

Xylanase may also improve the nutritional properties of agricultural silage and grain feed. The use of this enzyme in poultry diets showed that the decrease in weight gain and feed conversion efficiency in rye-fed broiler chicks has been associated with intestinal viscosity [66]. The incorporation of xylanase from *Trichoderma longibrachiatum* into the rye-based diet of broiler chickens reduced intestinal viscosity, thus, improving both the weight gain of the chicks and their feed conversion efficiency [67].

Xylanases can also be used in cereals as a pretreatment for arabinoxylan-containing substrates, as arabinoxylans are partly water soluble and result in a highly viscous aqueous solution. This high viscosity of cereal grain water extract may lead to brewing problems, by decreasing the rate of filtration or haze formation in beer. Additionally, it is unfavorable in the cereal grains used in animal feeding [68,69].

The enzymatic hydrolysis of xylan may also result in oligomers known as xylooligosaccharides (XOs), which may be used in pharmaceutical, agriculture and feed products. XOs have prebiotic effects, as they are neither hydrolyzed nor absorbed in the upper gastrointestinal tract, and they affect the host by selectively stimulating the growth or activity of one or a number of bacteria in the colon, thus improving health [70-72]. Among their key physiological advantages are the reduction of cholesterol, maintenance of gastrointestinal health, and improvement of the biological availability of calcium. They also inhibit starch retrogradation, improving the nutritional and sensory properties of food [73]. For the production of XOs, the enzyme complex must have low exoxylanase or β-xylosidase activity, to prevent the production of high amounts of xylose, which has inhibitory effects on XO production [74,75].

3.3. Xylanase assays

The xylanase activity is often assayed based on measurement of reducing sugar released during the course of hydrolysis of xylan, by DNS or Nelson-Somogyi methods. Due to absent of standardization, Bailey *et al.* [76] compared the measurement of xylanase activity by twenty different laboratories. According to the author, the major source of variation between apparent xylanase activities was probably the substrate chosen, although small differences in protocols were also significant. After standardization of substrate and method, the interlaboratory standard variation of the results decreased from 108% to 17% from the mean. Others researchers use the 4-o-methylglucuronoxylan covalently dyed with Remazol Brilliant Blue (RBB xylan) as substrate, and the xylanase is assayed based on the release of the dyed fragments [77]. There are also available some commercial methods for xylanase assays, as the fluorescence-based method EnzChek® Ultra Xylanase Assay Kit (Invitrogen, Carlsbad, CA) or the Xylazyme tablet (Megazyme, Bray, Ireland), which employs azurine-crosslinked arabinoxylan (AZCLArabinoxylan) as substrate and its hydrolysis by xylanase produces water soluble dyed fragments.

3.4. Producing microorganisms

Microorganisms, in particular, have been regarded as a good source of useful enzymes because they multiply at extremely high rates and synthesize biologically active products that can be controlled by humans. In recent years, there has been a phenomenal increase in the use of enzymes as industrial catalysts. These enzymes offer advantages over the use of conventional chemical catalysts for numerous reasons: they exhibit high catalytic activity and a high degree of substrate specificity, they can be produced in large amounts, they are highly biodegradable, they pose no threat to the environment and they are economically viable [4].

In this context, microbial xylanases are the preferred catalysts for xylan hydrolysis, due to their high specificity, mild reaction conditions, negligible substrate loss and side product generation. Xylanases derived from microorganisms have many potential applications in the food, feed, and paper pulp industries [10,12,78]. Complete xylanolytic enzyme systems, which including all of these activities, have been found to be widespread among fungi [20,24], actinomycetes [79] and bacteria [12], and some of the most important xylanolytic enzyme producers include *Aspergillus, Trichoderma, Streptomyces, Phanerochaetes, Chytridiomycetes, Ruminococcus, Fibrobacteres, Clostridia* and *Bacillus* [12,78,80,81]. The ecological niches of these microorganisms are diverse and widespread and typically include environments where plant materials accumulate and deteriorate, as well as in the rumen of ruminants [78,82,83].

Although there have been many reports on microbial xylanases since the 1960s, the prime focus has been on plant pathology related studies [84]. Only during the 1980's did the use of xylanases for biobleaching begin to be tested [85]. Since 1982, several microorganisms, including fungi and bacteria, have been reported to readily hydrolyze xylans by synthesizing 1,4-β-D endoxylanases (E.C. 3.2.18) and β-xylosidases (EC.3.2.1.37) [86]. Table 1 presents a list of some of the xylanase-producing microorganisms and their activities.

The production of xylanases must be improved by finding more potent fungal or bacterial strains or by inducing mutant strains to excrete greater amounts of the enzymes. Moreover, the level of microbial enzyme production is influenced by a variety of nutritional and physiological factors, such as the supply of carbon and nitrogen, physical circumstances and chemical conditions [98].

Microorganisms	Xylanases	Cultivation conditions	Media	Reference
Penicillium canescens	18,895 IU/g	pH 7.0; 30 °C	Soya oil cake and casein peptone	[87]
Streptomyces sp. P12–137	27.8 IU/mL	pH 7.2; 28 °C	Wheat bran and KNO_3	[88]
Thermomyces lanuginosus SD-21	8,237 IU/g	pH 6.0; 40 °C	Corn cob and wheat bran and $(NH_4)_2SO_4$	[89]
Penicillium fellutanum	39.7 IU/mL	30 °C	Oat spelt xylan, urea, peptone and yeast extract	[90]
Penicillium clerotiorum	7.5 IU/mL	pH 6.5; 30 °C	Wheat bran	[91]
Acremonium furcatum	33.1 IU/mL	30 °C	Oat spelt xylan, urea, peptone and yeast extract	[90]
Aspergillus niger PPI	16.0 IU/mL	pH 5.0; 28 °C	Oat and urea	[92]
Neocallimastix sp. strain L2	1.13 IU/mL	50 °C	Avicel (PH 105) from Serva (Heidelberg, Germany)	[93]
Cochliobolus sativus Cs6	1,469.4 IU/g	pH 4.5; 30 °C	Wheat straw and $NaNO_3$	[94]
Bacillus circulans D1	8.4 IU/mL	pH 9.0; 45 °C	Bagasse hydrolysates	[95]
Streptomycessp. strain Ib 24D	1,447.0 IU/mL	pH 7.5; 28 °C	Tomato pomace	[96]
Paecilomyces themophila J18	18,580.0 IU/g	pH 6.9; 50 °C	Wheat straw and yeast extract	[97]

Table 1. Review of xylanases-producing microorganisms.

3.4.1. Fungi

Filamentous fungi are particularly interesting producers of xylanases and other xylan-degrading enzymes because they excrete the enzymes into the medium and their enzyme levels are much higher than those of yeast and bacteria. In addition to xylanases, fungi produce several auxiliary enzymes required for the degradation of substituted xylan [2].

The fungal genera *Trichoderma, Aspergillus, Fusarium,* and *Pichia* are considered great producers of xylanases [99]. White-rot fungi have also been shown to produce extracellular xylanases that act on a wide range of hemicellulosic materials, are useful as food sources [100] and produce metabolites of interest to the pharmaceutical, cosmetic, and food industries [78]. White-rot basidiomycetes normally secrete large amounts of these enzymes to degrade lignocellulosic materials. For example, *Phanerochaete chrysosporium* produces high levels of α-glucuronidase [101], and *Coriolus versicolor* produces a complex xylanolytic combination of enzymes [102]. Xylanase is also produced by *Cuninghamella subvermispora* when growing on plant cell-wall polysaccharides or on wood chips [103].

Fungal xylanases are generally associated with celluloses [104]. On cellulose these strains produce both cellulase and xylanase, which may be due to traces of hemicellulose present in the cellulosic substrates [105]; however, selective production of xylanase may be possible using only xylan as the carbon source. The mechanisms that govern the formation of extracellular enzymes with regards to the carbon sources present in the medium are influenced by the availability of precursors for protein synthesis. Therefore, in some fungi, growing the cells on xylan uncontaminated by cellulose under a lower nitrogen/carbon ratio may be a possible strategy for producing xylanolytic systems free of cellulases [106]. Another major problem associated with fungi is the reduced xylanase yield in fermenter studies. Agitation is normally used to maintain the medium homogeneity, but the shearing forces in the fermenter can disrupt the fragile fungal biomass, leading to the reported low productivity. Higher rates of agitation may also lead to hyphal disruption, further decreasing the xylanase activity [50].

3.4.2. Bacteria

Xylanases have been reported in *Bacillus, Streptomyces* and other bacterial genera that do not have any role related to plant pathogenicity [86]. The extreme thermophile *Rhodothermus marinus* has been reported to produce α-L-arabinofuranosidase [107], and two different polypeptides with α-arabinofuranosidase activity from *Bacillus polymyxa* were characterized at the gene level for the production of α-arabinofuranosidases [108].

Bacteria, just like many other industrial enzymes, have fascinated researchers due to their alkaline-thermostable xylanase-producing trait [33]. The optimum pH of bacterial xylanases are, in general, slightly higher than the optimal pH of fungal xylanases [109], which is a suitable characteristic in most industrial applications, especially the paper and pulp industries. Noteworthy producers of high levels of xylanase activity at an alkaline pH and high temperature are *Bacillus* spp. [33]. When considering only temperature, a handful of xylanases that show optimum activity at higher temperatures have been reported from various microorganisms. These include *Geobacillus thermoleovorans, Streptomyces sp.* S27, *Bacillus firmus, Actinomadura sp.* strain Cpt20 and *Saccharopolyspora pathunthaniensis* S582, all of which produce xylanases that show activity between 65 and 90 °C [8]. One xylanase, reported from *Thermotoga sp.* [110], has been shown to exhibit a temperature optima between 100 and 105 °C.

3.5. Production of xylanases under SSF and SmF

Xylanases are produced by either solid-state or submerged fermentation [5]. Although most xylanase manufacturers produce these enzymes using submerged fermentation (SmF) techniques (nearly for 90% of the total xylanase sales worldwide) [2], the enzyme productivity via solid-state fermentation (SSF) is normally much higher than that of submerged fermentation [5]. The growing interest in using solid-state fermentation (SSF) techniques to produce a wide variety of enzymes, including xylanases from fungal origins, is primarily due to the economic and engineering advantages of this process [111].

The advantages of SSF processes over SmF include a low cultivation cost for the fermentation, lower risk of contamination [1], improved enzyme stability, mimicking the natural habit of the fungus, production of enzymes with higher specific activities, generation of a protein-enriched byproduct, and easier downstream processing of the enzymes produced [112]. SSF conditions are especially suitable for the growth of fungi, as these organisms are able to grow at relatively low water activities, contrary to most bacteria and yeast, which will not proliferate under these culture conditions [113].

On the contrary, submerged fermentation allows better control of the conditions during fermentation [114]. The submerged fermentation of aerobic microorganisms is a well-known and widely used method for the production of cellulase and xylanase [115]. In general, SmF is the preferred method of production when the preparations require more purified enzymes, whereas synergistic effects from a battery of xylan-degrading enzymes can easily be found in preparations obtained by SSF using complex substrates, though the latter is commonly sought in applications aimed at improving animal feed [113].

The choice of the substrate is of great importance for the selection of the fermentation process and the successful production of xylanases. In this context, purified xylans can be excellent substrates because the low molecular weight compounds derived from them are the best xylanase inducers. The use of such substrates has led to increased yields of xylanase production and a selective induction of xylanases, with concomitantly low cellulase activity in a number of microorganisms. However, for large-scale processes other alternatives have to be considered due to the cost of such substrates. Some lignocellulolytic substrates such as barley husk, corn cobs, hay, wheat bran or straw have been compared in relation to pure substrates, and many have performed significantly better than isolated xylans (or celluloses) with respect to the yields of xylanase in large-scale production processes. Solid-state fermentation processes are practical for complex substrates, including agricultural, forestry and food processing residues and wastes, which are used as inducing carbon sources for the production of xylanases [113]. The use of abundantly available and cost-effective agricultural residues, such as wheat bran, corn cobs, rice bran, rice husks, and other similar substrates, to achieve higher xylanase yields via SSF allows the reduction of the overall manufacturing cost of biobleached paper. This has facilitated the use of this environmentally friendly technology in the paper industry [1].

3.6. Cloning and expression of xylanases

To meet specific industrial needs, an ideal xylanase should have specific properties, such as stability over a wide range of pH values and temperatures, high specific activity, and strong resistance to metal cations and chemicals [116]. Other specifications include cost-effectiveness, eco-friendliness, and ease of use [32]. Therefore, most of the reported xylanases do not possess all of the characteristics required by industry [8].

Native enzymes are not sufficient to meet the demand, due to low yields and incompatibility of the standard industrial fermentation processes [35]. Therefore, molecular approaches must be implemented to design xylanases with the required characteristics [8]. Heterologous expression is the main tool for the production of xylanases at the industrial level [35]. Protein engineering (alteration or modification of existing proteins) by recombinant DNA technology could be useful in improving the specific characteristics of existing xylanases [8]. Genetic engineering and recombinant DNA technology allow the large-scale expression of xylanases in homologous or heterologous protein-expression hosts. As industrial applications require cheaper enzymes, the elevation of expression levels and efficient secretion of xylanases are vital for ensuring the viability of the process [23].

An increasing number of publications have described numerous xylanases from several sources and the cloning, sequencing, mutagenesis and crystallographic analysis of these enzymes [12]. The available amino acid sequence data, X-ray crystallographic data, molecular dynamics and computational design of xylanases provide information that authenticates the relationship between the structure and function of xylanases. All of these methods aid in the design of xylanases that are required in industrial processes, such as improvement of the stability of xylanases at higher temperatures and alkaline pHs [8].

To attempt these processes for commercial purposes, genes encoding several xylanases have been cloned in homologous and heterologous hosts [12,48]. Recombinant xylanases have shown equivalent or better properties than the native enzymes, and the xylanase genes from anaerobic microorganisms have also been expressed successfully in hosts that can be employed in the fermentation industry [35].

3.6.1. Expression in bacteria

Escherichia coli is known for its ease of manipulation, inexpensive growth conditions, simple techniques required for transformation and accumulation of high levels of product in the cell cytoplasm; therefore, this organism has become the most widely used expression host [117]. Despite *Escherichia coli's* use as a good cloning host for recombinant proteins, it does not provide efficient and functional expression of many xylanases [23,24], and not all genes are easily expressed in *E. coli* [117]. This problem may be due to the repetitive appearance of rare codons and the requirement for specific translational modifications, such as disulfide-bond formation and glycosylation [23]. Therefore, this microorganism is useful for the detailed study of xylanase gene structure and for the improvement of the enzymes via protein engineering [35].

Lactobacillus species and *Bacillus subitilis* have been attractive hosts for the production of heterologous proteins, obtaining higher expression levels than *E. coli* [23,118]. *B. subtilis* and *Lactobacillus* are gram-positive and perform N-glycosilation [119]. Their primary interest in industry and research, is due to the fact that are non-toxic and are generally recognized as safe (GRAS) [23,118]. Members of the genus *Bacillus*, unlike *E. coli*, do not contain endotoxins (lipopolysaccharides), which are difficult to remove from many proteins during the purification process. The secretory production could also be advantageous in industrial production [33].

3.6.2. Expression in yeast

Heterologous protein expression in yeast systems is highly attractive because they provide additional benefits over bacterial expression systems. Among these benefits are the ability to perform eukaryotic post-translational modifications, the ability to grow to very high cell densities and the ability to secrete proteins into the fermentation media. Moreover, yeast are free of toxins and the majority have GRAS status [23].

Saccharomyces cerevisiae secretes high amounts of xylanases into the culture medium. Because it has already been established as an industrial microorganism, it can be used conveniently for the industrial production of xylanases at low costs [35]. *Pichia pastoris* has also emerged as an excellent host for the commercial production of xylanases due to very high expression under its own promoters [35]. However, the success of this methylotrophic yeast, similar to *Hansenula polymorpha*, is reached with the promoters of alcohol oxidase, an enzyme involved in the methanol-utilization pathway [23]. Therefore, these promoters have limited use at the large scale due to the health and fire hazards of methanol [35].

3.6.3. Expression in filamentous fungi

Filamentous fungi are capable producers of xylanases, via both heterologous and homologous gene expression, and reach high expression yields with their own promoters [35]. Filamentous fungi have already undergone intricate strain improvement for high-level protein secretion and are feasible when using the native xylanase-expressing machinery for functional expression of foreign xylanases from remote sources. The xylanase gene from *P. griseofulvum* has been successfully expressed in *A. oryzae* [120].

4. Conclusion

Xylan, the major hemicellulose component, requires the synergistic action of several hemicellulase enzymes for its complete hydrolysis to monomer sugars. The principle enzyme in this processes is endo-1,4-β-xylanase, which cleaves the glycosidic bonds between xylosides, generating short xylooligosaccharides. The majority of the studied xylanases have been classified into the GH10 or GH11 families, whereas studies of the xylanases in families 5, 7, 8 and 43 are still emerging.

The conversion of xylan to useful products represents part of our efforts to strengthen the overall economics of the processing of lignocellulosic biomass and to develop new means of energy production from renewable resources. Among these products are xylanases, enzymes that have a wide range of important industrial applications. Therefore, in the future, new methods will be developed for easier and cheaper production of these enzymes to fulfill the demands of various industries. In this context, the use of lignocellulosic agricultural waste for the production of these enzymes by either submerged or solid-sate fermentation has been very attractive, in addition to molecular techniques that are being tested to improve the enzyme's characteristics and increase its expression rates. Moreover, as the native enzyme does not fulfill all of the process requirements, bioprospecting for new genes, rational engineering and directed evolution of known genes are powerful tools that can be used to improve these enzymes.

Acknowledgements

The authors acknowledge the financial support of Fapesp (Fundação de Amparo à Pesquisa do Estado de São Paulo).

http://dx.doi.org/10.5772/53544

Author details

F. L. Motta[1*], C. C. P. Andrade[2] and M. H. A. Santana[1]

*Address all correspondence to: flopesmotta@gmail.com

1 Development of Biotechnological Processes Laboratory, School of Chemical Engineering, University of Campinas, Campinas, Brazil

2 Bioprocess Engineering Laboratory, Food Engineering Department, University of Campinas, Campinas, Brazil

References

[1] Beg Q K, Kapoor M, Mahajan L, Hoondal G S. Microbial xylanases and their industrial applications: a review. Applied Microbiology and Biotechnology 2001;56: 326–338.

[2] Polizeli M L T M, Rizzatti C S, Monti R, Terenzi H F, Jorge J, Amorim D S. Xylanases from fungi: properties and industrial applications. Applied Microbiology and Biotechnology 2005;67: 577–91.

[3] Haltrich D, Nidetzky B, Kulbe K D, Steiner W, Župančič S. Production of fungal xylanases. Bioresource Technology 1996;58: 137–161.

[4] Gote M. Isolation, purification and characterization of thermostable-galactosidase from *Bacillus stearothermophilus* (NCIM-5146). University of Pune; 2004.

[5] Agnihotri S, Dutt D, Tyagi C H, Kumar A, Upadhyaya J S. Production and biochemical characterization of a novel cellulase-poor alkali-thermo-tolerant xylanase from *Coprinellus disseminatus* SW-1 NTCC 1165. World Journal of Microbiology and Biotechnology 2010;26: 1349–1359.

[6] Collins T, Meuwis M A, Stals I, Claeyssens M, Feller G, Gerday C. A novel family 8 xylanase, functional and physicochemical characterization. The Journal of Biological Chemistry 2002;277: 35133–35139.

[7] Pellerin P, Gosselin M, Lepoutre J P, Samain E, Debeire P. Enzymatic production of oligosaccharides from corncob xylan. Enzyme and Microbial Technology 1991;13: 617–621.

[8] Verma D, Satyanarayana T. Molecular approaches for ameliorating microbial xylanases. Bioresource Technology 2012;17: 360–367.

[9] Schulze E. Information regarding chemical composition of plant cell membrane. Ber Dtsch Chem Ges 1891; 24:2277–2287.

[10] Collins T, Gerday C, Feller G. Xylanases, xylanase families and extremophilic xylanases. FEMS Microbiology Reviews 2005;29: 3–23.

[11] Shallom D, Shoham Y. Microbial hemicellulases. Current Opinion in Microbiology 2003; 6:219–228.

[12] Kulkarni N, Shendye A, Rao M. Molecular and biotechnological aspects of xylanases. FEMS Microbiology Reviews 1999;23: 411–456.

[13] Li K, Azadi P, Collins R, Tolan J, Kim J, Eriksson K. Relationships between activities of xylanases and xylan structures. Enzyme and Microbial Technology 2000;27: 89–94.

[14] Singh S, Madlala A M, Prior B A. *Thermomyces lanuginosus*: properties of strains and their hemicellulases. FEMS Microbiology Reviews 2003;27: 3–16.

[15] Chanda S K, Hirst E L, Jones J K N, Percival E G V. The constitution of xylan from esparto grass. J. Chem. Soc. 1950;12889–12897.

[16] Eda S, Ohnishi A, Kato K. Xylan isolated from the stalk of *Nicotiana tabacum*. Agric. Biol. Chem. 1976;40: 359–364.

[17] Barry V, Dillon T. Occurrence of xylans in marine algae. Nature 1940;146: 620-620.

[18] Nunn J R, Parolis H, Russel I. Polysaccharides of the red algae *Chaetangium erinaceum*. Part I: Isolation and characterization of the water-soluble xylan. Carbohydrate Research. 1973;26: 169–180.

[19] Percival E G V, Chanda S K. The xylan of *Rhodymenia palmata,* Nature 1950;166: 787–788.

[20] Sunna A, Antranikian G. Xylanolytic enzymes from fungi and bacteria. Critical Reviews in Biotechnology 1997;17: 39–67.

[21] Voragen A G J, Gruppen H, Verbruggen M A, Vietor R J. Characterization of cereals arabinoxylans, in: Xylan and Xylanases, Elsevier, Amsterdam; 1992.

[22] Ferreira-Filho E X. The xylan-degrading enzyme system. Brazilian Journal of Medical and Biological Research 1994;27: 1093–1109.

[23] Juturu V, Wu J C. Microbial xylanases: Engineering, production and industrial applications. Biotechnology Advances 2011; doi:10.1016/j.biotechadv.2011.11.006.

[24] Belancic A, SCARPA J, Peirano A, Diaz r, Steiner J, Eyzayuirre J. *Penicillium purpurogenum* produces several xylanases: purification and properties of two of the enzymes. Journal of Biotechnology 1995;41: 71–79.

[25] Wong K K Y, Saddler J N. Applications of hemicellulases in the food, feed, and pulp and paper industries, in: M.P. Coughlan, G.P. Hazlewood (Eds.), Hemicelluloses and Hernicellulases, Portland Press, London; 1993.

[26] Coughlan G P, Hazlewood M P. β-1,4-D-xylan-degrading enzyme system: Biochemistry, molecular biology, and applications. Biotechnol. Appl. Biochem. 1993;17: 259–289.

[27] Hatanaka K. Incorporation of Fluorous Glycosides to Cell Membrane and Saccharide Chain Elongation by Cellular Enzymes. Top Curr Chem 2012;308: 291–306.

[28] Wong K K, Tan L U, Saddler J N. Multiplicity of β-1,4-xylanase in microorganisms: functions and applications. Microbiological Reviews 1988; 52: 305–17.

[29] Jeffries T W. Biochemistry and genetics of microbial xylanases. Current Opinion in Biotechnology. 1996;7: 337–342.

[30] Henrissat B, Coutinho P M. Classification of glycoside hydrolases and glycosyltransferases from hyperthermophiles. Methods in Enzymology 2001;330: 183–201.

[31] Cantarel B L, Coutinho P M, Rancurel C, Bernard T, Lombard V, Henrissat B. The Carbohydrate-Active EnZymes database (CAZy): an expert resource for Glycogenomics. Nucleic Acids Research 2009; 37:233–238.

[32] Taibi Z, Saoudi B, Boudelaa M, Trigui H, Belghith H, Gargouri A. Purification and biochemical characterization of a highly thermostable xylanase from *Actinomadura sp.* strain Cpt20 isolated from poultry compost. Applied Biochemistry and Biotechnology 2012; 166:663–679.

[33] Subramaniyan S, Prema P. Biotechnology of microbial xylanases: enzymology, molecular biology, and application. Critical Reviews in Biotechnology 2002;22: 33–64.

[34] Coutinho P M, Henrissat B. Carbohydrate-active enzymes Server; 1999.

[35] Ahmed S, Riaz S, Jamil A. Molecular cloning of fungal xylanases: an overview. Applied Microbiology and Biotechnology 2009;84: 19–35.

[36] Buchert J, Tenkanen M, Kantelinen A, Viikari L. Application of xylanases in the pulp and paper industry. Bioresource Technology 1995;50: 65–72.

[37] Biely P, Vrsanská M, Tenkanen M, Kluepfel D. Endo-β-1,4-xylanase families: differences in catalytic properties. Journal of Biotechnology 1997;57: 151–166.

[38] Lo Leggio L, Kalogiannis S, Bhat M K, Pickersgill R W. High resolution structure and sequence of *T. aurantiacus* xylanase I: implications for the evolution of thermostability in family 10 xylanases and enzymes with (β)alpha-barrel architecture. Proteins 1999;36: 295–306.

[39] Paës G, Berrin J G, Beaugrand J. GH11 xylanases: Structure/function/properties relationships and applications. Biotechnology Advances 2012;30: 564–592.

[40] Faulds C B, Mandalari G, Lo Curto R B, Bisignano G, Christakopoulos P, Waldron K W. Synergy between xylanases from glycoside hydrolase family 10 and family 11 and

a feruloyl esterase in the release of phenolic acids from cereal arabinoxylan. Applied Microbiology and Biotechnology 2006;71: 622–629.

[41] Davies G J, Wilson K S, Henrissat B. Nomenclature for sugar-binding subsites in glycosyl hydrolases. Biochemical Journal 1997;321: 557–559.

[42] Maslen S L, Goubet F, Adam A, Dupree P, Stephens E. Structure elucidation of arabinoxylan isomers by normal phase HPLC-MALDI-TOF/TOF-MS/MS. Carbohydrate Research 2007; 342: 724–35.

[43] Dodd D, Cann I K O. Enzymatic deconstruction of xylan for biofuel production. Global Change Biology Bioenergy 2009;1: 2–17.

[44] Ryttersgaard C, Lo Leggio L, Coutinho P M, Henrissat B, Larsen S. *Aspergillus aculeatus* β-1,4-galactanase: substrate recognition and relations to other glycoside hydrolases in clan GH-A. Biochemistry 2002;41: 15135–15143.

[45] Larson S B, Day J, Paulina A, De B, Keen N T, Mcpherson A. First Crystallographic Structure of a Xylanase from Glycoside Hydrolase Family 5: Implications for Catalysis 2003;8411–8422.

[46] Viikari L, Kantelinen A, Sundquist J, Linko M. Xylanases in bleaching: From an idea to the industry. FEMS Microbiology Reviews 1994;13: 335–350.

[47] Bajpai P. Biotechnology for Pulp and Paper Processing. Springer US, Boston, MA; 2012.

[48] Pérez J, Muñoz-Dorado J, de la Rubia T, Martínez J. Biodegradation and biological treatments of cellulose, hemicellulose and lignin: an overview., International Microbiology: the Official Journal of the Spanish Society for Microbiology 2002;5: 53–63.

[49] Onysko K. Biological bleaching of chemical pulps: a review. Biotechnology Advances 1993;11: 179–198.

[50] Subramaniyan S, Prema P. Cellulase-free xylanases from *Bacillus* and other microorganisms. FEMS Microbiology Letters 2000;183: 1–7.

[51] Chidi S B, Godana B, Ncube I, Rensburg E J V, Abotsi E K. Production , purification and characterization of celullase-free xylanase from *Aspergillus terreus* UL 4209, Journal of Biotechnology 2008;7: 3939–3948.

[52] Madlala A, Bissoon S, Singh S, Christov L. Xylanase-induced reduction of chlorine dioxide consumption during elemental chlorine-free bleaching of different pulp types. Biotechnology Letters 2001;23: 345–351.

[53] Chipeta Z A, Du Preez J C, Szakacs G, Christopher L. Xylanase production by fungal strains on spent sulphite liquor. Applied Microbiology and Biotechnology 2005;69: 71–78.

[54] Taherzadeh M J, Karimi K. Pretreatment of Lignocellulosic Wastes to Improve Ethanol and Biogas Production: A Review. International Journal of Molecular Sciences 2008;9: 1621–1651.

[55] Taherzadeh M J, Karimi K. Enzyme-based hydrolysis processes for ethanol from lignocellulosic materials: a review. BioResources 2007;2: 707–738.

[56] Viikari L, Vehmaanperä J, Koivula A. Lignocellulosic ethanol: From science to industry, Biomass and Bioenergy 2012;1–12.

[57] Olsson L, Hahn-Hägerdal B. Fermentation of lignocellulosic hydrolysates for ethanol production. Enzyme and Microbial Technology 1996;18: 312–331.

[58] Lee J. Biological conversion of lignocellulosic biomass to ethanol. Journal of Biotechnology 1997;56: 1–24.

[59] Chandrakant P, Bisaria V S. Simultaneous Bioconversion of Cellulose and Hemicellulose to Ethanol. Crit Rev Biotechnol 2008;18: 295-331.

[60] Sreenath H K, Santhanam K. The use of commercial enzymes in white grape juice clarification. Journal of Fermentation and Bioengineering 1992;73: 241–243.

[61] Parajó J C, Domínguez H, Domínguez J. Biotechnological production of xylitol. Part 1: Interest of xylitol and fundamentals of its biosynthesis. Bioresource Technology 1998;65: 191–201.

[62] Sirisansaneeyakul S. Screening of yeasts for production of xylitol from D-xylose. Journal of Fermentation and Bioengineering 1995;80: 565–570.

[63] Soleimani M, Tabil L, Panigrahi S, Alberta E. Bio-production of a Polyalcohol (Xylitol) from Lignocellulosic Resources: A Review Written for presentation at the CSBE/SCGAB 2006 Annual Conference, Society; 2006.

[64] Maat J, Roza M, Verbakel J, Stam H, DaSilra M, Egmond M. Xylanases and their application in bakery, in: J. Visser, G. Beldman, M. VanSomeren, A. Voragen (Eds.), Xylans and Xylanases, Elsevier, Amsterdam; 1992.

[65] Collins T, Hoyoux A, Dutron A, Georis J, Genot B, Dauvrin T. Use of glycoside hydrolase family 8 xylanases in baking. Journal of Cereal Science 2006;43: 79–84.

[66] Paridon P A, Boonman J C P, Selten G C M, Geerse C, Barug D, de Bot P H M, Hemke G. The application of fungal endoxylanase in poulty diets, in: Visser J, Beldman G, Kusters-van Someren M A, Voragen A G J, (Eds.), Xylans and Xylanases, Elsevier, Amsterdam; 1992.

[67] Bedford M, Classen H. The influence of dietary xylanase on intestinal viscosity and molecular weight distribution of carbohydrates in rye-fed broiler chick, in: J. Visser, G. Beldman, M. VanSomeren, A. Voragen (Eds.), Xylans and Xylanases, Elsevier, Amsterdam; 1992.

[68] Dervilly-pinel G, Saulnier L. Experimental evidence for a semi-flexible conformation 2001;330: 365–372.

[69] Dervilly G, Leclercq C, Zimmermann D, Roue C, Thibault J, Saulnier L. Isolation and characterization of high molar mass water-soluble arabinoxylans from barley and barley malt. Carbohydr. Polym 2002;47: 143–149.

[70] Roberfroid M B. Health benefits of non-digestible oligosaccharides. Advances in Experimental Medicine and Biology 1997;427: 211–219.

[71] Collins M D, Gibson G R. Probiotics, prebiotics, and synbiotics: approaches for modulating the microbial ecology of the gut. The American Journal of Clinical Nutrition 1999;69: 1052S–1057S.

[72] Vázquez M, Alonso J, Domínguez H, Parajó J. Xylooligosaccharides: manufacture and applications. Trends in Food Science & Technology 2000;11: 387–393.

[73] Voragen A G J. Technological aspects of functional carbohydrates. Trends in Food Science & Technology 1998;9: 328–335.

[74] Akpinar O, Erdogan K, Bostanci S. Enzymatic production of Xylooligosaccharide from selected agricultural wastes. Food and Bioproducts Processing 2009;87: 145–151.

[75] Vázquez M, Alonso J, Domínguez H, Parajó J. Enzymatic processing of crude xylooligomer solutions obtained by autohydrolysis of eucalyptus wood. Food Biotechnology 2002;16: 91–105.

[76] Bailey M J, Biely P, Poutanen K. Interlaboratory testing of methods for assay of xylanase activity. J. Biotechnology 1992;23: 257–271.

[77] Biely P, Mislovicová D, Toman R. Soluble chromatogenic substrates for the assay of endo-1,4-β-xylanases and endo-1,4-β-glucanases. Analytical Biochemistry 1985;144: 142–146.

[78] Qinnghe C, Xiaoyu Y, Tiangui N, Cheng J, Qiugang M. The screening of culture condition and properties of xylanase by white-rot fungus *Pleurotus ostreatus*. Process Biochemistry 2004;39: 1561–1566.

[79] Elegir G, Szakács G, Jeffries T W. Purification, Characterization, and Substrate Specificities of Multiple Xylanases from *Streptomyces sp.* Strain B-12-2. Applied and Environmental Microbiology 1994;60: 2609–2615.

[80] Wubah D A, Akin D E, Borneman W S. Biology, fiber-degradation, and enzymology of anaerobic zoosporic fungi. Critical Reviews in Microbiology 1993;19: 99–115.

[81] Matte A, Forsberg C W. Purification, characterization, and mode of action of endoxylanases 1 and 2 from *Fibrobacter succinogenes* S85. Applied and Environmental Microbiology 1992;58: 157–168.

[82] Prade R A. Xylanases: from biology to biotechnology. Biotechnology & Genetic Engineering Reviews 1996;13: 101–131.

[83] Krause D O, Denman S E, Mackie R I, Morrison M, Rae A L, Attwood G T. Opportunities to improve fiber degradation in the rumen: microbiology, ecology, and genomics. FEMS Microbiology Reviews 2003;27: 663–693.

[84] Lebeda A, Luhova L, Sedlarova M, Jancova D. The role of enzymes in plant-fungal pathogens interactions. Zeitschrift Für Pflanzenkrankheiten Und Pflanzenschutz 2001;108: 89–111.

[85] Viikari L, Ranua M, Kantelinen A, Sundiquist J, Linko M. Biotechnology in the pulp and paper industry, in: Bleaching with Enzymes, Proc. 3rd Int. Con. Stockholm; 1986.

[86] Esteban R, Villanueva I R, Villa T G. β-D-Xylanases of *Bacillus circulans* WL-12, Can. J. Microbiol. 1982;28: 733–739.

[87] Assamoi A A, Destain J, Thonart P. Xylanase Production by Penicillium canescens on Soya Oil Cake in Solid-State Fermentation. Applied Biochemistry and Biotechnology 2010; 160:50-62.

[88] Coman G, Bahrim G. Optimization of xylanase production by *Streptomyces sp.* P12-137 using response surface methodology and central composite design. Annals of Microbiology 2011;61: 773–779.

[89] Ge Y, Lili H, Fuliang Z. Study on the Solid-state Fermentation Conditions for Producing Thermostable Xylanase Feed in a Pressure Pulsation Bioreactor. Advanced Materials Research 2011;236-238: 72–76.

[90] Palaniswamy M, Vaikuntavasan B, Ramaswamy P. Isolation, identification and screening of potential xylanolytic enzyme from litter degrading fungi. African Journal of Biotechnology 2008;7: 1978–1982.

[91] Knob A, Carmona E C. Xylanase Production by *Penicillium sclerotiorum* and its Characterization. World Applied Sciences Journal 2008; 4: 277–283.

[92] Pandey P, Pandey A K. Production of cellulase-free thermostable xylanases by an isolated strain of *Aspergillus niger* PPI utilizing various lignocellulosic wastes. World Journal of Microbiology and Biotechnology 2002;18: 281–283.

[93] Dijkerman R, Ledeboer J, Camp H J M O D, Prins R A, Drift C V D. The Anaerobic Fungus *Neocallimastix sp.* Strain L2: Growth and Production of (Hemi)Cellulolytic Enzymes on a Range of Carbohydrate Substrates. Current Microbiology 1997;34: 91–96.

[94] Arabi M I E, Jawhar M, Bakri Y. Effect of Additional Carbon Source and Moisture Level on Xylanase Production by *Cochliobolus sativus* in Solid Fermentation. Microbiology 2011;80: 150–153.

[95] Bocchini D A, Oliveira O M M, Gomes E, Silva R D. Use of sugarcane bagasse and grass hydrolysates as carbon sources for xylanase production by *Bacillus circulans* D1 in submerged fermentation. Process Biochemistry 2005;40: 3653–3659.

[96] Rawashdeh R, Saadoun I, Mahasneh A. Effect of cultural conditions on xylanase production by *Streptomyces sp*. (strain Ib 24D) and its potential to utilize tomato pomace. African Journal of Biotechnology 2005;4: 251–255.

[97] Yang S Q, Yan Q J, Jiang Z Q, Li L T, Tian H M, Wang Y Z. High-level of xylanase production by the thermophilic *Paecilomyces themophila* J18 on wheat straw in solid-state fermentation. Bioresource Technology 2006;97: 1794–800.

[98] Nagar S, Gupta V K, Kumar D, Kumar L, Kuhad R C. Production and optimization of cellulase-free, alkali-stable xylanase by Bacillus *pumilus* SV-85S in submerged fermentation. Journal of Industrial Microbiology & Biotechnology 2010;37: 71–83.

[99] Adsul M G, Ghule J E, Shaikh H, Singh R, Bastawde K B, Gokhale D V. Enzymatic hydrolysis of delignified bagasse polysaccharides. Carbohydrate Polymers 2005;62: 6–10.

[100] Buswell J A, Chang S T. Biomass and extracellular hydrolytic enzyme production by six mushroom species grown on soybean waste. Biotechnology Letters 1994;16: 1317–1322.

[101] Castanares A, Hay A J, Gordon A H, McCrae S I, Wood T M. D-Xylan-degrading enzyme system from the fungus *Phanerochaete chrysosporium*: Isolation and partial characterisation of an α-(4-O-methyl)-D-glucuronidase. Journal of Biotechnology 1995;43: 183-194

[102] Abd El-Nasser N H, Helmy S M, El-Gammal A A. Formation of enzymes by biodegradation of agricultural wastes with white rot fungi. Polymer Degradation and Stability 1997;55: 249–255.

[103] de Souza-Cruz P B, Freer J, Siika-Aho M, Ferraz A. Extraction and determination of enzymes produced by *Ceriporiopsis subvermispora* during biopulping of *Pinus taeda* wood chips. Enzyme and Microbial Technology 2004;34: 228–234.

[104] Steiner W, Lafferty R M, Gomes I, Esterbauer H. Studies on a wild type strain of *Schizophyllum commune*: Cellulase and xylanase production and formation of the extracellular polysaccharide schizophyllan. Biotechnol. Bioeng. 1987;30: 169–178.

[105] Gilbert H J, Hazlewood G P. Bacterial cellulases and xylanases. J Gen Microbiol. 1993;139: 187–194.

[106] Biely P. Biotechnological potential and production of xylanolytic systems free of cellulases, ACS Symp. Ser. 1991;460: 408–416.

[107] Gomes J, Gomes I, Steiner W. Thermolabile xylanase of the Antarctic yeast *Cryptococcus adeliae*: production and properties. Extremophiles 2000;4: 227–235.

[108] Morales P, Sendra J M, Perez-Gonzalez J A. Purification and characterisation of an arabinofuranosidase from *Bacillus polymyxa* expressed in *Bacillus subtilis*. Appl. Microbiol. Biotechnol. 1995;44: 112–117.

[109] Khasin A, Alchanati I, Shoham Y. Purification and characterization of a thermostable xylanase from Bacillus *stearothermophilus* T-6. Appl. Environ. Microbiol. 1993;59: 1725–1730.

[110] Yoon H, Han N S, Kim C H. Expression of *Thermotoga maritima* Endo-β-1,4-xylanase Gene in *E. coli* and Characterization of the Recombinant Enzyme. Agric. Chem. Biotechnol. 2004;47: 157–160.

[111] Pandey A, Selvakumar P, Soccol C R, Nigam P. Solid state fermentation for the production of industrial enzymes. Current Science 1999;77: 149–162.

[112] Considine P J, Coughlan M P. Enzyme System for Lignocellulose Degradation, in: M.P. Coughlan (Ed.), Enzyme System for Lignocellulose Degradation, Elsevier Applied Science, London; 1989.

[113] Corral, O L, Villaseñor-Ortega F. Xylanases. Advances in Agricultural and Food Biotechnology 2006;305-322.

[114] Frost G M, Moss D M. Production of enzymes by fermentation, in: Rehm H J, Reed G (eds.,) Biotechnology; 1987.

[115] Garcia-Kirchner O, Muñoz-Aguilar M, Pérez-Villalva R, Huitrón-Vargas C. Mixed submerged fermentation with two filamentous fungi for cellulolytic and xylanolytic enzyme production. Applied Biochemistry and Biotechnology 2002;1105–1114.

[116] Qiu Z, Shi P, Luo H, Bai Y, Yuan T, Yang P. A xylanase with broad pH and temperature adaptability from *Streptomyces megasporus* DSM 41476, and its potential application in brewing industry. Enzyme and Microbial Technology 2010;46: 506–512.

[117] Jhamb K, Sahoo D K. Production of soluble recombinant proteins in *Escherichia coli*: effects of process conditions and chaperone co-expression on cell growth and production of xylanase. Bioresource Technology 2012; doi: http://dx.doi.org/10.1016/j.biortech.2012.07.011.

[118] Bron S, Bolhuis A, Tjalsma H, Holsappel S, Venema G, van Dijl J. Protein secretion and possible roles for multiple signal peptidases for precursor processing in bacilli. Journal of Biotechnology 1998;64: 3–13.

[119] Upreti R K, Kumar M, Shankar V. Review Bacterial glycoproteins: Functions, biosynthesis and applications. Proteomics 2003;3: 363–379.

[120] Nevalainen K M H, Te'o V S J, Bergquist P L. Heterologous protein expression in filamentous fungi. Trends in Biotechnology 2005;23: 468–74.